Solar Cell Nanotechnology

Solar Cell Nanotechnology

Edited by Anderson Sheilds

Larsen & Keller
www.larsen-keller.com

Solar Cell Nanotechnology
Edited by Anderson Sheilds
ISBN: 979-8-88836-080-4 (Hardback)

Larsen & Keller
Published by Larsen and Keller Education,
5 Penn Plaza,
19th Floor,
New York, NY 10001, USA

Cataloging-in-Publication Data

Solar cell nanotechnology / edited by Anderson Sheilds.
p. cm.
Includes bibliographical references and index.
ISBN 979-8-88836-080-4
1. Solar cells. 2. Nanotechnology. 3. Nanostructured materials. I. Sheilds, Anderson.
TK2960 .S65 2023
333.794--dc23

For more information regarding Larsen and Keller Education and its products, please visit the publisher's website www.larsen-keller.com

Table of Contents

	Preface	**VII**
Chapter 1	**Improving the Photocurrent in Quantum-Dot-Sensitized Solar Cells by Employing Alloy $Pb_xCd_{1-x}S$ Quantum Dots as Photosensitizers** Chunze Yuan, Lin Li, Jing Huang, Zhijun Ning, Licheng Sun and Hans Ågren	**1**
Chapter 2	**The Influence of Fluorination on Nano-Scale Phase Separation and Photovoltaic Performance of Small Molecular/$PC_{71}BM$ Blends** Zhen Lu, Wen Liu, Jingjing Li, Tao Fang, Wanning Li, Jicheng Zhang, Feng Feng and Wenhua Li	**14**
Chapter 3	**Titanium Dioxide-Coated Zinc Oxide Nanorods as an Efficient Photoelectrode in Dye-Sensitized Solar Cells** Qiang Zhang, Shengwen Hou and Chaoyang Li	**26**
Chapter 4	**Nano-Photonic Structures for Light Trapping in Ultra-Thin Crystalline Silicon Solar Cells** Prathap Pathi, Akshit Peer and Rana Biswas	**35**
Chapter 5	**Correlation between CdSe QD Synthesis, Post-Synthetic Treatment and BHJ Hybrid Solar Cell Performance** Michael Eck and Michael Krueger	**51**
Chapter 6	**Transparent Platinum Counter Electrode Prepared by Polyol Reduction for Bifacial, Dye-Sensitized Solar Cells** Alvien Ghifari, Dang Xuan Long, Seonhyoung Kim, Brian Ma and Jongin Hong	**63**
Chapter 7	**Influence of Nitrogen Doping on Device Operation for TiO_2-Based Solid-State Dye-Sensitized Solar Cells: Photo-Physics from Materials to Devices** Jin Wang, Kosti Tapio, Aurélie Habert, Sebastien Sorgues, Christophe Colbeau-Justin, Bernard Ratier, Monica Scarisoreanu, Jussi Toppari, Nathalie Herlin-Boime and Johann Bouclé	**72**
Chapter 8	**Enhancing the Photocurrent of Top-Cell by Ellipsoidal Silver Nanoparticles: Towards Current-Matched GaInP/GaInAs/Ge Triple-Junction Solar Cells** Yiming Bai, Lingling Yan, Jun Wang, Lin Su, Zhigang Yin, Nuofu Chen and Yuanyuan Liu	**91**
Chapter 9	**Arrays of Plasmonic Nanostructures for Absorption Enhancement in Perovskite Thin Films** Tianyi Shen, Qiwen Tan, Zhenghong Dai, Nitin P. Padture and Domenico Pacifici	**99**
Chapter 10	**Numerical Study of Complementary Nanostructures for Light Trapping in Colloidal Quantum Dot Solar Cells** Jue Wei, Qiuyang Xiong, Seyed Milad Mahpeykar and Xihua Wang	**112**
Chapter 11	**Enhancing Solar Cell Efficiency using Photon Upconversion Materials** Yunfei Shang, Shuwei Hao, Chunhui Yang and Guanying Chen	**121**

Chapter 12 **Wavelength-Conversion-Material-Mediated Semiconductor Wafer Bonding for Smart Optoelectronic Interconnects** **143**
Kodai Kishibe, Soichiro Hirata, Ryoichi Inoue, Tatsushi Yamashita and Katsuaki Tanabe

Chapter 13 **Investigating the Effect of Carbon Nanotube Diameter and Wall Number in Carbon Nanotube/Silicon Heterojunction Solar Cells** **151**
Tom Grace, LePing Yu, Christopher Gibson, Daniel Tune, Huda Alturaif, Zeid Al Othman and Joseph Shapter

Chapter 14 **Graphene and Carbon Quantum Dot-Based Materials in Photovoltaic Devices: From Synthesis to Applications** **164**
Sofia Paulo, Emilio Palomares and Eugenia Martinez-Ferrero

Chapter 15 **Improvement of Exciton Collection and Light-Harvesting Range in Ternary Blend Polymer Solar Cells based on Two Non-Fullerene Acceptors** **184**
Yanbin Wang, Changlong Zhuang, Yawen Fang, Hyung Do Kim, Huang Yu, Biaobing Wang and Hideo Ohkita

Chapter 16 **Ultraviolet Plasmonic Aluminium Nanoparticles for Highly Efficient Light Incoupling on Silicon Solar Cells** **196**
Yinan Zhang, Boyuan Cai and Baohua Jia

Chapter 17 **Morphology-Controlled High-Efficiency Small Molecule Organic Solar Cells without Additive Solvent Treatment** **207**
Il Ku Kim, Jun Hyung Jo and Jung-Ho Yun

Chapter 18 **Ag Nanoparticle–Functionalized Open-Ended Freestanding TiO_2 Nanotube Arrays with a Scattering Layer for Improved Energy Conversion Efficiency in Dye-Sensitized Solar Cells** **214**
Won-Yeop Rho, Myeung-Hwan Chun, Ho-Sub Kim, Hyung-Mo Kim, Jung Sang Suh and Bong-Hyun Jun

Permissions

List of Contributors

Index

Preface

A solar cell refers to a type of electrical device that uses the photovoltaic effect to transform light into electricity. The utilization of nanotechnology and nanostructures in photovoltaics has the potential for increasing the efficiency of solar cells, either through the application of novel physical mechanisms or by developing solar cells with efficiencies that are closer to their theoretical maximum. Nanostructured solar cells are significant for generating renewable energy. They are also environmentally friendly. The nano-grating structures have a progressive change in refractive index. This functions as a multilayer antireflective coating, which helps in decreasing light reflection losses across the broadband range of incidence angles and wavelengths. The nanostructured solar cells are balanced between efficiency and cost. This book contains some path-breaking studies on solar cell nanotechnology. It elucidates the concepts and innovative models around prospective developments with respect to nanostructured solar cells. This book will serve as a reference to a broad spectrum of readers.

This book is a result of research of several months to collate the most relevant data in the field.

When I was approached with the idea of this book and the proposal to edit it, I was overwhelmed. It gave me an opportunity to reach out to all those who share a common interest with me in this field. I had 3 main parameters for editing this text:

1. Accuracy – The data and information provided in this book should be up-to-date and valuable to the readers.

2. Structure – The data must be presented in a structured format for easy understanding and better grasping of the readers.

3. Universal Approach – This book not only targets students but also experts and innovators in the field, thus my aim was to present topics which are of use to all.

Thus, it took me a couple of months to finish the editing of this book.

I would like to make a special mention of my publisher who considered me worthy of this opportunity and also supported me throughout the editing process. I would also like to thank the editing team at the back-end who extended their help whenever required.

Editor

Improving the Photocurrent in Quantum-Dot-Sensitized Solar Cells by Employing Alloy $Pb_xCd_{1-x}S$ Quantum Dots as Photosensitizers

Chunze Yuan [1,†,‡], Lin Li [2,†], Jing Huang [1], Zhijun Ning [1], Licheng Sun [2,*] and Hans Ågren [1,*]

1 Department of Theoretical Chemistry and Biology, School of Biotechnology, Royal Institute of Technology, 10691 Stockholm, Sweden; chunze@stanford.edu (C.Y.); jinghuang@theochem.kth.se (J.H.); zhijunning@gmail.com (Z.N.)

2 Center of Molecular Devices, Department of Chemistry, School of Chemical Science and Engineering, Royal Institute of Technology, 10044 Stockholm, Sweden; lin3@kth.se

* Correspondence: lichengs@kth.se (L.S.); agren@theochem.kth.se (H.Å.)

† These authors contributed equally to the paper.

‡ Present address: Department of Chemistry, Stanford University, Stanford, CA 94305, USA.

Academic Editor: Guanying Chen

Abstract: Ternary alloy $Pb_xCd_{1-x}S$ quantum dots (QDs) were explored as photosensitizers for quantum-dot-sensitized solar cells (QDSCs). Alloy $Pb_xCd_{1-x}S$ QDs ($Pb_{0.54}Cd_{0.46}S$, $Pb_{0.31}Cd_{0.69}S$, and $Pb_{0.24}Cd_{0.76}S$) were found to substantially improve the photocurrent of the solar cells compared to the single CdS or PbS QDs. Moreover, it was found that the photocurrent increases and the photovoltage decreases when the ratio of Pb in $Pb_xCd_{1-x}S$ is increased. Without surface protecting layer deposition, the highest short-circuit current density reaches 20 mA/cm^2 under simulated AM 1.5 illumination (100 mW/cm^2). After an additional CdS coating layer was deposited onto the $Pb_xCd_{1-x}S$ electrode, the photovoltaic performance further improved, with a photocurrent of 22.6 mA/cm^2 and an efficiency of 3.2%.

Keywords: quantum dot-sensitized solar cells; photocurrent; alloy; PbS

1. Introduction

Quantum-dot-sensitized solar cells (QDSCs), which are similar to dye-sensitized solar cells (DSCs) in terms of configuration and working mechanism [1,2], have been intensively investigated in the last decade due to the excellent properties of quantum dots (QDs), *i.e.*, size-dependent spectral tunability, high molar extinction coefficients, and low fabrication cost [3–7]. Especially, QD-based solar cells have shown promising conversion efficiency in the infrared region, where light cannot be effectively harvested by DSCs. In addition, the unique multi-electron generation effect of QDs proposes the theoretically maximum efficiency of QDSCs to be as high as 44% [8]. Therefore, QDSCs have been studied as a promising alternative or supplement to DSCs [9–11]. However, this efficiency is still far from the theoretically highest efficiency [12,13]. To enable the photovoltaic performance of QDSCs to compete with DSCs or other type solar cells, many efforts have been made, e.g., exploring new kinds of QDs with better light harvesting capability, electron injection efficiency, electrolyte and electrode materials in order to alleviate the series resistance between the electrolyte and cathode, as well as the carrier recombination between anode and electrolyte [14–19]. The extension of the light-harvesting region of the photosensitizers to the near-infrared (NIR) region has received more and more attention owing to its critical role to further improve the general solar cell performance [20–22]. Thus, although currently silicon solar cells can harvest light well in the visible region, there is lack of high-performance materials that can efficiently harvest infrared light.

So far, few types of QDs with absorption spectra stretching down to the NIR area have been proven to be effective for QDSCs. The broadening of the absorption spectrum has motivated the development of type-II core/shell-structured QDs, the bandgap of which is the difference between the conduction band (CB) edge of the component with deeper conduction band and the valence band (VB) edge of the component with a smaller VB [23,24]. QDs alloyed with multiple elements serve as another possible approach to broaden the absorption spectrum since they may present lower bandgaps than the corresponding binary QDs by the effect of optical bowing [25]. Another popular approach is to apply some traditional QDs with narrow bandgaps, such as PbS, PbSe, and Ag_2S [26–28]. PbS QDs have been intensely studied for the solar cell application due to the narrow bandgap (0.41 eV for bulk PbS) and potential multiple exciton collection [29–35]. The Bohr radius of PbS is as large as 18 nm [34], which makes it possible to tune its bandgap in a wide range by modifying the size of QDs.

PbS QDs have been successfully applied in heterojunction solar cells, showing remarkable conversion efficiency up to 10% [36–41]. However, PbS QDSCs show much lower device performance due to the low CB level, heavy charge recombination, and low stability with the presently available liquid electrolytes [30,33]. Coating of CdS or CdPbS shells onto PbS QDs have been explored, showing increased device performance compared with PbS only QDs, as the recombination and corrosion with the electrolyte are then inhibited [30,42]. Alloy QDs have been proposed as a good strategy for improving the optical and electronic properties of QDs by controlling the proportion of their component elements [25,43].

Solution phase *in-situ* ion adsorption by the method of successive ionic-layer adsorption and reaction (SILAR) [5,44] has been widely used for QDSCs and shows promising device performance, due to its convenient fabrication process. SILAR-adsorbed PbS QDs has been explored for QDSCs, albeit giving quite low device performance. Mixed Cd and Pb ions by using $Pb(NO_3)_2$ as lead salts for alloy $Pb_xCd_{1-x}S$ adsorption have been used; however, the device performance remains still low. It still remains unclear how the ratio of Pb:Cd of $Pb_xCd_{1-x}S$ QDs can influence the photovoltaic performance of QDSCs [45]. Since their bandgaps and CB levels are in between those of PbS and CdS [42], $Pb_xCd_{1-x}S$ QDs may present a higher CB edge than PbS and a wider absorption spectrum than CdS. By tuning the Pb:Cd ratio of $Pb_xCd_{1-x}S$, the bandgap and CB level of the alloy QDs can be adjusted, which should affect the light-harvesting ability, carrier injection, and carrier recombination in the QDSCs. On the other hand, it has been shown that the addition of a certain amount of Cd in PbS QDs can significantly limit the carrier recombination by reducing the trap density without affecting the carrier extraction, which brings much improved device performance [36]. Herein, in this report, we systematically study the effect of the ratio of Pb:Cd. This ratio is varied by controlling the Pb^{2+} ion concentration in the cationic precursor solutions, which is used to prepare three kinds of $Pb_xCd_{1-x}S$ QDs by the SILAR process.

2. Results and Discussion

Firstly, a kind of $Pb_xCd_{1-x}S$ QDs, termed PbCdS-1, was investigated as a photosensitizer compared to the CdS and PbS QDs. PbCdS-1 was prepared by a mixed precursor solution containing 0.004 M $PbCl_2$ and 0.1 M $Cd(NO_3)_2$. More details are given in the experimental section. The CdS, PbS, and PbCdS-1 were deposited on the TiO_2 electrodes by the SILAR process. The absorption spectra of the CdS, PbS, and PbCdS-1 QD-sensitized TiO_2 electrodes after three SILAR cycles are shown in the inset of Figure 1a. The CdS-sensitized electrode shows the narrowest absorption range (<500 nm), due to its wide bandgap, and the lowest absorbance intensity. In contrast, the PbS electrode shows the highest absorbance intensity and the widest absorption range over 800 nm, due to the considerably narrow bandgap and high molar extinction coefficient. For the PbCdS-1 electrode, the absorbance intensity is slightly lower than for PbS. The absorption performance of the PbCdS-1 QDs is similar to that of the PbS QDs, indicating that PbCdS-1 can harvest similar photon energies as PbS.

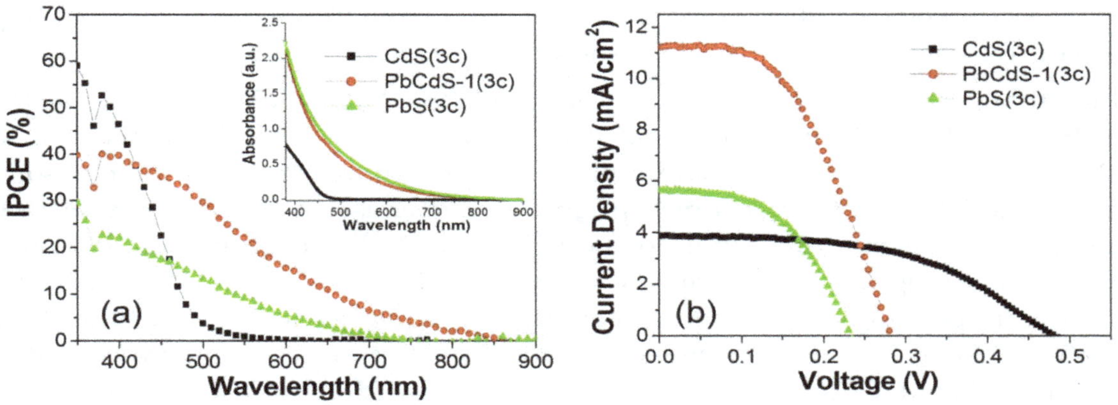

Figure 1. Incident photon to current efficiency (IPCE) spectra (**a**) and current–voltage (I–V) curves (**b**) of quantum-dot-sensitized solar cells (QDSCs) employing three successive ionic-layer adsorption and reaction (SILAR) cycles of CdS, PbS, and PbCdS-1 as photosensitizers. Inset of (**a**) is the absorption spectra of three-SILAR-cycle QD-sensitized TiO_2 films. "3C" means three SILAR cycles, and the bare TiO_2 film was used as blank.

Figure 1a shows the incident photon to current efficiency (IPCE) spectra of QDSCs employing CdS, PbS, and PbCdS-1 with three SILAR cycles as photosensitizers. From the absorption spectra, the sizes of three-cycle CdS, PbS, and PbCdS-1 can be estimated to be ~3 nm, ~2 nm, and 2–4 nm, respectively [45–48]. The CdS QDSCs show the highest IPCE and narrowest region (only up to 550 nm). The highest IPCE is due to the high CB energy level of CdS, which leads to a high driving force for electron injection from CB of CdS to TiO_2 [5]. The narrow photocurrent response range of the CdS cell gives an agreement with its absorption spectrum, which is limited by the wide bandgap of CdS. In contrast, the PbS QDSCs present a narrower IPCE range (<700 nm) than their absorption spectra (up to 800 nm), possibly due to the non-radiative decay through deep energy level traps or low CB energy levels. For PbCdS-1 QDSCs, it is clear that the range of photocurrent response extends up to 800 nm, which is in concordance with its absorption spectrum. In addition, Cd cation may be bound to the unpassivated S in PbS, thus removing valence-band-associated trap states [36]. The possible reasons of the much improved efficiency in the infrared region are the higher CB edge and the much reduced trap density of states for the alloy QDs. Compared to PbS, a remarkably improved overall photocurrent response was observed with a maximum IPCE close to 40%. The IPCE is almost double to that of PbS.

Figure 1b shows the current–voltage (I–V) curves of QDSCs employing CdS, PbS, and PbCdS-1 with three SILAR cycles under simulated one sun illumination (AM 1.5, 100 mW/cm^2). The corresponding parameters of performance are shown in Table S1. Among the three QDs, CdS gives the lowest photocurrent and the highest voltage due to its wide bandgap and high CB edge. PbS offers higher current and much lower voltage than CdS. The relative high current is due to the strong ability of PbS to harvest photons. The low voltage stems from that PbS suffers a serious carrier recombination due to low CB level and high trap densities. When PbCdS-1 was applied, the highest current (11.2 mA/cm^2) was obtained, while the voltage is larger than that of PbS. The highest current stems mainly from the high IPCE of PbCdS-1 as discussed above. In addition to an efficient photon harvesting ability, there might be other reasons that contribute to the high current and voltage. According to recent publications, Hg- or Cu-doped PbS could push the CB position to a higher energy level, which further promotes the electron injection and leads to a high current [34,49]. In another recent research work, the introduction of Cd into PbS was found to bring a similar result [42,45]. Therefore, one can confirm that the CB edge of PbCdS-1 is much higher than for PbS, which means that the introduction of Cd into PbS is favorable for the electron injection. This favorable electron injection could reduce electron recombination between PbCdS-1 and TiO_2, which contributes to a higher voltage.

Another reason might be the reduced defects by alloying Cd into PbS that may decrease trap density and restrain the recombination between TiO_2 and QDs.

To study the effect of the proportions of Pb and Cd in the $Pb_xCd_{1-x}S$ QDs, two more $Pb_xCd_{1-x}S$ QDs—PbCdS-2 and PbCdS-3—were investigated as photosensitizers for QDSC applications. PbCdS-1, -2, and -3 were prepared with gradually decreased Pb^{2+} concentrations of precursor solution, as shown in Table S2. Inductively coupled plasma-atomic emission spectrometry (ICP-AES) was carried out to measure the three kinds of $Pb_xCd_{1-x}S$ QD electrodes prepared by five SILAR cycles, directly revealing the concentrations of Pb and Cd of $Pb_xCd_{1-x}S$ QDs listed in Table S3. Therefore, the corresponding chemical formula of PbCdS-1, -2, and -3 can be deduced as $Pb_{0.54}Cd_{0.46}S$, $Pb_{0.31}Cd_{0.69}S$, and $Pb_{0.24}Cd_{0.76}S$, respectively. Although these formulas may not be quite exact for other cycles of $Pb_xCd_{1-x}S$ QDs, the proportion of Pb in the $Pb_xCd_{1-x}S$ QDs can be confirmed to be in the order: PbCdS-1 > PbCdS-2 > PbCdS-3.

The absorption spectra of the five-SILAR-cycle QD-sensitized electrodes are shown in Figure 2a. Compared to the inset of Figure 1a, the five-SILAR-cycle CdS, PbS, and PbCdS-1 show stronger absorbance intensities and broader absorption ranges than the corresponding three-SILAR-cycle QDs. Increasing the number of cycles makes the QDs to grow bigger in size, leading to narrower bandgap and stronger and broader absorption. The effect of the proportion of Pb in the $Pb_xCd_{1-x}S$ QD alloys on the optical properties is illustrated by comparing the absorption spectra of PbCdS-1, PbCdS-2, and PbCdS-3. It is clearly indicated that the absorbance intensity follows the order: PbCdS-1 > PbCdS-2 > PbCdS-3. Moreover, derived from the absorption spectra, the Tauc plots of $(h\nu\alpha)^2$ *vs.* hν, drawn in Figure 2b, can be employed to estimate the optical bandgap related by the Tauc equation [48].

$$(h\nu\alpha)^2 \propto (h\nu - E_g^{op}) \quad (1)$$

where E_g^{op} is optical bandgap, ν is light frequency, and α is the absorption coefficient. Therefore, the bandgap is determined by the intersection of the base line and the tangent line.

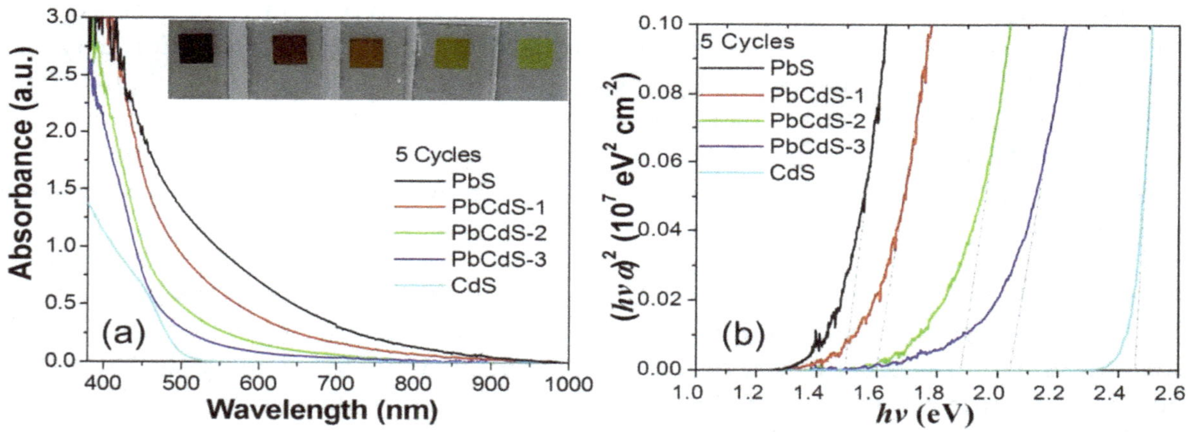

Figure 2. Absorption spectra (**a**) and Tauc plots (**b**) of five-SILAR-cycle PbS, PbCdS-1, -2, -3, and CdS-sensitized TiO_2 films. The bare TiO_2 film was used as blank in the absorption measurement. Inset images of (**a**) show the PbS, PbCdS-1, -2, -3, and CdS electrodes in turn from left to right. The doted tangent lines of the linear region of plots in (**b**) show the linear fit for the bandgap energy of QDs.

As discussed above, the CB energy level is an important factor affecting the performance of QDs solar cells. In this study, cyclic voltammogram (CV) was carried out in a 0.1 M KCl aqueous solution at pH 6.9, shown in Figure 3. The CB edge can be determined by the onset potential of the reduction current [50,51]. Compared to the CV performance of bare FTO and TiO_2 electrodes (Figure S1), the onset parts of the reduction current surrounded by the dotted line should result from QDs. Therefore, combining the obtained bandgaps and CB edges, the relationship of energy levels for different QDs and TiO_2 is shown in Scheme 1. The CB edge of TiO_2 was −4.21 eV (*vs.* vacuum) taken from the literature [52]. The CB energy value *vs.* normal hydrogen electrode (NHE) obtained from the CV was converted to the vacuum level by the relation that 0 V *vs.* NHE is equal to −4.5 eV *vs.* vacuum [53]. Clearly, as shown in Scheme 1, the bandgap of $Pb_xCd_{1-x}S$ QDs is wider than for PbS and narrower than for CdS under the same SILAR cycles, and the higher proportion of Pb in $Pb_xCd_{1-x}S$ presents a lower bandgap. In other words, increasing the proportion of Pb leads to a higher absorbance intensity and a concomitant shift of the absorption feature towards the higher wavelength because the larger proportion of Pb decreases the bandgap of the QDs. Moreover, it is confirmed that the CB positions of $Pb_xCd_{1-x}S$ QDs are between PbS and CdS QDs, and PbS QDs have a low CB edge (−4.15 eV) close to the TiO_2 CB edge. Therefore, a higher CB level causes a faster electron injection from QDs into TiO_2.

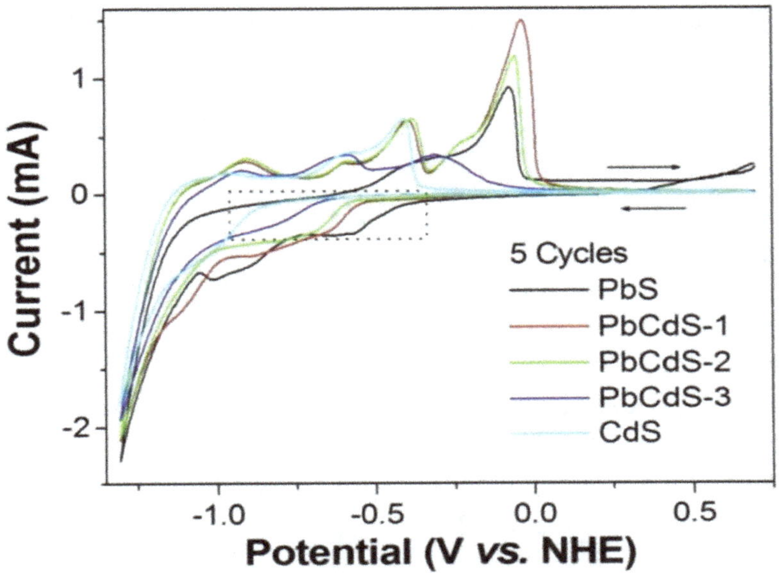

Figure 3. Cyclic voltammograms (CVs) of five-SILAR-cycle PbS, PbCdS-1, -2, -3, and CdS-sensitized TiO_2 electrode. CVs were measured with the electrolyte of 0.1 M KCl in deionized water (pH 6.9), and the scan rate was 10 mV/s. The arrows indicate the scan direction. The dotted part in the CV figures is to show the onset potentials of the reduction current varied at the different electrodes.

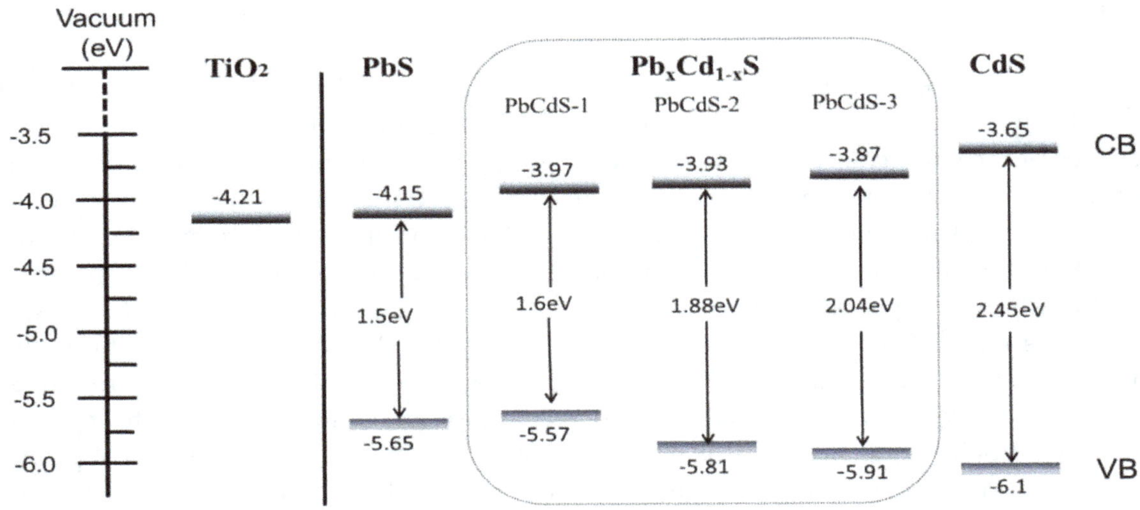

Scheme 1. Schematic diagram of edge energy levels of the conduction band (CB) edge and valence band (VB) edge for TiO_2 and five-SILAR-cycle PbS, CdS, and three kinds of $Pb_xCd_{1-x}S$.

Generally, the major factors limiting the conversion efficiency of QDSC are the relatively slow electron transportation within the TiO_2 film and the slow hole transfer, the main consequence of which is a large carrier loss due to the recombination between the TiO_2 film and electrolyte [54]. Moreover, the trap states of QDs in the middle position between the TiO_2 CB edge and the redox level of the electrolyte [3], which are generated by the interfacial effect and inner defects, can aggravate the recombination through back electron transfer via trap states. Therefore, it is vital to reduce the trap density of QDs for improving the performance of QDSCs. As discussed above, the presence of Cd in PbS may reduce the trap states. In order to verify the effect of Cd, dark I–V measurements were carried out, which is a good method to investigate the carrier recombination between TiO_2 and electrolyte. As shown in Figure 4, the onset of dark current of the devices based on $Pb_xCd_{1-x}S$ occurred at a much higher bias-voltage than that of PbS, indicating that the recombination in $Pb_xCd_{1-x}S$-based QDSCs is remarkably reduced. It also suggests that the increase of Cd in $Pb_xCd_{1-x}S$ can further reduce the recombination degree. The result proved that the utilization of $Pb_xCd_{1-x}S$ can effectively decrease the trap density.

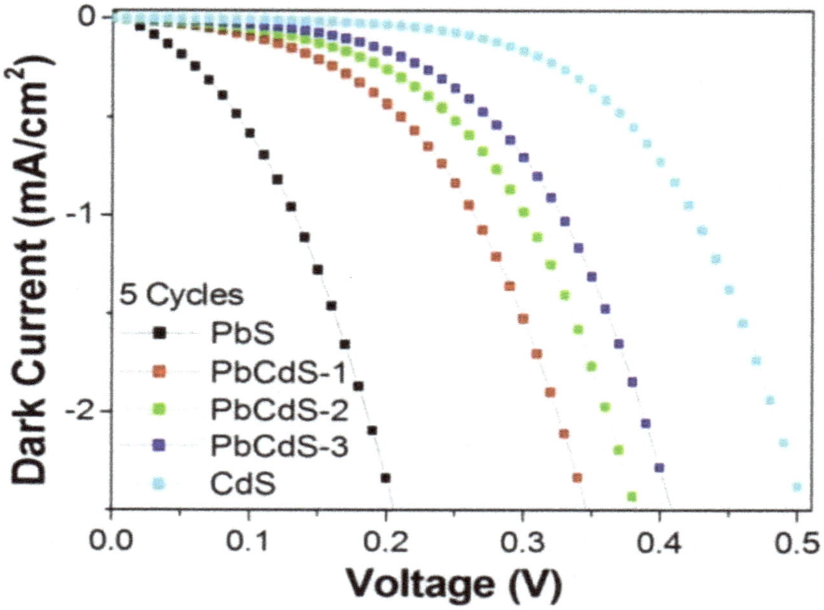

Figure 4. Dark I–V curves of QDSCs employing five-SILAR-cycle PbS, PbCdS-1, -2, -3, and CdS.

The optimization of the SILAR cycles is a key factor to boost the performance of the solar cells. Herein, we discuss the effect of SILAR cycles on the photovoltaic performance of the solar cells. The I–V curves of the solar cells of PbS, CdS, and $Pb_xCd_{1-x}S$ under one sun illumination are depicted in Figure S2. The photovoltaic parameters are listed in Table S4. In order to analyze the effect of the SILAR cycles, the photovoltaic parameters were presented as a function of the number of such cycles as shown in Figure 5. It was observed that the performance of the PbS solar cells decreases with the increase of SILAR cycles, probably due to the narrower bandgap and the lower CB energy level of the PbS QDs when the SILAR cycles increases after three cycles. The photocurrents of the three kinds of $Pb_xCd_{1-x}S$ and CdS are enhanced by increasing SILAR cycles until reaching a number of certain cycles. To further optimize the performance by increasing the cycle number, it was found that 7-cycle PbCdS-1, 7-cycle PbCdS-2, 9-cycle PbCdS-3, and 9-cycle CdS QDSCs provide the best photocurrent. Moreover, the trend of the efficiency is similar to that of the photocurrent. The comparison of the parameters between the different QD solar cells indicates that the QDs with the higher proportion of Pb show the higher short-circuit current and the lower open-circuit voltage.

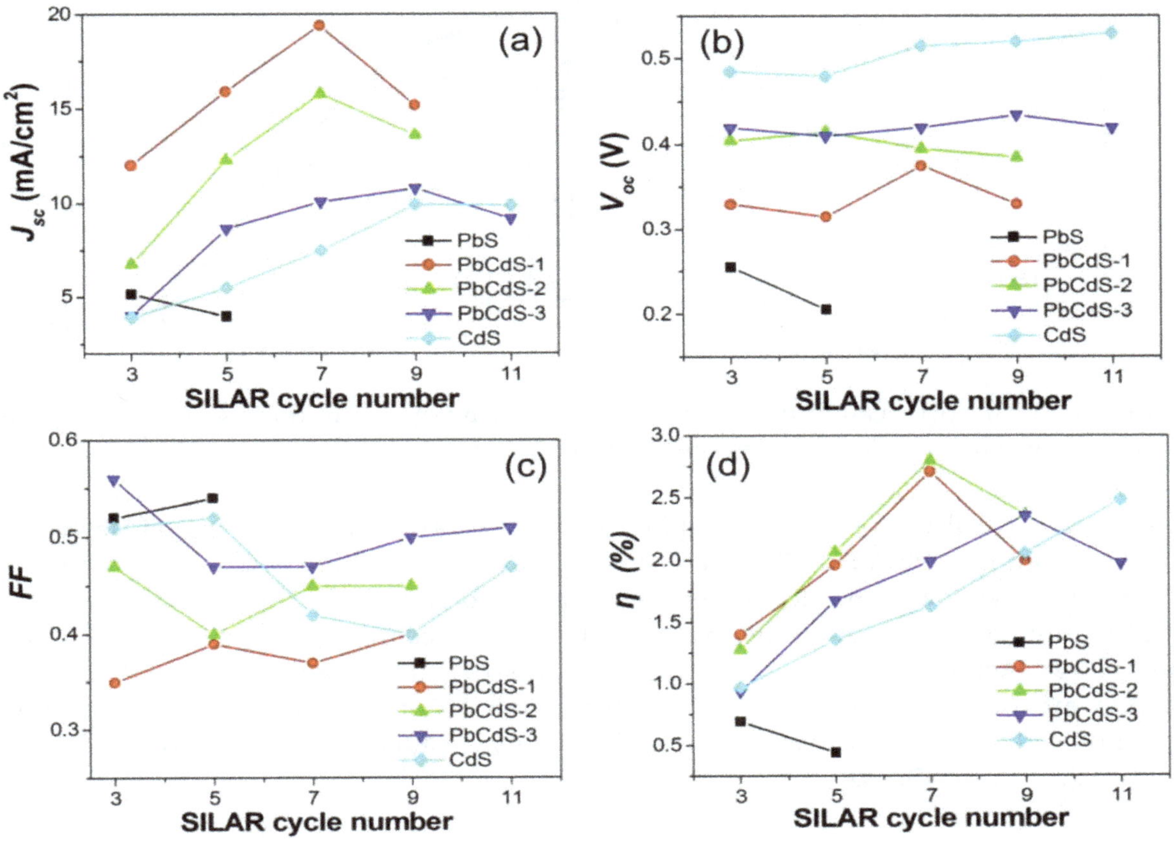

Figure 5. The photovoltaic parameters of QDSCs employing CdS, PbS, and three $Pb_xCd_{1-x}S$ QDs in terms of the number of SILAR cycles. (**a**) short-circuit current density (J_{sc}); (**b**) open-circuit voltage (V_{oc}); (**c**) fill factor (*FF*); (**d**) efficiency (η).

After analyzing the optimization of solar cell parameters, some characteristic trends can be unraveled. In particular, the proportion of Pb and Cd in the $Pb_xCd_{1-x}S$ alloys will affect these parameters. In order to better understand the effect, the optimized performance of the solar cells based on $Pb_xCd_{1-x}S$ QDs are re-illustrated and compared in Figure 6; these are the PbCdS-1 (seven SILAR cycles), PbCdS-2 (seven SILAR cycles), and PbCdS-3 (nine SILAR cycles) QDSCs. These three samples show a similar maximum IPCE close to 60%. However, a clear broadening of the IPCE features is observed for the PbCdS-1 solar cell which presents an IPCE range up to more than 900 nm, while the PbCdS-3 only reaches 800 nm. In other words, increasing the proportion of Pb (and thus decreasing the proportion of Cd) in the $Pb_xCd_{1-x}S$ alloys leads to a significant broadening of the IPCE features. This result is consistent with the absorption performance of the five cycles $Pb_xCd_{1-x}S$. Moreover, the IPCE trend also reflects the photocurrent in the I–V curves of the QDSCs. It is clear that the photocurrent follows the trend: PbCdS-1 > PbCdS-2 > PbCdS-3, something that can be mainly attributed to the light-harvesting ability of the $Pb_xCd_{1-x}S$ alloys. The highest photocurrent J_{sc} obtained by the PbCdS-1 QDSCs reaches almost 20 mA/cm^2. However, the voltage follows the reverse trend: PbCdS-1 < PbCdS-2 < PbCdS-3. Increasing the proportion of Pb leads to a lower voltage, which is caused by the effect of a lower CB level and higher trap density, as discussed above. PbCdS-2 QDSCs exhibit the efficiency of 2.8%, referring to their moderate photocurrent of 15.8 mA/cm^2 and voltage of 0.4 V. Therefore, controlling the ratio between Pb and Cd in the $Pb_xCd_{1-x}S$ QDs alloys is very important for the photovoltaic performance of the QDSCs, especially for controlling the photocurrent and voltage.

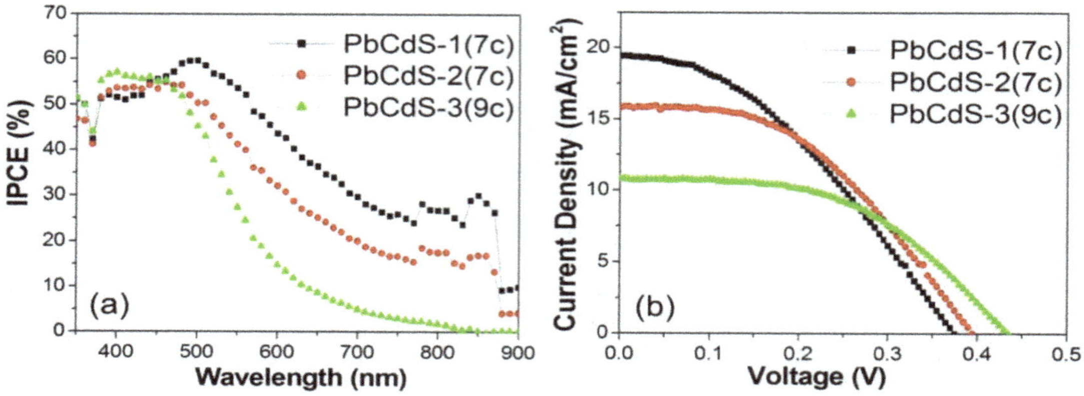

Figure 6. (**a**) IPCE spectra and (**b**) I–V curves of QDSCs employing PbCdS-1 (seven SILAR cycles), PbCdS-2 (seven SILAR cycles), and PbCdS-3 (nine SILAR cycles).

Furthermore, we also investigated the photovoltaic performance after an extra CdS coating layer was introduced onto the PbCdS-1 QDs, shown in Figure 7. The CB edge of CdS is higher than that of $Pb_xCd_{1-x}S$, which will favor the electron injection of CdS to TiO_2 via PbCdS-1. Therefore, it could be observed that the IPCE of seven-cycle PbCdS-1 QDSCs after coating three-cycle CdS QDs improve, especially at wavelengths larger than 500 nm. In addition, CdS coating could serve as a protection layer to prevent the inner layered QDs from the corrosion of electrolyte and reduce the charge recombination [30]. This improvement gives us finally a photocurrent of 22.6 mA/cm^2 and an efficiency of 3.2% for the QDSC.

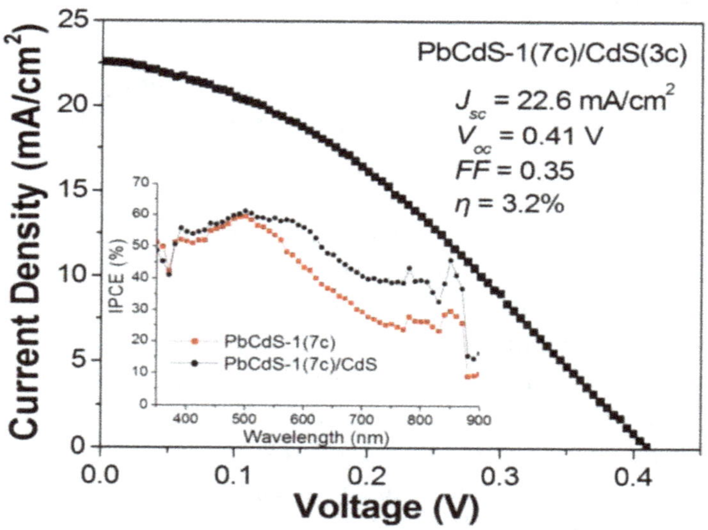

Figure 7. I–V curve of QDSC employing a hybrid QDs of PbCdS-1 (seven SILAR cycles) and CdS (three SILAR cycles), where PbCdS-1 was sensitized onto a TiO_2 film, followed by subsequently depositing CdS onto PbCdS-1. Inset: IPCE spectrum of the hybrid QDs compared to the corresponding PbCdS-1 QDs.

3. Experimental Section

3.1. Preparation of the QDs-Sensitized Solar Cells

Mesoporous TiO_2 films with a triple-layer structure, containing the compact, transparent, and scattering layer, were prepared according to the procedure reported in [55]. First, fluorine-doped tin oxide (FTO, Solaronix TCO22-7) glass was sequentially cleaned in saturated sodium hydroxide isopropanol solution, absolute ethanol, and deionized (DI) water for 15 min under ultrasonic treatment. To deposit the first layer, *i.e.*, compact layer, onto the FTO substrate, a freshly aqueous $TiCl_4$ solution (40 mM) was used to treat the cleaned FTO glass for 30 min at 70 °C. Subsequently, the transparent and scattering layers were successively deposited by screen printing with TiO_2 pastes (Solaronix, Ti-Nanoxide T/SP and R/SP), followed by sintering at 500 °C for 30 min in a muffle furnace to remove organic components. The sintered film was post-treated with an aqueous $TiCl_4$ solution. The produced TiO_2 films provided a thickness of 11 μm (7 μm for transparent layer and 4 μm for scattering layer) and a working area of 5 mm × 5 mm.

QD-sensitized TiO_2 photoelectrodes were fabricated by growing QDs directly onto a TiO_2 film according to the method of successive ionic layer adsorption and reaction (SILAR). In this work, PbS, CdS, and $Pb_xCd_{1-x}S$ QD alloys were sensitized onto TiO_2 films to produce the photoanode. In order to sensitize these QDs, several necessary precursor solutions containing different ions should be prepared: 0.1 M $Cd(NO_3)_2$ aqueous solution as a Cd^{2+} source, 0.004 M $PbCl_2$ aqueous solution as a Pb^{2+} source, and 0.1 M Na_2S solution in methanol/DI water (1:1 by volume) as a S^{2-} source. The aqueous ionic solution of mixed Pb^{2+} and Cd^{2+} made by adding 1%–4% $PbCl_2$ into 0.1 M $Cd(NO_3)_2$ solution was used as a cationic source to grow $Pb_xCd_{1-x}S$ QDs upon reacting with an anionic S^{2-} source. In this work, three kinds of $Pb_xCd_{1-x}S$ QDs, termed as PbCdS-1, PbCdS-2, and PbCdS-3, were prepared by the mixed precursor solution containing Pb^{2+} of 0.004 M, 0.002 M, and 0.001 M, respectively. The typical SILAR procedure is described by using PbS QDs as an example. The TiO_2 film was alternately immersed into the Pb^{2+} source for 2 min and the S^{2-} source for another 2 min. The films were washed thoroughly with DI water and methanol to remove excess precursor between each immersion step. Such a single procedure is denoted as one growth cycle, and repeating the cycle to deposit more PbS onto the TiO_2. The precursor solutions containing Pb^{2+} were replaced by a fresh identical solution for a further immersion process every two SILAR cycles. After QDs sensitization, the surface of the QDs was passivated by depositing a layer coating of ZnS in order to optimize the performance of solar cells. The ZnS coating was prepared by two SILAR cycles using a 0.1 M methanol solution of a $Zn(NO_3)_2$ as Zn^{2+} source.

The preparation process of Cu_2S counter electrode is similar to that in the previous report [56]. Briefly, a tailored brass foil was immersed in HCl solution for 5 min at 70 °C, cleaned by DI water, and subsequently dipped into an aqueous polysulfide solution containing 1 M S and 1 M Na_2S to generate a layer of fresh Cu_2S onto the brass substrate. The sandwich structure of QDSCs was fabricated by assembling a QD-sensitized TiO_2 film and Cu_2S counter electrode with a 50-μm Surlyn film as the sealing thermoplastic material. Polysulfide electrolyte, containing Na_2S (2 M) and S (3 M) in water/methanol (7:3 by volume), was injected to empty space between the two electrodes through a hole on the counter electrode.

3.2. Measurement and Equipment

UV-Vis spectra of the QD-deposited TiO_2 films were measured by a Lambda 750 UV-Vis spectrometer (PerkinElmer, Waltham, MA, USA). The current–voltage (I–V) characteristics of the solar cell were performed by using a Keithley source-meter under AM 1.5 illumination (100 $mW\cdot cm^{-2}$) from a Newport 300 W solar simulator (Newport Corporation, Irvine, CA, USA). The I–V measurement setup was calibrated by using an IR-filtered silicon solar cell (Fraunhofer ISE, Freiburg, Germany). Incident photon-to-current conversion efficiencies (IPCEs) were determined using a measurement system including a 300 W xenon lamp, a 1/8 m monochromator, a Keithley source-meter, and a power meter with a 818-UV detector head. IPCEs were measured by monitoring photocurrent from low to high wavelength every 10 nm. The measured IPCE at higher wavelengths was lower than the actual value because PbS QDSCs are not stable and have slow photoelectric response. The cells with strong photoelectric response at above ~750 nm exhibit a similar fluctuation in IPCE. This should be an equipment problem, possibly resulting from the xenon lamp in the IPCE system. To weaken the light refection, a black mask (6 mm × 6 mm) was used for photovoltaic measurements. The inductively coupled plasma-atomic emission spectrometry (ICP-AES) was measured by a Perkin Elmer Model

Optima 7300DV ICP AEOS (PerkinElmer, Waltham, MA, USA). Under each condition, at least three samples were prepared for the measurements in order to obtain a medium value as the final data. Cyclic voltammogram (CV) was carried out at ambient temperature with a electrochemical workstation (model 660A, CH Instruments, Austin, TX, USA) using a regular three-electrode electrochemical cell. The measured sample films were used as the working electrode, a Pt sheet as the counter electrode, and a Ag/AgCl electrode as the reference electrode. The potential conversion of reference from the Ag/AgCl electrode to normal hydrogen electrode by E (*vs.* NHE) = E (*vs.* Ag/AgCl) + 0.197 V [57].

4. Conclusions

This work was motivated by the fact that, while PbS quantum dots (QDs) have been successfully applied in heterojunction solar cells, showing remarkable conversion efficiency, corresponding PbS QD-sensitized solar cells (QDSCs) still have shown low device performance. One way to improve such QDSCs and utilize the high underlying quantum efficiency is alloying other metals to give a higher quality QDs. We have therefore fabricated ternary alloy $Pb_xCd_{1-x}S$ photosensitizers for quantum-dot-sensitized solar cells (QDSCs) by the process of successive ionic layer adsorption and reaction (SILAR). The photovoltaic performance of the QDSCs based on three kinds of $Pb_xCd_{1-x}S$ QDs were explored by the comparison with the corresponding PbS and CdS QDs. Firstly, we found that the three-SILAR-cycle PbCdS-1 ($Pb_{0.54}Cd_{0.46}S$) presents a much higher photocurrent compared to PbS and CdS. Then, by investigation of the absorption spectrum, cyclic voltammogram (CV), and dark I–V current of five-SILAR-cycle QDs, it is suggested that $Pb_xCd_{1-x}S$ QDs have a wider absorption range compared to the CdS QDs and a higher conduction band (CB) edge and reduced trap density compared to the PbS QDs. This indicates that the $Pb_xCd_{1-x}S$ QD alloys can overcome the shortcomings of CdS and PbS for QDSC applications. Furthermore, by comparing the PbCdS-1, PbCdS-2 ($Pb_{0.31}Cd_{0.69}S$), and PbCdS-3 ($Pb_{0.24}Cd_{0.76}S$) QDSCs, we found that the solar cells based on the $Pb_xCd_{1-x}S$ alloy with a higher proportion of Pb exhibit a larger photocurrent and a lower photovoltage. As a result, the PbCdS-1 solar cells present a significant short-circuit current density (J_{sc}), up to 20 mA/cm^2, by the optimization of the SILAR cycles. This indicates that the employment of $Pb_xCd_{1-x}S$ QD alloys can be a strategy to improve the photocurrent for the PbS and CdS QDSCs. Indeed, the control of the ratio of Pb:Cd in the $Pb_xCd_{1-x}S$ alloys is very critical for the photovoltaic performance of QDSCs based on $Pb_xCd_{1-x}S$ QDs. A good proportion and balance of Pb and Cd is thus required for achieving an optimum current and voltage of the solar cells. Finally, a coating layer of CdS deposited onto $Pb_xCd_{1-x}S$ photoelectrode gives enhancements in the photocurrent to 22.6 mA/cm^2 and in the efficiency to 3.2%. It is reasonable to anticipate the remarkable enhancement of QDSCs by improving QD quality with less or even no defects and developing more effective passivation materials.

Supplementary Materials
Table S1: The photovoltaic parameters of QDSCs employing three SILAR cycles of CdS, PbS, and PbCdS-1 as photosensitizers. Table S2: The concentrations of precursor solutions of cationic sources used in this work. Table S3: The concentrations of Pb and Cd in $Pb_xCd_{1-x}S$ QDs were assayed by inductively coupled plasma-atomic emission spectrometry (ICP-AES). Table S4: The photovoltaic parameters of QDSCs depending on the number of SILAR cycles, employing CdS, PbS, and $Pb_xCd_{1-x}S$ as photosensitizers; short-circuit current (J_{sc}), open-circuit voltage (V_{oc}), fill factor (*FF*), efficiency (η). Figure S1: Cyclic voltammograms of the bare FTO glass and TiO_2 electrode measured under the same condition with Figure 3. Figure S2: I–V curves of QDSCs depending on the number of SILAR cycles, employing (a) CdS, (b) PbS, (c) PbCdS-1, (d) PbCdS-2, (e) PbCdS-3 as photosensitizers.

Acknowledgments: We acknowledge support from the Swedish Science Research Council, contract nr. C0334701, the Swedish Energy Agency, K & A Wallenberg Foundation, and National Natural Science Foundation of China (21120102036, 91233201).

Author Contributions: C.Y. and L.L. conceived, designed, and performed the experiments, as well as analyzed the data and wrote the paper; Z.N. measured ICP-AES, analyzed data, and provided helpful suggestions; J.H. participated in the electrode fabrication and electrochemistry measurements; L.S. and H.Å. contributed to the discussion, the analysis of the data, and the writing of the paper.

References

1. Robel, I.; Subramanian, V.; Kuno, M.; Kamat, P.V. Quantum Dot Solar Cells. Harvesting Light Energy with CdSe Nanocrystals Molecularly Linked to Mesoscopic TiO_2 Films. *J. Am. Chem. Soc.* **2006**, *128*, 2385–2393. [CrossRef] [PubMed]
2. O'Regan, B.; Grätzel, M. A Low-Cost, High-Efficiency Solar Cell Based on Dye-Sensitized Colloidal Titanium Dioxide Films. *Nature* **1991**, *353*, 737–740. [CrossRef]
3. Mora-Sero, I.; Gimenez, S.; Fabregat-Santiago, F.; Gomez, R.; Shen, Q.; Toyoda, T.; Bisquert, J. Recombination in Quantum Dot Sensitized Solar Cells. *Acc. Chem. Res.* **2009**, *42*, 1848–1857. [CrossRef] [PubMed]
4. Hossain, M.A.; Jennings, J.R.; Shen, C.; Pan, J.H.; Koh, Z.Y.; Mathews, N.; Wang, Q. CdSe-Sensitized Mesoscopic TiO_2 Solar Cells Exhibiting >5% Efficiency: Redundancy of CdS Buffer Layer. *J. Mater. Chem.* **2012**, *22*, 16235–16242. [CrossRef]
5. Lee, Y.-L.; Lo, Y.-S. Highly Efficient Quantum-Dot-Sensitized Solar Cell Based on Co-Sensitization of CdS/CdSe. *Adv. Funct. Mater.* **2009**, *19*, 604–609. [CrossRef]
6. Yu, X.-Y.; Liao, J.-Y.; Qiu, K.-Q.; Kuang, D.-B.; Su, C.-Y. Dynamic Study of Highly Efficient CdS/CdSe Quantum Dot-Sensitized Solar Cells Fabricated by Electrodeposition. *ACS Nano* **2011**, *5*, 9494–9500. [CrossRef] [PubMed]
7. Zhang, X.; Huang, X.; Yang, Y.; Wang, S.; Gong, Y.; Luo, Y.; Li, D.; Meng, Q. Investigation on New $CuInS_2$/Carbon Composite Counter Electrodes for CdS/CdSe Cosensitized Solar Cells. *ACS Appl. Mater. Interfaces* **2013**, *5*, 5954–5960. [CrossRef] [PubMed]
8. Ellingson, R.J.; Beard, M.C.; Johnson, J.C.; Yu, P.; Micic, O.I.; Nozik, A.J.; Shabaev, A.; Efros, A.L. Highly Efficient Multiple Exciton Generation in Colloidal PbSe and PbS Quantum Dots. *Nano Lett.* **2005**, *5*, 865–871. [CrossRef] [PubMed]
9. Nozik, A.; Beard, M.; Luther, J.; Law, M. Semiconductor Quantum Dots and Quantum Dot Arrays and Applications of Multiple Exciton Generation to Third-Generation Photovoltaic Solar Cells. *Chem. Rev.* **2010**, *110*, 6873–6890. [CrossRef] [PubMed]
10. Yang, Z.; Chen, C.Y.; Roy, P.; Chang, H.T. Quantum Dot-Sensitized Solar Cells Incorporating Nanomaterials. *Chem. Commun.* **2011**, *47*, 9561–9571. [CrossRef] [PubMed]
11. McDaniel, H.; Fuke, N.; Makarov, N.S.; Pietryga, J.M.; Klimov, V.I. An Integrated Approach to Realizing High-Performance Liquid-Junction Quantum Dot Sensitized Solar Cells. *Nat. Commun.* **2013**, *4*. [CrossRef] [PubMed]
12. Wang, G.; Wei, H.; Luo, Y.; Wu, H.; Li, D.; Zhong, X.; Meng, Q. A Strategy to Boost the Cell Performance of $CdSe_xTe_{1-x}$ Quantum Dot Sensitized Solar Cells over 8% by Introducing Mn Modified CdSe Coating Layer. *J. Power Sources* **2016**, *302*, 266–273. [CrossRef]
13. Wang, J.; Li, Y.; Shen, Q.; Izuishi, T.; Pan, Z.; Zhao, K.; Zhong, X. Mn Doped Quantum Dot Sensitized Solar Cells with Power Conversion Efficiency Exceeding 9%. *J. Mater. Chem. A* **2016**, *4*, 877–886. [CrossRef]
14. Ning, Z.; Tian, H.; Yuan, C.; Fu, Y.; Sun, L.; Ågren, H. Pure Organic Redox Couple for Quantum-Dot-Sensitized Solar Cells. *Chem. Eur. J.* **2011**, *17*, 6330–6333. [CrossRef] [PubMed]
15. Lightcap, I.V.; Kamat, P.V. Fortification of CdSe Quantum Dots with Graphene Oxide. Excited State Interactions and Light Energy Conversion. *J. Am. Chem. Soc.* **2012**, *134*, 7109–7116. [CrossRef] [PubMed]
16. Farrow, B.; Kamat, P.V. CdSe Quantum Dot Sensitized Solar Cells. Shuttling Electrons through Stacked Carbon Nanocups. *J. Am. Chem. Soc.* **2009**, *131*, 11124–11131. [CrossRef] [PubMed]
17. Zhang, Q.; Chen, G.; Yang, Y.; Shen, X.; Zhang, Y.; Li, C.; Yu, R.; Luo, Y.; Li, D.; Meng, Q. Toward Highly Efficient CdS/CdSe Quantum Dots-Sensitized Solar Cells Incorporating Ordered Photoanodes on Transparent Conductive Substrates. *Phys. Chem. Chem. Phys.* **2012**, *14*, 6479–6486. [CrossRef] [PubMed]
18. Salant, A.; Shalom, M.; Tachan, Z.; Buhbut, S.; Zaban, A.; Banin, U. Quantum Rod-Sensitized Solar Cell: Nanocrystal Shape Effect on the Photovoltaic Properties. *Nano Lett.* **2012**, *12*, 2095–2100. [CrossRef] [PubMed]
19. Ning, Z.; Yuan, C.; Tian, H.; Hedstrom, P.; Sun, L.; Ågren, H. Quantum Rod-Sensitized Solar Cells. *ChemSusChem* **2011**, *4*, 1741–1744. [CrossRef] [PubMed]
20. Nazeeruddin, M.K.; Pechy, P.; Renouard, T.; Zakeeruddin, S.M.; Humphry-Baker, R.; Comte, P.; Liska, P.; Cevey, L.; Costa, E.; Shklover, V.; *et al.* Engineering of Efficient Panchromatic Sensitizers for Nanocrystalline TiO_2-Based Solar Cells. *J. Am. Chem. Soc.* **2001**, *123*, 1613–1624. [CrossRef] [PubMed]

21. Kinoshita, T.; Dy, J.T.; Uchida, S.; Kubo, T.; Segawa, H. Wideband Dye-Sensitized Solar Cells Employing a Phosphine-Coordinated Ruthenium Sensitizer. *Nat. Photonics* **2013**, *7*, 535–539. [CrossRef]
22. Yuan, C.Z.; Chen, G.Y.; Prasad, P.N.; Ohulchanskyy, T.Y.; Ning, Z.J.; Tian, H.N.; Sun, L.C.; Ågren, H. Use of Colloidal Upconversion Nanocrystals for Energy Relay Solar Cell Light Harvesting in the near-Infrared Region. *J. Mater. Chem.* **2012**, *22*, 16709–16713. [CrossRef]
23. Kim, S.; Fisher, B.; Eisler, H.-J.; Bawendi, M. Type-II Quantum Dots: CdTe/CdSe(Core/Shell) and CdSe/ZnTe(Core/Shell) Heterostructures. *J. Am. Chem. Soc.* **2003**, *125*, 11466–11467. [CrossRef] [PubMed]
24. Ning, Z.; Tian, H.; Yuan, C.; Fu, Y.; Qin, H.; Sun, L.; Ågren, H. Solar Cells Sensitized with Type-II ZnSe-CdS Core/Shell Colloidal Quantum Dots. *Chem. Commun.* **2011**, *47*, 1536–1538. [CrossRef] [PubMed]
25. Pan, Z.; Zhao, K.; Wang, J.; Zhang, H.; Feng, Y.; Zhong, X. Near Infrared Absorption of $CdSe_xTe_{1-x}$ Alloyed Quantum Dot Sensitized Solar Cells with More Than 6% Efficiency and High Stability. *ACS Nano* **2013**, *7*, 5215–5222. [CrossRef] [PubMed]
26. Tubtimtae, A.; Wu, K.-L.; Tung, H.-Y.; Lee, M.-W.; Wang, G.J. Ag_2S Quantum Dot-Sensitized Solar Cells. *Electrochem. Commun.* **2010**, *12*, 1158–1160. [CrossRef]
27. Parsi Benehkohal, N.; González-Pedro, V.; Boix, P.P.; Chavhan, S.; Tena-Zaera, R.; Demopoulos, G.P.; Mora-Seró, I. Colloidal PbS and PbSeS Quantum Dot Sensitized Solar Cells Prepared by Electrophoretic Deposition. *J. Phys. Chem. C* **2012**, *116*, 16391–16397. [CrossRef]
28. Long, R.; Prezhdo, O.V. *Ab Initio* Nonadiabatic Molecular Dynamics of the Ultrafast Electron Injection from a PbSe Quantum Dot into the TiO_2 Surface. *J. Am. Chem. Soc.* **2011**, *133*, 19240–19249. [CrossRef] [PubMed]
29. Sambur, J.B.; Novet, T.; Parkinson, B.A. Multiple Exciton Collection in a Sensitized Photovoltaic System. *Science* **2010**, *330*, 63–66. [CrossRef] [PubMed]
30. Braga, A.; Giménez, S.; Concina, I.; Vomiero, A.; Mora-Seró, I. Panchromatic Sensitized Solar Cells Based on Metal Sulfide Quantum Dots Grown Directly on Nanostructured TiO_2 electrodes. *J. Phys. Chem. Lett.* **2011**, *2*, 454–460. [CrossRef]
31. Ma, W.; Luther, J.M.; Zheng, H.; Wu, Y.; Alivisatos, A.P. Photovoltaic Devices Employing Ternary PbS_xSe_{1-x} Nanocrystals. *Nano Lett.* **2009**, *9*, 1699–1703. [CrossRef] [PubMed]
32. Lee, H.; Leventis, H.C.; Moon, S.-J.; Chen, P.; Ito, S.; Haque, S.A.; Torres, T.; Nüesch, F.; Geiger, T.; Zakeeruddin, S.M.; *et al.* PbS and CdS Quantum Dot-Sensitized Solid-State Solar Cells: "Old Concepts, New Results". *Adv. Funct. Mater.* **2009**, *19*, 2735–2742. [CrossRef]
33. Gonzalez-Pedro, V.; Sima, C.; Marzari, G.; Boix, P.P.; Gimenez, S.; Shen, Q.; Dittrich, T.; Mora-Sero, I. High Performance PbS Quantum Dot Sensitized Solar Cells Exceeding 4% Efficiency: The Role of Metal Precursors in the Electron Injection and Charge Separation. *Phys. Chem. Chem. Phys.* **2013**, *15*, 13835–13843. [CrossRef] [PubMed]
34. Lee, J.W.; Son, D.Y.; Ahn, T.K.; Shin, H.W.; Kim, I.Y.; Hwang, S.J.; Ko, M.J.; Sul, S.; Han, H.; Park, N.G. Quantum-Dot-Sensitized Solar Cell with Unprecedentedly High Photocurrent. *Sci. Rep.* **2013**, *3*. [CrossRef] [PubMed]
35. Niu, G.; Wang, L.; Gao, R.; Ma, B.; Dong, H.; Qiu, Y. Inorganic Iodide Ligands in *Ex Situ* PbS Quantum Dot Sensitized Solar Cells with I^-/I^{3-} Electrolytes. *J. Mater. Chem.* **2012**, *22*, 16914–16919. [CrossRef]
36. Ip, A.H.; Thon, S.M.; Hoogland, S.; Voznyy, O.; Zhitomirsky, D.; Debnath, R.; Levina, L.; Rollny, L.R.; Carey, G.H.; Fischer, A.; *et al.* Hybrid Passivated Colloidal Quantum Dot Solids. *Nat. Nanotechnol.* **2012**, *7*, 577–582. [CrossRef] [PubMed]
37. Tang, J.; Kemp, K.W.; Hoogland, S.; Jeong, K.S.; Liu, H.; Levina, L.; Furukawa, M.; Wang, X.; Debnath, R.; Cha, D.; *et al.* Colloidal-Quantum-Dot Photovoltaics Using Atomic-Ligand Passivation. *Nat. Mater.* **2011**, *10*, 765–771. [CrossRef] [PubMed]
38. Ning, Z.; Zhitomirsky, D.; Adinolfi, V.; Sutherland, B.; Xu, J.; Voznyy, O.; Maraghechi, P.; Lan, X.; Hoogland, S.; Ren, Y.; *et al.* Graded Doping for Enhanced Colloidal Quantum Dot Photovoltaics. *Adv. Mater.* **2013**, *25*, 1719–1723. [CrossRef] [PubMed]
39. Etgar, L.; Moehl, T.; Gabriel, S.; Hickey, S.G.; Eychmuller, A.; Grätzel, M. Light Energy Conversion by Mesoscopic PbS Quantum Dots/TiO_2 Heterojunction Solar Cells. *ACS Nano* **2012**, *6*, 3092–3099. [CrossRef] [PubMed]
40. Kim, G.H.; Garcia de Arquer, F.P.; Yoon, Y.J.; Lan, X.; Liu, M.; Voznyy, O.; Yang, Z.; Fan, F.; Ip, A.H.; Kanjanaboos, P.; *et al.* High-Efficiency Colloidal Quantum Dot Photovoltaics via Robust Self-Assembled Monolayers. *Nano Lett.* **2015**, *15*, 7691–7696. [CrossRef] [PubMed]

41. Lan, X.; Voznyy, O.; Kiani, A.; Garcia de Arquer, F.P.; Abbas, A.S.; Kim, G.H.; Liu, M.; Yang, Z.; Walters, G.; Xu, J.; *et al.* Passivation Using Molecular Halides Increases Quantum Dot Solar Cell Performance. *Adv. Mater.* **2016**, *28*, 299–304. [CrossRef] [PubMed]
42. Kim, J.; Choi, H.; Nahm, C.; Kim, C.; Ik Kim, J.; Lee, W.; Kang, S.; Lee, B.; Hwang, T.; Hejin Park, H.; *et al.* Graded Bandgap Structure for PbS/CdS/ZnS Quantum-Dot-Sensitized Solar Cells with a $Pb_xCd_{1-x}S$ Interlayer. *Appl. Phys. Lett.* **2013**, *102*. [CrossRef]
43. Santra, P.K.; Kamat, P.V. Tandem-Layered Quantum Dot Solar Cells: Tuning the Photovoltaic Response with Luminescent Ternary Cadmium Chalcogenides. *J. Am. Chem. Soc.* **2013**, *135*, 877–885. [CrossRef] [PubMed]
44. Lee, H.J.; Wang, M.; Chen, P.; Gamelin, D.R.; Zakeeruddin, S.M.; Grätzel, M.; Nazeeruddin, M.K. Efficient CdSe Quantum Dot-Sensitized Solar Cells Prepared by an Improved Successive Ionic Layer Adsorption and Reaction Process. *Nano Lett.* **2009**, *9*, 4221–4227. [CrossRef] [PubMed]
45. Shu, T.; Zhou, Z.-M.; Wang, H.; Liu, G.-H.; Xiang, P.; Rong, Y.-G.; Zhao, Y.-D.; Han, H.-W. Efficient CdPbS Quantum Dots-Sensitized TiO_2 Photoelectrodes for Solar Cell Applications. *J. Nanosci. Nanotechnol.* **2011**, *11*, 9645–9649. [CrossRef] [PubMed]
46. Singh, N.; Mehra, R.M.; Kapoor, A.; Soga, T. ZnO Based Quantum Dot Sensitized Solar Cell Using CdS Quantum Dots. *J. Renew. Sustain. Energy* **2012**, *4*. [CrossRef]
47. Gocalińska, A.; Saba, M.; Quochi, F.; Marceddu, M.; Szendrei, K.; Gao, J.; Loi, M.A.; Yarema, M.; Seyrkammer, R.; Heiss, W.; *et al.* Size-Dependent Electron Transfer from Colloidal PbS Nanocrystals to Fullerene. *J. Phys. Chem. Lett.* **2010**, *1*, 1149–1154. [CrossRef]
48. Tauc, J.; Menth, A.; Wood, D.L. Optical and Magnetic Investigations of the Localized States in Semiconducting Glasses. *Phys. Rev. Lett.* **1970**, *25*, 749–752. [CrossRef]
49. Huang, Z.; Zou, X.; Zhou, H. A Strategy to Achieve Superior Photocurrent by Cu-Doped Quantum Dot Sensitized Solar Cells. *Mater. Lett.* **2013**, *95*, 139–141. [CrossRef]
50. Haram, S.K.; Kshirsagar, A.; Gujarathi, Y.D.; Ingole, P.P.; Nene, O.A.; Markad, G.B.; Nanavati, S.P. Quantum Confinement in CdTe Quantum Dots: Investigation through Cyclic Voltammetry Supported by Density Functional Theory. *J. Phys. Chem. C* **2011**, *115*, 6243–6249. [CrossRef]
51. Lee, M.S.; Cheon, I.C.; Kim, Y.I. Photoelectrochemical Studies of Nanocrystalline TiO_2 Film Electrodes. *Bull. Korean Chem. Soc.* **2003**, *24*, 1155–1162.
52. Xu, Y.; Schoonen, M.A.A. The Absolute Energy Positions of Conduction and Valence Bands of Selected Semiconducting Minerals. *Am. Mineral.* **2000**, *85*, 543–556. [CrossRef]
53. Grätzel, M. Photoelectrochemical Cells. *Nature* **2001**, *414*, 338–344. [CrossRef] [PubMed]
54. Kamat, P.V. Quantum Dot Solar Cells. The Next Big Thing in Photovoltaics. *J. Phys. Chem. Lett.* **2013**, *4*, 908–918. [CrossRef] [PubMed]
55. Ito, S.; Murakami, T.N.; Comte, P.; Liska, P.; Grätzel, C.; Nazeeruddin, M.K.; Grätzel, M. Fabrication of Thin Film Dye Sensitized Solar Cells with Solar to Electric Power Conversion Efficiency over 10%. *Thin Solid Films* **2008**, *516*, 4613–4619. [CrossRef]
56. Gimenez, S.; Mora-Sero, I.; Macor, L.; Guijarro, N.; Lana-Villarreal, T.; Gomez, R.; Diguna, L.J.; Shen, Q.; Toyoda, T.; Bisquert, J. Improving the Performance of Colloidal Quantum-Dot-Sensitized Solar Cells. *Nanotechnology* **2009**, *20*. [CrossRef] [PubMed]
57. Esswein, A.J.; Surendranath, Y.; Reece, S.Y.; Nocera, D.G. Highly Active Cobalt Phosphate and Borate Based Oxygen Evolving Catalysts Operating in Neutral and Natural Waters. *Energy Environ. Sci.* **2011**, *4*, 499–504. [CrossRef]

2

The Influence of Fluorination on Nano-Scale Phase Separation and Photovoltaic Performance of Small Molecular/PC$_{71}$BM Blends

Zhen Lu [1], Wen Liu [1], Jingjing Li [1], Tao Fang [2], Wanning Li [1], Jicheng Zhang [2], Feng Feng [1,*] and Wenhua Li [2,*]

1 College of Chemistry and Environmental Engineering, ShanXi DaTong University, Datong 037009, China; luzhen0313@aliyun.com (Z.L.); liuwen19701021@163.com (W.L.); lijingjing8150@163.com (J.L.); liwanning2016@163.com (W.L.)

2 Beijing Key Laboratory of Energy Conversion and Storage Materials, College of Chemistry, Beijing Normal University, Beijing 100875, China; fangtao@bnu.edu.cn (T.F.); zhangjichengbnu@hotmail.com (J.Z.)

* Correspondence: feng-feng64@263.net (F.F.); liwenhua@bnu.edu.cn (W.L.)

Academic Editor: Guanying Chen

Abstract: To investigate the fluorination influence on the photovoltaic performance of small molecular based organic solar cells (OSCs), six small molecules based on 2,1,3-benzothiadiazole (BT), and diketopyrrolopyrrole (DPP) as core and fluorinated phenyl (DFP) and triphenyl amine (TPA) as different terminal units (DFP-BT-DFP, DFP-BT-TPA, TPA-BT-TPA, DFP-DPP-DFP, DFP-DPP-TPA, and TPA-DPP-TPA) were synthesized. With one or two fluorinated phenyl as the end group(s), HOMO level of BT and DPP based small molecular donors were gradually decreased, inducing high open circuit voltage for fluorinated phenyl based OSCs. DFP-BT-TPA and DFP-DPP-TPA based blend films both displayed stronger nano-scale aggregation in comparison to TPA-BT-TPA and TPA-DPP-TPA, respectively, which would also lead to higher hole motilities in devices. Ultimately, improved power conversion efficiency (PCE) of 2.17% and 1.22% was acquired for DFP-BT-TPA and DFP-DPP-TPA based devices, respectively. These results demonstrated that the nano-scale aggregation size of small molecules in photovoltaic devices could be significantly enhanced by introducing a fluorine atom at the donor unit of small molecules, which will provide understanding about the relationship of chemical structure and nano-scale phase separation in OSCs.

Keywords: small molecule; fluorinated phenyl (DFP) groups; organic solar cell; solution process; nanoscale phase separation

1. Introduction

Recently, organic solar cells (OSCs) have received great attention due to their advantages of solution processability, light weight, low cost, and flexibility [1–4]. Generally, bulk-heterojunction architecture was adopted with the electron-efficient conjugated polymer or small molecule as the donor and electron-deficient fullerene derivative such as (6,6)-phenyl-C$_{71}$-butyric acid methyl ester (PC$_{71}$BM) as the acceptor [5–7]. In comparison to widely investigated polymeric counterparts, small molecule-based OSCs have distinct advantages of well-defined chemical structures, easy purification, and high purity without batch-to-batch variation [5,6,8,9]. Therefore, small molecule-based materials are more suitable for mass production compared to polymer based ones. Driven by the developing of high efficiency small molecular donors and the investigation of nano-scale phase separation at donor/acceptor interfaces, power conversion efficiency (PCE) of nearly 10% for small molecule-based single-junction OSCs has been achieved [10,11].

Due to the strong electron-withdrawing capability, the introducing of fluorine atoms onto the conjugated backbones of small molecules could reduce the HOMO energy levels, resulting in higher open circuit voltage (V_{oc}) in OSCs [12–19]. Furthermore, the intra-molecular interaction of F–S and F–H could also induce closed packing properties and lead to superior hole mobiles [17,20–22]. Fluorinated molecules also showed good thermal and electrochemical stability, which would be helpful in the future commercial application. Although fluorinated small molecules have been broadly studied, fluorine atoms were mostly introduced to the acceptor unit of donor materials [18,19,23–25], works focus on the investigation of small molecules with a fluorine atom at the donor unit were still rare. Triphenylamine (TPA) based small molecules were common donor materials in organic semiconductor devices [8,26–30]. Nevertheless, the inferior planarity of TPA unit would induce weak intramolecular packing of small molecules and lead to a relatively low hole mobility in OSCs [31–33]. Conjugating a fluorine unit could therefore enhance the aggregation of TPA based small molecules and result in an appropriate nano-scale phase separation when blended with $PC_{71}BM$. As a result, higher hole motilities and PCE were expected.

In this contribution, six small molecules (DFP-BT-DFP, DFP-BT-TPA, TPA-BT-TPA, DFP-DPP-DFP, DFP-DPP-TPA, and TPA-DPP-TPA) based on 2,1,3-benzothiadiazole (BT) or diketopyrrolopyrrole (DPP) (Chart 1) as the core and TPA or fluorinated phenyl (DFP) groups as the flanks were synthesized and applied as the donors in OSCs [34–36]. BT or DPP groups were chosen as the acceptor units due to their uniquely planarity and remarkably electron-withdrawing capabilities, and various DFP groups were conjugated as the end groups to investigate the influence of fluorinated donor unit on the performance of small molecule based OSCs. With one or two DFP as the end group(s), the HOMO level of BT-based small molecular donors TPA-BT-TPA, DFP-BT-TPA, and DFP-BT-DFP were gradually decreased; DPP-based small molecular donors also exhibited similar tendency, which would be beneficial to achieve high V_{oc} in OSCs. Due to the inferior solubility of DFP-BT-DFP and DFP-DPP-DFP, only TPA-BT-TPA, TPA-DPP-TPA, DFP-BT-TPA, and DFP-DPP-TPA were used as donor materials in OSCs. DFP-BT-TPA and DFP-DPP-TPA based blend films both displayed stronger nano-scale aggregation in comparison to TPA-BT-TPA and TPA-DPP-TPA, respectively, leading to higher hole motilities in devices. Ultimately, a PCE of 2.17% with a V_{oc} of 0.90 V and a PCE of 1.22% with a V_{oc} of 0.78 V was acquired for DFP-BT-TPA and DFP-DPP-TPA-based devices, respectively. Our results demonstrated that the nano-scale aggregation size of small molecules in photovoltaic devices could be significantly enhanced by introducing a fluorine atom at the donor unit of small molecules, which provided useful information in the further design of high efficiency small molecular donors for OSCs.

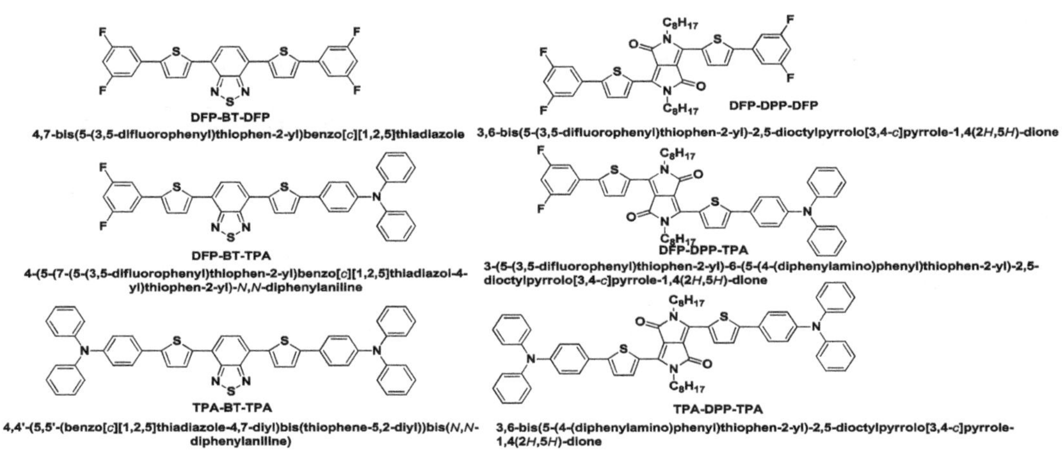

Chart 1. Chemical structures of small molecules (DFP-BT-DFP, DFP-BT-TPA, TPA-BT-TPA, DFP-DPP-DFP, DFP-DPP-TPA, and TPA-DPP-TPA). DFP: fluorinated phenyl; BT: 2,1,3-benzothiadiazole; DPP: diketopyrrolopyrrole; TPA: triphenyl amine.

2. Results and discussions

2.1. Synthesis

The synthesis route of six small molecules (DFP-BT-DFP, DFP-BT-TPA, TPA-BT-TPA, DFP-DPP-DFP, DFP-DPP-TPA, and TPA-DPP-TPA) is outlined in Scheme 1. All small molecules were synthesized by using the Suzuki cross-coupling reaction between a boronic acid ester and a brominated aromatic compound in toluene and K_2CO_3 aqueous solution under N_2 with $Pd(PPh_3)_4$ as the catalyst procedure. DFP-BT-TPA, TPA-BT-TPA, DFP-DPP-TPA, and TPA-DPP-TPA show good solubility in common organic solvents such as chloroform ($CHCl_3$), tetrahydrofuran (THF), and chlorobenzene (CB). However, DFP-BT-DFP and DFP-DPP-DFP showed poor solubility in these solvents, which might be ascribed to the fluorinated and symmetric molecular structure. As shown in Figure S1, DSC images demonstrated that all these small molecules were crystalline, which might be beneficial for the closed packing properties when blended with $PC_{71}BM$. As shown in Figure S2 and Table 1, DFP-BT-DFP, DFP-BT-TPA, TPA-BT-TPA, DFP-DPP-DFP, DFP-DPP-TPA, and TPA-DPP-TPA all exhibited good thermal stability with a 5% weight loss of 350, 450, 461, 374, 400, and 408 °C, respectively. Our results revealed TPA-based small molecules possess better thermal stability than DFP based small molecules, which is beyond our expectation. The 5% weight loss temperature of DFP-BT-TPA and TPA-BT-TPA could reach to around 450 °C, making them very stable in the application of OSCs.

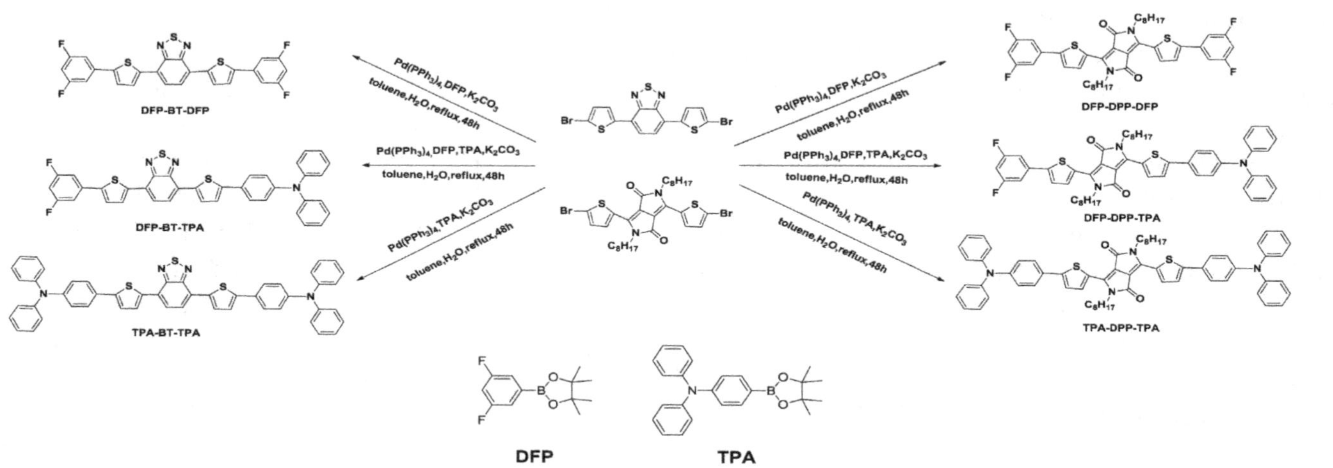

Scheme 1. Synthetic route of six small molecules (DFP-BT-DFP, DFP-BT-TPA, TPA-BT-TPA, DFP-DPP-DFP, DFP-DPP-TPA, and TPA-DPP-TPA).

Table 1. Optical, electrochemical, and physical properties of small molecules.

Donor	λ_{abs}(S) (nm) [a]	λ_{abs}(F) (nm) [b]	λ_{onset} (nm)	$E_{g,\,opt}$(ev)	HOMO [c]	LUMO [d]	T_m (°C)	T_g (°C)
DFP-BT-DFP	488	505	636	1.95	−5.73	−3.78	246.6	350
DFP-BT-TPA	525	530	643	1.93	−5.34	−3.41	248.8	450
TPA-BT-TPA	544	547	659	1.88	−5.17	−3.29	254.4	461
DFP-DPP-DFP	598	616, 663	715	1.73	−5.59	−3.86	216.9	374
DFP-DPP-TPA	619	583, 627	746	1.66	−5.13	−3.47	195.3	400
TPA-DPP-TPA	634	588, 656	760	1.63	−5.07	−3.44	217.3	408

[a] S stands for in solutions; [b] F stands for as films; [c] HOMO means highest occupied molecular orbital; [d] LUMO means lowest unoccupied molecular orbital.

2.2. Optical Properties

The normalized ultraviolet-visible spectroscopy (UV-VIS) absorption spectra of six small molecules in dilute CB solution and as thin films at 25 °C are shown in Figure 1. These small molecules exhibit a broad absorption in the range from 300 to 700 nm with two distinguishable absorption bands, which could be attributed to the π-π* transition and the internal charge transfer (ICT) interaction between the donor and the acceptor units [4,21,28]. Upon going from solution to the solid state, the absorption spectra become broader and redshift 17 nm for DFP-BT-DFP, 5 nm for DFP-BT-TPA, and 5 nm for TPA-BT-TPA, respectively, which can be attributed to the closer π-π stacking in solid state. The onsets of film absorption spectra are 636, 643, and 659 nm, respectively, for these small molecules, and optical band gaps ($E_{g,\ opt}$) were calculated to be 1.95, 1.93, and 1.88 eV, respectively. Similar to BT based small molecules, DFP-DPP-DFP, DFP-DPP-TPA, TPA-DPP-TPA show broader absorption over the range from 300 to 750 nm with two distinguishable absorption bands. Specifically, the absorption for DFP-DPP-DFP showed stronger shoulder peaks, revealing more ordered molecular stacking capabilities in the solid state [37,38]. From the absorption onsets, the optical band gaps were estimated to be 1.73, 1.66, and 1.63 eV for DFP-DPP-DFP, DFP-DPP-TPA, and TAP-DPP-TPA, respectively.

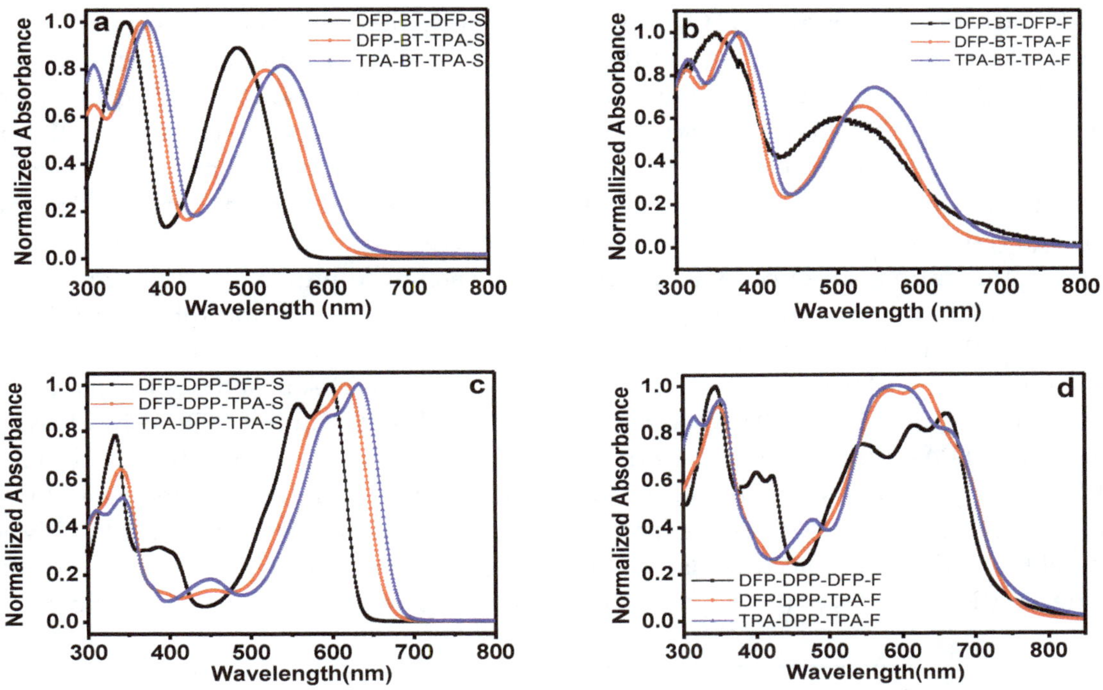

Figure 1. Normalized ultraviolet-visible spectroscopy (UV-VIS) absorption spectra of the six small molecules. (**a**,**c**) in dilute chlorobenzene (CB) solution; (**b**,**d**) as thin film on a quartz substrate.

2.3. Cyclic Voltammetry

Cyclic voltammetry (CV) experiments were carried out to evaluate electrochemical characteristics of these small molecules. As shown in Figure 2, the onset oxidation potentials (E_{ox}) were 1.02, 0.63, 0.46 V for DFP-BT-DFP, DFP-BT-TPA, and TPA-BT-TPA, respectively, According to the equation $E_{HOMO} = -(4.71 + E_{ox})$ (eV), highest occupied molecular orbital (HOMO) energy levels of these small molecules could be determined to be −5.73, −5.34, and −5.17 eV. With the combination of the optical band gap and the equation $E_{LUMO} = E_{HOMO} + E_{g,\ opt}$, [39] the lowest unoccupied molecular orbital (LUMO) energy levels were calculated to be −3.78, −3.41, and −3.29 eV for DFP-BT-DFP, DFP-BT-TPA, and TPA-BT-TPA, respectively. Similarly, the HOMO and LUMO energy levels were determined to be −5.59, −5.13, −5.07 and −3.86, −3.47, −3.44 eV for DFP-DPP-DFP, DFP-DPP-TPA, and TAP-DPP-TPA, respectively. These results demonstrated that the incorporation of fluorine atom could significantly decrease the HOMO energy levels of small molecules, which would be beneficial to achieve high V_{oc} in OSCs. The data are also summarized in Table 1.

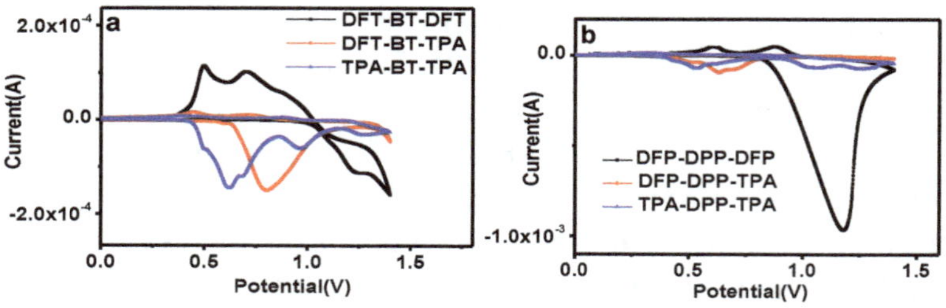

Figure 2. Cyclic voltammograms of small molecules. (**a**) DFP-BT-DFP, DFP-BT-TPA, TPA-BT-TPA; and (**b**) DFP-DPP-DFP, DFP-DPP-TPA, TPA-DPP-TPA.

2.4. Photovoltaic Properties

Photovoltaic properties of these small molecules were investigated by a general device structure of indium tin oxides (ITO)/poly(3,4-ethylenedioxythiophene):poly(styrenesulfonate) (PEDOT:PSS)/donor:$PC_{71}BM$/LiF/Al. Different spin-coating speeds and weight ratios of small molecules to $PC_{71}BM$ in a CB solution were screened to optimize the photovoltaic performance. Due to the inferior solubility of DFP-BT-DFP and DFP-DPP-DFP, typical cluster structures with boosted big domains could be easily observed by naked eyes. As a result, photovoltaic performance of these two small molecules based devices would not be discussed. For other small molecules, the current density-voltage (J-V) curves are shown in Figure 3a and device characteristics are summarized in Table 2. After optimization, superior PCE values could be achieved with a weight ratio of 1:4 (w/w) for all devices, and a PCE of 1.95% with a V_{oc} of 0.82 V, J_{sc} of 5.97 mA$\cdot$ cm^{-2} and fill factor (FF) of 0.40 was achieved for TPA-BT-TPA devices. When one TPA unit was replaced by one DFP group, increased V_{oc} of 0.90 V and J_{sc} of 6.12 mA$\cdot$ cm^{-2} were obtained for DFP-BT-TPA based device, leading to higher PCE of 2.17%. Similarly, DFP-DPP-TPA based solar cells exhibited a higher PCE of 1.22% with a V_{oc} of 0.78 V than that of TPA-DPP-TPA based devices. This improvement could be attributed to the increase of V_{oc} and J_{sc}. Since V_{oc} is related to the offset between HOMO level of donor materials and LUMO level of $PC_{71}BM$, the decrease of HOMO levels for DFP based small molecules would lead to higher V_{oc} [3]. Meanwhile, DFP based small molecules also possesses bigger nano-scale aggregates, and resulting higher hole mobilities in the active layer (*vide infra*), which would also lead to higher J_{sc} in devices.

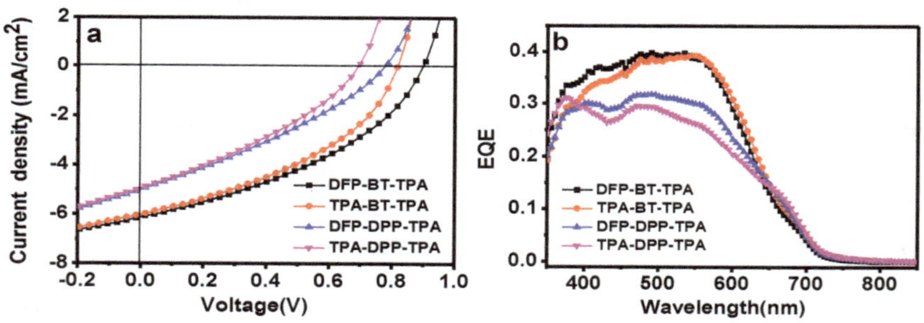

Figure 3. J-V characteristics (**a**) and external quantum efficiencie (EQE) curves (**b**) of devices fabricated from the blend of small molecule: $PC_{71}BM$.

Table 2. Photovoltaic performance and hole mobilities of small molecule-based devices. FF: fill factor; PCE: power conversion efficiency; SCLC: space charge limited current.

Donor Molecule	V_{oc} (V)	J_{sc} (mA$\cdot$ cm^{-2})	FF	PCE (max/ave) [a] (%)	Thickness (nm)	SCLC (cm$^2\cdot$ V$^{-1}\cdot$ s^{-1})
DFP-BT-TPA	0.90	6.12	0.39	2.17/2.08	76	3.81×10^{-4}
TPA-BT-TPA	0.82	5.97	0.40	1.95/1.85	82	4.42×10^{-5}
DFP-DPP-TPA	0.78	4.95	0.32	1.22/1.17	70	9.35×10^{-5}
TPA-DPP-TPA	0.70	4.90	0.34	1.15/1.02	85	3.26×10^{-5}

[a] Average value recorded over 20 devices.

To verify the J_{sc} obtained from *J-V* measurement, the external quantum efficiencies (EQEs) of devices were measured. As shown in Figure 3b, the EQE images of devices all displayed a broad response in the range from 350 to 600 nm, and the maximum EQE value of DFP-BT-TPA and TPA-BT-TPA-based devices are both 40%, which are higher than that of DFP-DPP-TPA and TPA-DPP-TPA. J_{sc} values integrated from EQE curves all agreed approximately with that obtained from *J-V* measurement.

The hole mobilities of these devices were measured by the space charge limited current (SCLC) method. Hole-only devices were fabricated with a structure of ITO/PEDOT:PSS/donor:$PC_{71}BM$/Au and resulting hole mobility value (μ) was calculated from the dark *J-V* experiments. Dark *J-V* curves were fitted by using the Mott–Gurney equation: $J = 9\varepsilon_o\varepsilon_r\mu V^2/8L^3$, where J is the space charge limited current, ε_o is the vacuum permittivity, ε_r is the permittivity of the active layer, μ is the hole mobility of small molecules, and L is the thickness of the active layer. μ of devices based on DFP-BT-TPA and DFP-DPP-TPA were evaluated to be 3.81×10^{-4} and 9.35×10^{-5} $cm^2 \cdot V^{-1} \cdot s^{-1}$, respectively, and μ of TPA-BT-TPA and TPA-DPP-TPA based devices were determined to be 4.42×10^{-5} and 3.26×10^{-5} $cm^2 \cdot V^{-1} \cdot s^{-1}$, respectively. μ of DFP based devices were obviously higher than that of TPA based ones, which might be attributed to their planar chemical structure and resulting bigger nano-scale aggregation in the blend films.

2.5. Film Morphology

Atomic force microscopy (AFM) experiments were carried out to investigate the surface morphology of the active layer. Well-mixed blend films were in favor of exciton dissociation in devices, whereas appropriate nano-scale aggregation of small molecules was also important for the charge transport. Therefore, suitable nano-scale phase separation is critical to obtain a superior photovoltaic performance [40,41]. As shown in Figure 4, blend films based on TPA-BT-TPA and TPA-DPP-TPA were both uniform and smooth with a root-mean-square (RMS) roughness value of 0.78 and 0.65 nm, which might be ascribed to the inferior packing properties of TPA group. For devices based on DFP-BT-TPA and DFP-DPP-TPA, rougher blend film surface with a RMS roughness value of 2.35 nm and 0.84 nm were acquired, demonstrating the domain size of DFP-BT-TPA and DFP-DPP-TPA were higher than TPA-BT-TPA and TPA-DPP-TPA in devices. Our results demonstrated that incorporating a DFT group as the end groups could enhance the nano-scale aggregation in the devices, which would be beneficial to the charge transport in the blend films, and lead to higher J_{sc} and PCE.

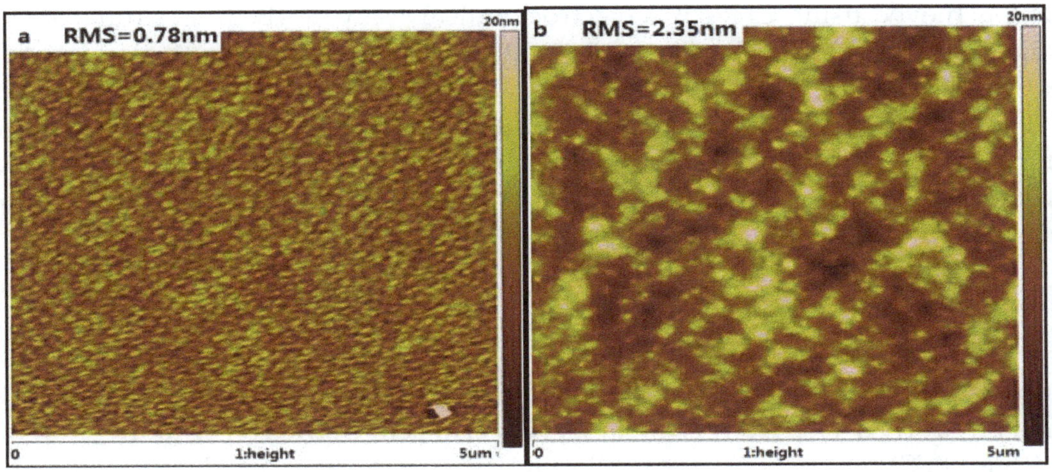

Figure 4. *Cont.*

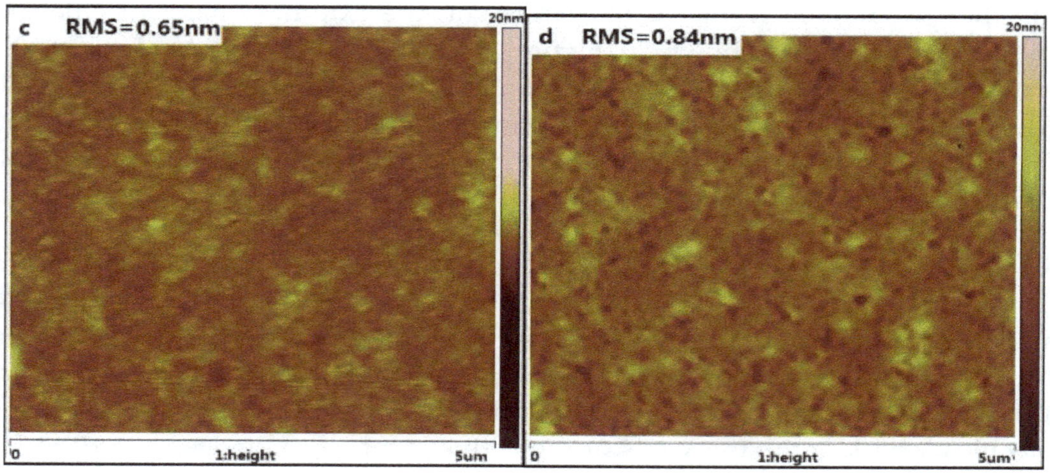

Figure 4. Atomic force microscopy (AFM) (5 × 5 μm^2) height images of blend films. (**a**) TPA-BT-TPA:$PC_{71}BM$; (**b**) DFP-BT-TPA:$PC_{71}BM$; (**c**) TPA-DPP-TPA:$PC_{71}BM$; and (**d**) DFP-DPP-TPA:$PC_{71}BM$.

3. Experimental Section

3.1. Materials

All chemicals were purchased from commercial sources and used without further purification, and all solvents were purified by standard techniques. Additionally, all reactions were carried out under the nitrogen atmosphere unless otherwise stated. The catalyst procedure $Pd(PPh_3)_4$ [34], 4,7-bis(5-bromothiophen-2-yl)benzo[c][1,2,5]thiadiazole [35] and 3,6-bis(5-bromothiophen-2-yl)-2,5-dioctylpyrrolo[3,4-c]pyrrole-1,4-dione [36] were synthesized according to the literature procedures.

3.2. Measurements and Characterization

All reactions were monitored by thin layer chromatography (TLC) on silica gel 60 F254 (Merck, Beijing, China, 0.2 mm), and column chromatography was performed on silica gel (200~300 mesh). ^{1}H and ^{13}C NMR spectra were recorded on a Bruker AV 400 spectrometer (Saarbrücken, Germany) using $CDCl_3$ as the solvents. Differential scanning calorimetry (DSC) measurements were performed on the Perkin-Elmer Diamond DSC instrument (Shanghai, China) under a nitrogen atmosphere with a heating rate of 20 °C/min. UV-VIS absorption spectra was recorded on the PerkinElmer UV-VIS spectrometer model Lambda 750 (Shanghai, China). The electrochemical behavior of small molecules was achieved using CHI 630A electrochemical analyzer (Beijing, China) with a standard three-electrode electrochemical cell in 0.1 M Bu_4NPF_6 in CH_3CN solution. A glassy carbon working electrode, a Pt wire counter electrode, and an $Ag/AgNO_3$ (0.01 M in CH_3CN) reference electrode were used. Elemental analyses were performed on a Flash EA 1112 analyzer (Lakewood, CO, USA), and atomic force microscopy (AFM) measurements were carried out in the tapping mode using a Digital Instrument Multimode Nanoscope IIIA (Plainview, TX, USA). The thickness of the active layers was determined by the Dektak 6 M surface profilometer (Beijing, China).

3.3. Fabrication and Characterization of SCLC

Devices used to measure the space charge limited current (SCLC) were fabricated with a configuration of ITO/PEDOT:PSS/small molecule:$PC_{71}BM$/Au. The conductivity of ITO was 20 Ω. PEDOT:PSS is Baytron Al 4083 from H.C.Starck (Leverkusen, Germany) and filtered with a 0.45 μm polyethersulfone (PES) film before use. The PEDOT:PSS was spin-coated on top of a ITO substrate from a speed of 3000 rpm/s for 60 s and dried at 130 °C for 15 min on a hotplate. The thickness of the PEDOT:PSS layer was about 40 nm. Small molecule and $PC_{71}BM$ was dissolved in chlorobenzene

(CB) at 110 °C overnight, and then spin-coated onto the PEDOT:PSS layer. The top electrode was subsequently thermally evaporated with a 100 nm of gold under a pressure of 10^{-4} Pa by a shadow mask. Dark current-voltage characteristics were recorded using an Agilent B2902A Source (Taiwan) under darkness in a range of 0 V to 5.0 V.

3.4. Fabrication and Characterization of PSCs

PSCs were fabricated with the device configuration of ITO/PEDOT:PSS/small molecule:$PC_{71}BM$/LiF/Al. ITO glasses with a conductivity of 20 Ω were cleaned before use. PEDOT:PSS is Baytron Al 4083 from H.C.Starck, and filtered with a 0.45 μm PES film before use. The thin layer of PEDOT:PSS was spin-coated on top of a cleaned ITO substrate at 3000 rpm/s for 60 s and dried subsequently at 130 °C for 15 min on a hotplate. The thickness of the PEDOT:PSS layer was about 40 nm. The active layer was subsequently prepared by spin-coating the blend solution of small molecule and $PC_{71}BM$ on the top of ITO/PEDOT:PSS. The top electrode was thermally evaporated with a 0.6 nm LiF layer, and followed by 100 nm of aluminum under a pressure of 10^{-4} Pa through a shadow mask. Five cells were fabricated on one substrate with an effective area of 0.04 cm^2. The current-voltage (*I-V*) curves of devices were completed by a computer-controlled Keithley 236 Source Measure Unit. An AM 1.5G AAA class solar simulator (model XES-301S, SAN-EI) in an intensity of 100 mW/cm^2 was used as the simulative sun light source, and the light intensity of the light was calibrated under a standard single-crystal Si photovoltaic cell.

3.5. General Procedure for the Synthesis of Small Molecules

The general procedure for the synthesis were shown in Scheme 1, A solution of BT or DPP, DFP or TPA, K_2CO_3, toluene (50 mL), and H_2O (5 mL) was carefully gas and degassed before and after $Pd(PPh_3)_4$ was added, and the mixture was stirring at 100 °C for two days. After the reaction, the mixture was poured into water (100 mL) and extracted with $CHCl_3$. The organic layer was washed with water three times, and then dried by $MgSO_4$. After evaporating to dryness, the pure product was obtained from column chromatography by a silica gel.

4,7-bis(5-(3,5-difluorophenyl)thiophen-2-yl)benzo[c][1,2,5]thiadiazole (DFP-BT-DFP): BT (0.96 g, 2.1 mmol), DFP (1.01 g, 4.2 mmol), K_2CO_3 (1.38 g, 10.0 mmol), and $Pd(PPh_3)_4$ (127.1 mg, 0.11 mmol) were used. Yield: 0.73 g, 66% purple crystal. ^{1}H NMR (400 MHz, $CDCl_3$) δ. 8.12 (d, 1H), 7.94 (s, 1H), 7.44 (d, 1H) 7.52 (m, 2H), 6.77 (m, 1H). ^{13}C NMR (100 MHz, $CDCl_3$) δ. 155.61, 151.55, 148.01, 143.27, 142.00, 137.55, 131.73, 120.63, 116.91. Analyze (Anal.) Calcd for $C_{26}H_{12}F_4N_2S_3$: C, 59.53; H, 2.31; N, 5.34; Found: C, 59.42; H, 2.40; N, 5.26. MALDI-TOF: calculated for $C_{26}H_{12}F_4N_2S_3$ 524.6, found 523.8.

4-(5-(7-(5-(3,5-difluorophenyl)thiophen-2-yl)benzo[c][1,2,5]thiadiazol-4-yl)thiophen-2-yl)-*N*,*N*-diphenylaniline (DFP-BT-TPA): BT (1.15 g, 2.5 mmol), DFP (0.60 g, 2.5 mmol), TPA (0.93 g, 2.5 mmol), K_2CO_3 (1.38 g, 10.0 mmol), and $Pd(PPh_3)_4$ (144.3 mg, 0.11 mmol) were used. Yield: 0.67 g, 41% purple crystal. ^{1}H NMR (400 MHz, $CDCl_3$) δ. 8.14 (d, 1H), 8.08 (d, 1H), 7.90 (t, 2H) 7.57 (d, 2H), 7.43 (d, 1H), 7.34 (d, 1H), 7.29 (t, 3H), 7.22 (d, 2H), 7.16–7.00 (m, 8H), 6.75 (t, 1H). ^{13}C NMR (100 MHz, $CDCl_3$) δ. 161.66, 158.56, 153.96, 145.33, 143.04, 139.03, 132.72, 129.49, 127.99, 124.80, 123.41, 123.32, 115.51. Anal. Calcd for $C_{38}H_{23}F_2N_3S_3$: C, 69.60; H, 3.54; N, 6.41 Found: C, 69.45; H, 3.52; N, 6.40. MALDI-TOF: calculated for $C_{38}H_{23}F_2N_3S_3$ 655.8, found 654.9.

4,4′-(5,5′-(benzo[c][1,2,5]thiadiazole-4,7-diyl)bis(thiophene-5,2-diyl))bis(*N*,*N*-diphenylaniline) (TPA-BT-TPA): BT (1.01 g, 2.2 mmol), TPA (1.63 g, 4.4 mmol), K_2CO_3 (1.22 g, 8.8 mmol), and $Pd(PPh_3)_4$ (127.1 mg, 0.11 mmol) were used. Yield: 1.30 g, 75% purple crystal. ^{1}H NMR (400 MHz, $CDCl_3$) δ. 8.11 (s, 2H), 7.85 (s, 2H), 7.56 (m, 4H), 7.29 (m, 10H), 7.08 (m, 16H). ^{13}C NMR (100 MHz, $CDCl_3$) δ. 152.57, 147.58, 147.39, 129.35, 128.67, 128.03, 126.60, 125.61, 125.14, 124.64, 123.46, 123.23. Anal. Calcd for $C_{50}H_{34}N_4S_3$: C, 76.30; H, 4.35; N, 7.12. Found: C, 76.17; H, 4.36; N, 7.02. MALDI-TOF: calculated for $C_{50}H_{34}N_4S_3$ 787.0, found 785.9.

3,6-bis(5-(3,5-difluorophenyl)thiophen-2-yl)-2,5-dioctylpyrrolo[3,4-c]pyrrole-1,4(2H,5H)-dione(DFP-DPP-DFP): DPP (1.37 g, 2.0 mmol) and DFP (0.96 g, 4.0 mmol), K_2CO_3 (1.11 g, 8.0 mmol), and $Pd(PPh_3)_4$ (115.5 mg, 0.11 mmol) were used. Yield: 0.91 g, 66% gray solid. ^1H NMR (400 MHz, $CDCl_3$) δ. 8.94 (d, 2H), 7.48 (d, 2H), 7.19 (d, 4H), 6.82 (t, 2H), 4.11 (t, 4H), 1.77 (t, 4H), 1.46–1.28 (m, 20H), 0.86 (t, 6H) ^{13}C NMR (100 MHz, $CDCl_3$) δ. 161.26, 155.77, 155.15, 149.36, 146.58, 139.32, 136.46, 129.90, 125.89, 109.15, 108.85, 42.32, 31.79, 30.01, 29.28, 29.21, 26.90, 22.63, 14.07. Anal. Calcd for $C_{42}H_{44}F_4N_2O_2S_2$: C, 67.36; H, 5.92; N, 3.74. Found: C, 67.12; H, 5.92; N, 3.73. MALDI-TOF: calculated for $C_{42}H_{44}F_4N_2O_2S_2$ 748.9, found 748.1.

3-(5-(3,5-difluorophenyl)thiophen-2-yl)-6-(5-(4-(diphenylamino)phenyl)thiophen-2-yl)-2,5-dioctylpyrrolo pyrrole-1,4(2H,5H)-dione(DFP-DPP-TPA): DPP (1.91 g, 2.8 mmol), DFP (0.67 g, 2.8 mmol) and TPA (1.04 g, 2.8 mmol), K_2CO_3 (3.33 g, 14.0 mmol), and $Pd(PPh_3)_4$ (161.7 mg, 0.14 mmol) were used. Yield: 0.76 g, 31% gray solid. ^1H NMR (400 MHz, $CDCl_3$) δ. 9.04 (s, 1H), 8.87 (s, 1H), 7.52 (s, 2H), 7.44 (s, 1H), 7.36 (s, 1H), 7.30 (t, 4H), 7.14 (d, 6H), 7.08 (dd, 4H), 6.78 (s, 1H), 4.09 (t, 4H), 1.76 (t, 4H), 1.45–1.26 (m, 20H), 0.85 (t, 6H). ^{13}C NMR (100 MHz, $CDCl_3$) δ. 148.74, 147.02, 137.71, 129.47, 126.99, 126.30, 125.09, 123.80, 123.64, 122.58, 109.05, 108.77, 42.32, 31.82, 30.04, 29.27, 29.22, 26.94, 22.64, 14.12. Anal. Calcd for $C_{55}H_{59}F_2N_3O_2S_2$: C, 73.71; H, 6.64; N, 4.69. Found: C, 73.25; H, 6.35; N, 4.73. MALDI-TOF: calculated for $C_{55}H_{59}F_2N_3O_2S_2$ 896.2, found 895.3.

3,6-bis(5-(4-(diphenylamino)phenyl)thiophen-2-yl)-2,5-dioctylpyrrolo[3,4-c]pyrrole-1,4(2H,5H)-dione (TPA-DPP-TPA): DPP (1.30 g, 1.9 mmol) and TPA (1.41 g, 3.8 mmol), K_2CO_3 (1.31 g, 9.5 mmol), and $Pd(PPh_3)_4$ (115.5 mg, 0.10 mmol) were used. Yield: 1.35 g, 70% gray solid. ^1H NMR (400 MHz, $CDCl_3$) δ. 7.30 (t, 15H), 7.10 (d, 17H), 4.10 (t, 4H), 1.80 (t, 4H), 1.46–1.26 (m, 20H), 0.84 (t, 6H). ^{13}C NMR (100 MHz, $CDCl_3$) δ. 162.54, 160.20, 155.76, 148.66, 147.40, 132.11, 129.44, 126.93, 124.82, 123.60, 123.69, 122.72, 110.77, 42.28, 31.81, 30.03, 29.24, 29.22, 26.93, 22.63, 14.10. Anal. Calcd for $C_{66}H_{66}N_4O_2S_2$: C, 78.38; H, 6.58; N, 5.54. Found: C, 78.17; H, 6.61; N, 5.38. MALDI-TOF: calculated for $C_{66}H_{66}N_4O_2S_2$ 1011.4, found 1010.1.

4. Conclusions

In conclusion, six small molecules (DFP-BT-DFP, DFP-BT-TPA, TPA-BT-TPA, DFP-DPP-DFP, DFP-DPP-TPA, and TPA-DPP-TPA) based on 2,1,3-benzothiadiazole (BT) or diketopyrrolopyrrole (DPP) as the core and TPA or fluorinated phenyl (DFP) as the end groups were designed and synthesized as the donors in OSCs. To investigate the influence of fluorinated donor unit on the photovoltaic performance of devices, various DFP groups were conjugated. With one or two DFP as the end group(s), HOMO level of TPA-BT-TPA, DFP-BT-TPA, and DFP-BT-DFP was gradually decreased; similar tendency could also be observed for DPP based small molecular donors, inducing high V_{oc} for DFP based OSCs. DFP-BT-TPA and DFP-DPP-TPA based blend films both displayed stronger nano-scale aggregation in comparison to TPA-BT-TPA and TPA-DPP-TPA, respectively, which would also lead to higher hole motilities in devices. Ultimately, a PCE of 2.17% with a V_{oc} of 0.90 V was acquired for DFP-BT-TPA based devices. Our results demonstrated that the nano-scale aggregation size of small molecules in photovoltaic devices could be significantly enhanced by introducing a fluorine atom at the donor unit of small molecules. Although the PCE is not attractive enough, this work will provide understanding about the relationship of chemical structure and nano-scale phase separation in OSCs.

Acknowledgments: Financial support from PhD Research Startup Foundation of Shanxi Datong University (2013-B-01), School foundation of Shanxi Datong University (XDC2014110), the NSF of China (91233205), the NSF of China (51403021), the NSF of China (21375083) and Beijing Natural Science Foundation (2152017).

Author Contributions: The experimental design was planned by Feng Feng and Wenhua Li. The experimental work and data analysis were performed by Zhen Lu, Wen Liu, Jingjing Li, Tao Fang, Wannin Li and Jicheng Zhang.

References

1. Liang, Y.; Xu, Z.; Xia, J.; Tsai, S.-T.; Wu, Y.; Li, G.; Ray, C.; Yu, L. For the Bright Future—Bulk Heterojunction Polymer Solar Cells with Power Conversion Efficiency of 7.4%. *Adv. Mater.* **2010**, *22*, E135–E138. [CrossRef] [PubMed]
2. Lu, L.; Zheng, T.; Wu, Q.; Schneider, A.M.; Zhao, D.; Yu, L. Recent Advances in Bulk Heterojunction Polymer Solar Cells. *Chem. Rev.* **2015**, *115*, 12666–12731. [CrossRef] [PubMed]
3. Chen, J.; Cao, Y. Development of Novel Conjugated Donor Polymers for High-Efficiency Bulk-Heterojunction Photovoltaic Devices. *Acc. Chem. Res.* **2009**, *42*, 1709–1718. [CrossRef] [PubMed]
4. Li, Y. Molecular Design of Photovoltaic Materials for Polymer Solar Cells: Toward Suitable Electronic Energy Levels and Broad Absorption. *Acc. Chem. Res.* **2012**, *45*, 723–733. [CrossRef] [PubMed]
5. Lin, Y.; Li, Y.; Zhan, X. Small molecule semiconductors for high-efficiency organic photovoltaics. *Chem. Soc. Rev.* **2012**, *41*, 4245–4272. [CrossRef] [PubMed]
6. Cui, C.; Guo, X.; Min, J.; Guo, B.; Cheng, X.; Zhang, M.; Brabec, C.J.; Li, Y. High-Performance Organic Solar Cells Based on a Small Molecule with Alkylthio-Thienyl-Conjugated Side Chains without Extra Treatments. *Adv. Mater. (Deerfield Beach Fla.)* **2015**, *27*, 7469–7475. [CrossRef] [PubMed]
7. Liu, C.; Yi, C.; Wang, K.; Yang, Y.L.; Bhatta, R.S.; Tsige, M.; Xiao, S.Y.; Gong, X. Single-Junction Polymer Solar Cells with Over 10% Efficiency by a Novel Two-Dimensional Donor-Acceptor Conjugated Copolymer. *ACS Appl. Mater. Interfaces* **2015**, *7*, 4928–4935. [CrossRef] [PubMed]
8. Shang, H.; Fan, H.; Liu, Y.; Hu, W.; Li, Y.; Zhan, X. A Solution-Processable Star-Shaped Molecule for High-Performance Organic Solar Cells. *Adv. Mater.* **2011**, *23*, 1554–1557. [CrossRef] [PubMed]
9. Sun, Y.; Welch, G.C.; Leong, W.L.; Takacs, C.J.; Bazan, G.C.; Heeger, A.J. Solution-processed small-molecule solar cells with 6.7% efficiency. *Nat. Mater.* **2012**, *11*, 44–48. [CrossRef] [PubMed]
10. Zhang, Q.; Kan, B.; Liu, F.; Long, G.; Wan, X.; Chen, X.; Zuo, Y.; Ni, W.; Zhang, H.; Li, M.; *et al.* Small-molecule solar cells with efficiency over 9%. *Nat. Photonics* **2015**, *9*, 35–41. [CrossRef]
11. Kan, B.; Zhang, Q.; Li, M.; Wan, X.; Ni, W.; Long, G.; Wang, Y.; Yang, X.; Feng, H.; Chen, Y. Solution-Processed Organic Solar Cells Based on Dialkylthiol-Substituted Benzodithiophene Unit with Efficiency near 10%. *J. Am. Chem. Soc.* **2014**, *136*, 15529–15532. [CrossRef] [PubMed]
12. Ni, W.; Li, M.; Kan, B.; Zuo, Y.; Zhang, Q.; Long, G.; Feng, H.; Wan, X.; Chen, Y. Open-circuit voltage up to 1.07 V for solution processed small molecule based organic solar cells. *Organic Electron.* **2014**, *15*, 2285–2294. [CrossRef]
13. Carsten, B.; Szarko, J.M.; Son, H.J.; Wang, W.; Lu, L.; He, F.; Rolczynski, B.S.; Lou, S.J.; Chen, L.X.; Yu, L. Examining the Effect of the Dipole Moment on Charge Separation in Donor-Acceptor Polymers for Organic Photovoltaic Applications. *J. Am. Chem. Soc.* **2011**, *133*, 20468–20475. [CrossRef] [PubMed]
14. Chen, H.Z.; Ling, M.M.; Mo, X.; Shi, M.M.; Wang, M.; Bao, Z. Air stable *n*-channel organic semiconductors for thin film transistors based on fluorinated derivatives of perylene diimides. *Chem. Mater.* **2007**, *19*, 816–824. [CrossRef]
15. Li, K.; Li, Z.; Feng, K.; Xu, X.; Wang, L.; Peng, Q. Development of Large Band-Gap Conjugated Copolymers for Efficient Regular Single and Tandem Organic Solar Cells. *J. Am. Chem. Soc.* **2013**, *135*, 13549–13557. [CrossRef] [PubMed]
16. Schroeder, B.C.; Huang, Z.; Ashraf, R.S.; Smith, J.; D'Angelo, P.; Watkins, S.E.; Anthopoulos, T.D.; Durrant, J.R.; McCulloch, I. Silaindacenodithiophene-Based Low Band Gap Polymers—The Effect of Fluorine Substitution on Device Performances and Film Morphologies. *Adv. Funct. Mater.* **2012**, *22*, 1663–1670. [CrossRef]
17. Shewmon, N.T.; Watkins, D.L.; Galindo, J.F.; Zerdan, R.B.; Chen, J.; Keum, J.; Roitberg, A.E.; Xue, J.; Castellano, R.K. Enhancement in Organic Photovoltaic Efficiency through the Synergistic Interplay of Molecular Donor Hydrogen Bonding and π-Stacking. *Adv. Funct. Mater.* **2015**, *25*, 5166–5177. [CrossRef]
18. Wang, L.; Yin, L.; Ji, C.; Li, Y. Tuning the photovoltaic performance of BT-TPA chromophore based solution-processed solar cells through molecular design incorporating of bithiophene unit and fluorine-substitution. *Dyes Pigments* **2015**, *118*, 37–44. [CrossRef]
19. Sonar, P.; Ng, G.-M.; Lin, T.T.; Dodabalapur, A.; Chen, Z.-K. Solution processable low bandgap diketopyrrolopyrrole (DPP) based derivatives: Novel acceptors for organic solar cells. *J. Mater. Chem.* **2010**, *20*, 3626–3636. [CrossRef]
20. Cui, R.; Fan, L.; Yuan, J.; Jiang, L.; Chen, G.; Ding, Y.; Shen, P.; Li, Y.; Zou, Y. Effect of fluorination on the performance of poly(thieno[2,3-*f*]benzofuran-*co*-benzothiadiazole) derivatives. *RSC Adv.* **2015**, *5*, 30145–30152. [CrossRef]

21. Guo, S.; Ning, J.; Koerstgens, V.; Yao, Y.; Herzig, E.M.; Roth, S.V.; Mueller-Buschbaum, P. The Effect of Fluorination in Manipulating the Nanomorphology in PTB7:$PC_{71}BM$ Bulk Heterojunction Systems. *Adv. Energy Mater.* **2015**, *5*. [CrossRef]
22. Jo, J.W.; Jung, J.W.; Jung, E.H.; Ahn, H.; Shin, T.J.; Jo, W.H. Fluorination on both D and A units in D-A type conjugated copolymers based on difluorobithiophene and benzothiadiazole for highly efficient polymer solar cells. *Energy Environ. Sci.* **2015**, *8*, 2427–2434. [CrossRef]
23. Kim, H.G.; Kang, B.; Ko, H.; Lee, J.; Shin, J.; Cho, K. Synthetic Tailoring of Solid-State Order in Diketopyrrolopyrrole-Based Copolymers via Intramolecular Noncovalent Interactions. *Chem. Mater.* **2015**, *27*, 829–838. [CrossRef]
24. Wang, J.-L.; Wu, Z.; Miao, J.-S.; Liu, K.-K.; Chang, Z.-F.; Zhang, R.-B.; Wu, H.-B.; Cao, Y. Solution-Processed Diketopyrrolopyrrole-Containing Small-Molecule Organic Solar Cells with 7.0% Efficiency: In-Depth Investigation on the Effects of Structure Modification and Solvent Vapor Annealing. *Chem. Mater.* **2015**, *27*, 4338–4348. [CrossRef]
25. Cho, A.; Kim, Y.; Song, C.E.; Moon, S.-J.; Lim, E. Synthesis and Characterization of Fluorinated Benzothiadiazole-Based Small Molecules for Organic Solar Cells. *Sci. Adv. Mater.* **2014**, *6*, 2411–2415. [CrossRef]
26. Paek, S.; Cho, N.; Song, K.; Jun, M.J.; Lee, J.K.; Ko, J. Efficient Organic Semiconductors Containing Fluorine-Substituted Benzothiadiazole for Solution-Processed Small Molecule Organic Solar Cells. *J. Phys. Chem. C* **2012**, *116*, 23205–23213. [CrossRef]
27. Cho, N.; Song, K.; Lee, J.K.; Ko, J. Facile Synthesis of Fluorine-Substituted Benzothiadiazole-Based Organic Semiconductors and Their Use in Solution-Processed Small-Molecule Organic Solar Cells. *Chem. Eur. J.* **2012**, *18*, 11433–11439. [CrossRef] [PubMed]
28. Dutta, P.; Yang, W.; Eom, S.H.; Lee, S.-H. Synthesis and characterization of triphenylamine flanked thiazole-based small molecules for high performance solution processed organic solar cells. *Organic Electron.* **2012**, *13*, 273–282. [CrossRef]
29. Li, Z.; Dong, Q.; Li, Y.; Xu, B.; Deng, M.; Pei, J.; Zhang, J.; Chen, F.; Wen, S.; Gao, Y.; Tian, W. Design and synthesis of solution processable small molecules towards high photovoltaic performance. *J. Mater. Chem.* **2011**, *21*, 2159–2168. [CrossRef]
30. Lin, Y.; Cheng, P.; Li, Y.; Zhan, X. A 3D star-shaped non-fullerene acceptor for solution-processed organic solar cells with a high open-circuit voltage of 1.18 V. *Chem. Commun.* **2012**, *48*, 4773–4775. [CrossRef] [PubMed]
31. Vijay Kumar, C.; Cabau, L.; Koukaras, E.N.; Sharma, G.D.; Palomares, E. Efficient solution processed D_1-A-D_2-A-D_1 small molecules bulk heterojunction solar cells based on alkoxy triphenylamine and benzo[1,2-*b*:4,5-*b*′]thiophene units. *Organic Electron.* **2015**, *26*, 36–47. [CrossRef]
32. Patil, H.; Chang, J.; Gupta, A.; Bilic, A.; Wu, J.; Sonar, P.; Bhosale, S.V. Isoindigo-Based Small Molecules with Varied Donor Components for Solution-Processable Organic Field Effect Transistor Devices. *Molecules* **2015**, *20*, 17362–17377. [CrossRef] [PubMed]
33. Bagde, S.S.; Park, H.; Yang, S.-N.; Jin, S.-H.; Lee, S.-H. Diketopyrrolopyrrole-based narrow band gap donors for efficient solution-processed organic solar cells. *Chem. Phys. Lett.* **2015**, *630*, 37–43. [CrossRef]
34. Mikroyannidis, J.A.; Stylianakis, M.M.; Suresh, P.; Balraju, P.; Sharma, G.D. Low band gap vinylene compounds with triphenylamine and benzothiadiazole segments for use in photovoltaic cells. *Organic Electron.* **2009**, *10*, 1320–1333. [CrossRef]
35. Tolman, C.A.; Seidel, W.C.; Gerlach, D.H. Triarylphosphine and ethylene complexes of zerovalent nickel, palladium, and platinum. *J. Am. Chem. Soc.* **1972**, *94*, 2669–2676. [CrossRef]
36. Kato, S.-I.; Matsumoto, T.; Ishi-i, T.; Thiemann, T.; Shigeiwa, M.; Gorohmaru, H.; Maeda, S.; Yamashita, Y.; Mataka, S. Strongly red-fluorescent novel donor-π-bridge-acceptor-π-bridge-donor (D-π-A-π-D) type 2,1,3-benzothiadiazoles with enhanced two-photon absorption cross-sections. *Chem. Commun.* **2004**, *20*, 2342–2343. [CrossRef] [PubMed]
37. Zou, Y.; Gendron, D.; Badrou-Aïch, R.; Najari, A.; Tao, Y.; Leclerc, M. A High-Mobility Low-Bandgap Poly(2,7-carbazole) Derivative for Photovoltaic Applications. *Macromolecules* **2009**, *42*, 2891–2894. [CrossRef]
38. Huang, J.; Zhan, C.; Zhang, X.; Zhao, Y.; Lu, Z.; Jia, H.; Jiang, B.; Ye, J.; Zhang, S.; Tang, A.; *et al.* Solution-Processed DPP-Based Small Molecule that Gives High Photovoltaic Efficiency with Judicious Device Optimization. *ACS Appl. Mater. Interfaces* **2013**, *5*, 2033–2039. [CrossRef] [PubMed]
39. Uhrich, C.; Schueppel, R.; Petrich, A.; Pfeiffer, M.; Leo, K.; Brier, E.; Kilickiran, P.; Baeuerle, P. Organic Thin-Film Photovoltaic Cells Based on Oligothiophenes with Reduced Bandgap. *Adv. Funct. Mater.* **2007**, *17*, 2991–2999. [CrossRef]

40. Lu, Z.; Li, C.-H.; Du, C.; Gong, X.; Bo, Z.-S. 6,7-dialkoxy-2,3-diphenylquinoxaline based conjugated polymers for solar cells with high open-circuit voltage. *Chin. J. Polym. Sci.* **2013**, *31*, 901–911. [CrossRef]
41. Shang, Y.; Hao, S.; Liu, J.; Tan, M.; Wang, N.; Yang, C.; Chen, G. Synthesis of Upconversion beta-$NaYF_4$:Nd^{3+}/Yb^{3+}/Er^{3+} Particles with Enhanced Luminescent Intensity through Control of Morphology and Phase. *Nanomaterials* **2015**, *5*, 218–232. [CrossRef]

3

Titanium Dioxide-Coated Zinc Oxide Nanorods as an Efficient Photoelectrode in Dye-Sensitized Solar Cells

Qiang Zhang [1], Shengwen Hou [2] and Chaoyang Li [1,2,*]

1 School of Systems Engineering, Kochi University of Technology, Kami, Kochi 782-8502, Japan; 216005z@gs.kochi-tech.ac.jp
2 Center for Nanotechnology, Kochi University of Technology, Kami, Kochi 782-8502, Japan; 186003p@gs.kochi-tech.ac.jp
* Correspondence: li.chaoyang@kochi-tech.ac.jp

Abstract: Well-arrayed zinc oxide nanorods applied as photoelectrodes for dye-sensitized solar cells were synthesized on an aluminum-doped zinc oxide substrate by the multi-annealing method. In order to improve the chemical stability and surface-to-volume ratio of photoanodes in dye-sensitized solar cells, the synthesized zinc oxide nanorods were coated with pure anatase phase titanium dioxide film using a novel mist chemical vapor deposition method. The effects of the titanium dioxide film on the morphological, structural, optical, and photovoltaic properties of zinc oxide–titanium dioxide core–shell nanorods were investigated. It was found that the diameter and surface-to-volume ratio of zinc oxide nanorods were significantly increased by coating them with titanium dioxide thin film. The power conversion efficiency of dye-sensitized solar cells was improved from 1.31% to 2.68% by coating titanium dioxide film onto the surface of zinc oxide nanorods.

Keywords: titanium dioxide; zinc oxide; core–shell nanorods; mist chemical vapor deposition; dye-sensitized solar cell

---

1. Introduction

Since Grätzel et al. developed the titanium dioxide (TiO_2)-based dye-sensitized solar cell (DSSC) in 1991 [1], the DSSC has emerged as a promising photovoltaic device, due to its promising power conversion efficiency (PCE), low fabrication cost, and low toxicity [2–5]. Hitherto, it has been reported that TiO_2-based DSSCs achieved a notable PCE of over 14% [6]. However, further improvements in PCE are difficult to achieve due to some disadvantages in current TiO_2-based DSSCs, such as the low carrier transportation rate of TiO_2 resulting from its low electron mobility, as well as the difficulty in fabricating TiO_2 nanostructures with a large surface-to-volume ratio [7,8]. Recently, zinc oxide (ZnO) has been widely investigated in different types of solar cells [9–11]. As an alternative photoanode material of DSSCs, ZnO has attracted much attention because it exhibits a similar bandgap and electron injection process from excited dye molecules to TiO_2 [12,13]. Moreover, the electron mobility of ZnO (200~1000 $cm^2/(V{\cdot}s)$) is much higher than that of TiO_2 (0.1~4 $cm^2/(V{\cdot}s)$) [14], which will enhance electron transportation. Additionally, compared with TiO_2, it is much easier to fabricate ZnO as various nanostructures to enlarge the surface-to-volume ratio [15]. Therefore, ZnO-based nanostructures and nanocomposites have much potential for application as a photoanode material to improve the PCE of DSSC.

However, the poor chemical stability of ZnO in the acidic dye solution and electrolyte solution of DSSCs has hampered its wider applicability as a photoanode material in DSSCs [16]. Additionally, defects easily form in ZnO, which increases the Zn^{2+}/dye complex and the electron–hole recombination at the interface [17–20]. In order to overcome the shortcomings of ZnO-based photoanodes, one solution is to coat a chemically stable shell onto the surface of as-deposited ZnO. This core–shell structure can passivate ZnO's surface to reduce the complex and form an energy barrier, thereby reducing the electron–hole recombination [21]. Among different ZnO-based nanocomposites, one of the most promising structures is ZnO–TiO_2's core–shell nanostructure. According to the literature [22–25], the PCE of ZnO photoanode-based DSSCs can be improved by about one to five times by replacing the ZnO photoanode with a corresponding ZnO–TiO_2 core–shell nanostructure. It is reported that ZnO's nanostructure could be coated with TiO_2 thin film using the sol–gel method [26], solution method [27], and atomic layer deposition [28]. However, the difficulties that arise with the uniformity and also in controlling the thickness of the TiO_2 layer are still unsolved.

Based on our previous study, DSSCs with a high PCE could be achieved by controlling the vertical alignment of ZnO nanorods and the quality of transparent conductive substrates [29–31]. In addition, mist chemical vapor deposition (mist CVD) has been proven to be an effective method for modifying ZnO nanorods [32,33]. In this study, ZnO nanorods with vertical alignment were fabricated by a multi-annealing process in reducing ambient. Compared with ZnO nanorods fabricated by other methods, the ZnO nanorods fabricated by multi-annealing showed a higher concentration of oxygen vacancies. The oxygen vacancies were generated due to the effect of reducing ambient and they enhanced the conductivity of ZnO nanorods. However, the oxygen vacancies on the surface of ZnO nanorods will trigger the recombination of electrons. In order to solve this issue, the TiO_2 thin layer was coated on ZnO nanorods by the mist CVD method to prevent the recombination of electrons and enhance the chemical stability of electrodes. Compared with other methods, the combination of the multiple annealing process and mist CVD method is an effective method to fabricate ZnO–TiO_2 core–shell nanorods applied as photoelectrodes for DSSCs. Figure 1 shows the fabrication mechanism and working principle of ZnO–TiO_2 core–shell nanorods. The electrons are injected from excited dye molecules to the conduction band (CB) of TiO_2. Then, the electrons are transferred from the CB of TiO_2 to the CB of ZnO. The ZnO core has high electron mobility and the TiO_2 shell can protect the ZnO core from corrosion and suppress the recombination of electrons. After coating, the obtained ZnO–TiO_2 core–shell nanorods, as well as the as-deposited ZnO nanorods, were used to fabricate DSSCs for comparison. The effects of TiO_2 coating on the properties of ZnO–TiO_2 core–shell nanorods were investigated in detail.

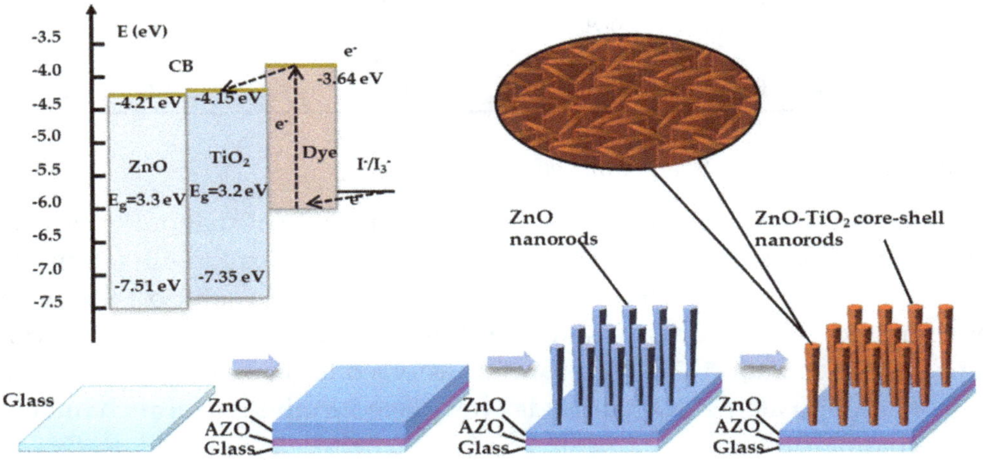

Figure 1. Fabrication mechanism and working principle of the zinc oxide–titanium dioxide (ZnO–TiO_2) core–shell nanorod.

2. Materials and Methods

2.1. Deposition of Thin Films

The aluminum-doped ZnO (AZO, 300 nm) thin films were deposited on alkali-free glass sheets (Eagle XG, Corning Inc., Corning, NY, USA) using a conventional radio frequency (RF, 13.56 MHz) magnetron sputtering system with an AZO target (2 wt.% Al_2O_3). Following the deposition of AZO films, ZnO films with a 500 nm thickness were deposited on AZO by the same sputtering system with a ZnO target (5N). Table 1 shows the deposition conditions of the AZO film and ZnO film. Argon was selected as the working gas, the flow rate of which was maintained at 30 sccm. During the deposition, the working distance and temperature were set and maintained at 60 mm and 150 °C, respectively. The pressure and RF power for AZO film deposition were maintained at 1 Pa and 60 W. For the deposition of ZnO, the pressure and RF power were held at 7 Pa and 180 W.

Table 1. Deposition conditions of AZO and ZnO films.

Film	AZO	ZnO
Target	AZO (2 wt.%)	ZnO (5N)
Working gas, Flow Rate (sccm)	Argon, 30	Argon, 30
Working distance (mm)	60	60
Deposition Temperature (°C)	150	150
Pressure (Pa)	1	7
RF Power (W)	60	180

2.2. Fabrication of ZnO Nanorods

After sputtering deposition, the fabricated ZnO films were treated using a multi-annealing process in a conventional annealing furnace. As shown in Table 2, the temperature was firstly kept at 300 °C for 2 h in a forming gas ambient (H_2:N_2 = 1.96%) to increase the density of zinc seeds on the surface. Then, the temperature was increased to 450 °C and kept at this level for 3 h for forming gas to produce the ZnO nanorods. Before the third forming gas annealing process, oxygen was introduced into the furnace for 40 min for surface oxidation to avoid an excessive reducing reaction. For safety considerations, nitrogen was introduced for 5 min between the forming gas and oxygen annealing processes.

Table 2. Annealing condition.

Step	Gas	Temperature (°C)	Time (min)
1	H_2 in N_2 (1.96%)	300	120
2	H_2 in N_2 (1.96%)	450	180
3	O_2	450	40
4	H_2 in N_2 (1.96%)	450	120

2.3. Fabrication of ZnO–TiO_2 Core–Shell Nanorods

Finally, TiO_2 film was coated onto the surface of the fabricated ZnO nanorods by a mist CVD system. Table 3 shows the deposition condition of the TiO_2 film. An ethanolic titanium tetraisopropoxide (TTIP, purity > 95.0%, Wako Pure Chemical Industries, Ltd., Osaka, Japan) solution with a concentration of 0.10 mol/L was prepared as the precursor solution. Mist droplets were generated from the precursor solution by ultrasonic atomization (2.4 MHz) and transferred to the reaction chamber by compressed air. The sample of as-deposited ZnO nanorods was placed in the reaction chamber and heated to 450 °C during the coating process.

Table 3. Deposition condition of TiO_2.

Solvent	Ethanol
Solute	TTIP
Concentration (mol/L)	0.10
Deposition Temperature (°C)	450
Carrier Gas, Flow Rate (L/min)	Compressed Air (dried), 2.5
Dilution Gas, Flow Rate (L/min)	Compressed Air (dried), 4.5

2.4. Fabrication of DSSC

The obtained $ZnO–TiO_2$ core–shell nanorods, as well as the as-deposited ZnO nanorods, were applied as photoanodes to fabricate DSSCs for comparison. N719 (Sigma Aldrich, St. Louis, MO, USA) was used as a dye sensitizer. The photoanodes were immersed in an ethanoic dye solution with a concentration of 5×10^{-4} mol/L for 12 h. A solution containing 0.10 mol/L lithium iodine and 0.05 mol/L iodine was used as the electrolyte. A platinum-coated indium-doped tin oxide film on glass was applied as the counter-electrode. Six samples of DSSCs were fabricated and investigated to confirm their reproducibility.

2.5. Characterization

The morphological properties of the AZO film, as-deposited ZnO nanorods, and $ZnO–TiO_2$ core–shell nanorods were evaluated using field emission scanning electron microscopy (FE-SEM, JSM-7400F, JEOL, Tokyo, Japan) and transmission emission microscopy (TEM, JEM 2100F, JEOL, Tokyo, Japan). The structural properties of the AZO film were measured by X-ray diffraction (XRD, ATX-G, Rigaku, Tokyo, Japan). The structural properties of the as-deposited ZnO nanorods and $ZnO–TiO_2$ core–shell nanorods were investigated by grazing incidence X-ray diffraction (GIXRD, ATX-G, Rigaku, Tokyo, Japan). The optical properties of the as-deposited ZnO nanorods and $ZnO–TiO_2$ core–shell nanorods were obtained using a spectrophotometer (U-4100, Hitachi, Tokyo, Japan). The fabricated DSSCs were characterized using a solar simulator (PEC-L01, AM 1.5 G, 100 mW/cm^2, Peccell Technologies Inc., Yokohama, Japan) and a source meter (Keithley 2400, Keithley Instruments Inc., Solon, OH, USA). All of the measurements were carried out at room temperature.

3. Results

The XRD pattern of the AZO film is shown in Figure 2. It was found that only the (002) diffraction peak was observed in the XRD pattern, which indicated that the AZO films had highly (002) preferred orientation with a c-axis perpendicular to the substrates. The insert image in Figure 2 shows the FE-SEM top view image of the AZO film. It is confirmed that an AZO film with a uniform surface was obtained after deposition.

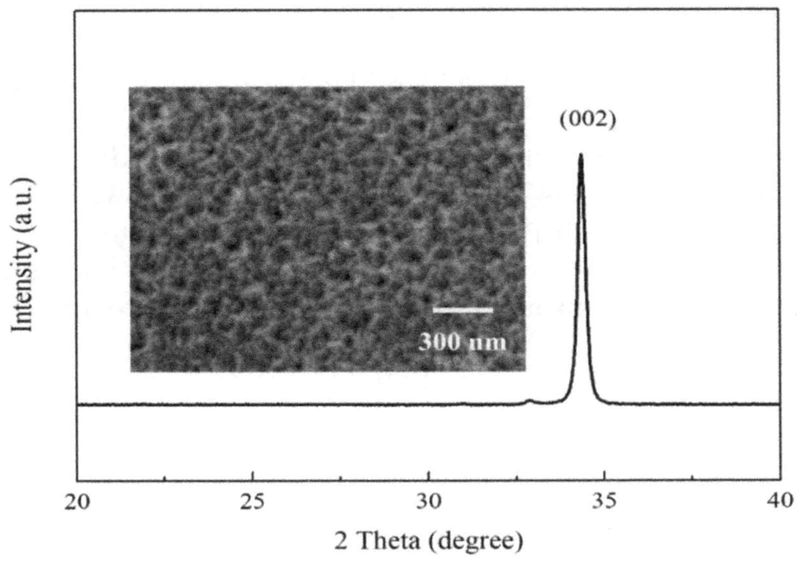

Figure 2. XRD pattern of AZO film (insert image shows the FE-SEM top view image of AZO film).

The FE-SEM images of the as-deposited ZnO nanorods and $ZnO–TiO_2$ core–shell nanorods are shown in Figure 3. The details of single nanorods are shown in the inset images. The as-deposited ZnO nanorods showed a well-arrayed hexagonal structure with a smooth surface. Compared with the as-deposited ZnO nanorods, the ZnO–TiO_2 core–shell nanorods had a higher surface roughness and a larger diameter. Intertwined TiO_2 nanosheets were observed on the surface of the ZnO–TiO_2 core–shell nanorods, indicating that the TiO_2 film was successfully coated onto the surface of the ZnO nanorods. Figure 3c shows the TEM image of a single ZnO–TiO_2 core–shell nanorod. It was confirmed that the thickness of the TiO_2 shell on the ZnO nanorods was around 15 nm.

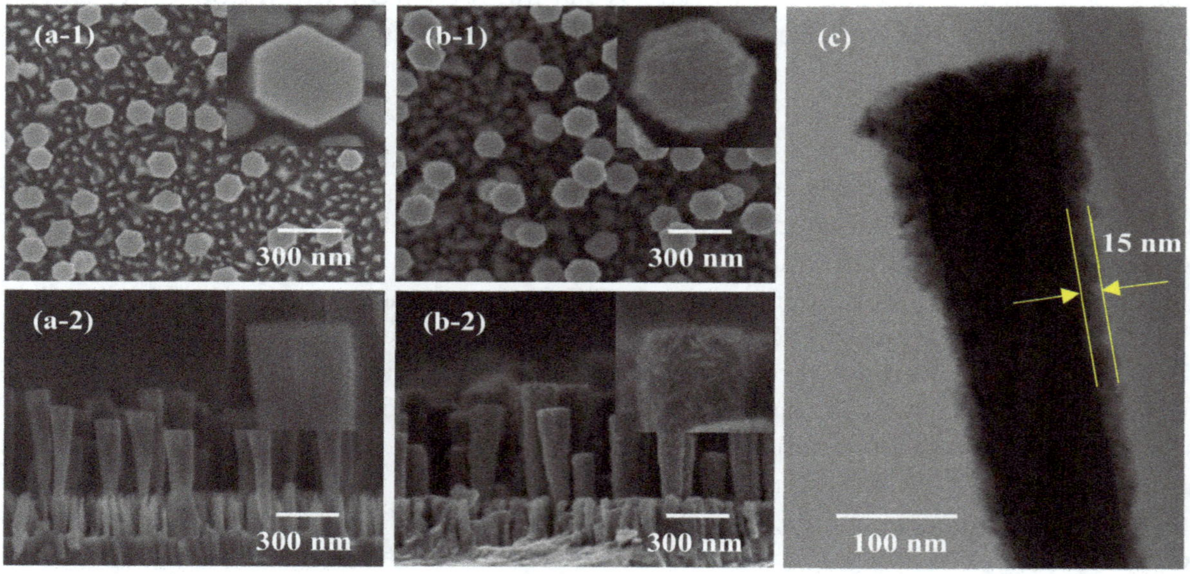

Figure 3. FE-SEM images of (**a**) as-deposited ZnO nanorods, (**b**) ZnO–TiO_2 core–shell nanorods, and (**c**) TEM image of single ZnO–TiO_2 core–shell nanorod ((**1**) Top view, (**2**) cross-section view).

The GIXRD patterns of the as-deposited ZnO nanorods and ZnO–TiO_2 core–shell nanorods are shown in Figure 4. It was found that only the (002) diffraction peak was observed in the GIXRD pattern of the as-deposited ZnO nanorods, suggesting that both the ZnO film and ZnO nanorods had highly (002) preferred orientation with a c-axis perpendicular to the substrates. This agrees well with the FE-SEM results. In the GIXRD pattern of the ZnO–TiO_2 core–shell nanorods, the observed peaks corresponded with the (101), (200), (211), (204), (220), and (215) diffraction peaks of the anatase phase TiO_2 and the (002) diffraction peak of ZnO. All of the diffraction peaks of TiO_2 were identified and corresponded with the anatase phase of TiO_2 (JCPDS 21-1272), indicating that the TiO_2 film coated on ZnO nanorods was pure anatase phase.

The optical transmission spectra of the as-deposited ZnO nanorods and ZnO–TiO_2 core–shell nanorods are shown in Figure 5. The as-deposited ZnO nanorods showed a high transmittance of 75% in visible range. After coating with TiO_2 film, the transmittance of the nanorods in visible range decreased to 55%, due to the scattering of TiO_2 nanosheets. It is well-known that the bandgap of material can be calculated from the transmittance data by the following equations [34,35]:

$$\alpha = \frac{1}{d}\ln\left(\frac{1}{T}\right) \tag{1}$$

$$(\alpha h\nu)^2 = A\left(h\nu - E_g\right) \tag{2}$$

where α is the absorption coefficient, d the thickness of material, T the transmittance, $h\nu$ the incident photon energy, A a constant, and E_g the bandgap. A plot of $(\alpha h\nu)^2$ as a function of $h\nu$ made to determine E_g by linear fitting is shown in Figure 6. After fitting, the bandgap of the as-deposited ZnO nanorods was determined as around 3.32 eV, corresponding with the bandgap of bulk ZnO (3.37 eV). The bandgap of the ZnO–TiO_2 core–shell nanorods was around 3.28 eV, corresponding with the bandgap of anatase phase TiO_2 (3.2 eV).

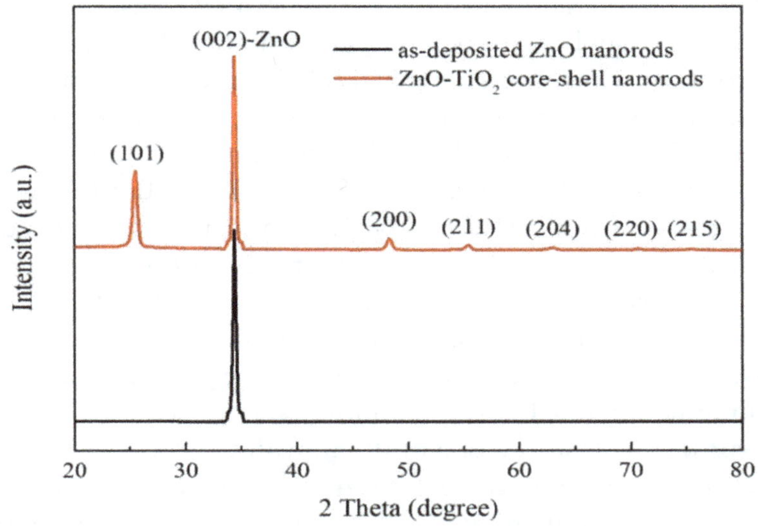

Figure 4. GIXRD patterns of as-deposited ZnO nanorods and ZnO–TiO_2 core–shell nanorods.

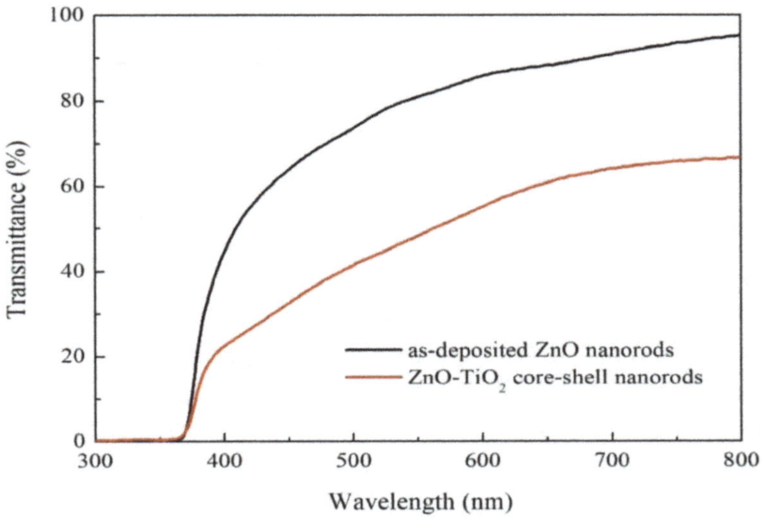

Figure 5. Optical transmission spectra of as-deposited ZnO nanorods and ZnO–TiO_2 core–shell nanorods.

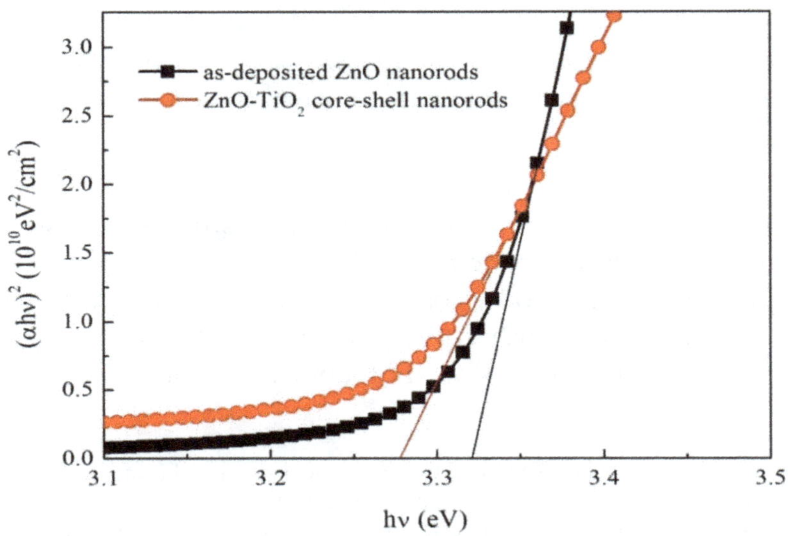

Figure 6. Variation of $(ahv)^2$ of the as-deposited ZnO nanorods and ZnO–TiO_2 core–shell nanorods as a function of the photon energy (hv).

Figure 7 shows the *J-V* characteristics of the demonstrated DSSCs applying as-deposited ZnO nanorods and ZnO–TiO_2 core–shell nanorods as photoanodes. Compared with the DSSCs using as-deposited ZnO nanorods, the DSSCs applying ZnO–TiO_2 core–shell nanorods showed higher open circuit voltage (V_{OC}), higher short circuit current density (J_{SC}), higher fill factor (*FF*), and higher PCE. After coating with TiO_2, the V_{OC} of the DSSCs increased from 0.60 V to 0.63 V, and the J_{SC} increased from 5.01 mA/cm^2 to 6.73 mA/cm^2. It was found that the *FF* increased from 43.41% to 63.13%, and the PCE increased from 1.31% to 2.68%. The results showed good reproducibility by checking all of the DSSCs samples. The significant improvement of the *FF* and PCE was due to the great improvement in the J_{SC}, which could be explained as follows: Firstly, the TiO_2 shell increased the surface-to-volume ratio of the ZnO nanorods. Therefore, more dye molecules were absorbed onto the surface of the nanorods, which enhanced their light harvesting. Secondly, the TiO_2 shell has a much lower electron–hole recombination rate than ZnO nanorods, which could greatly improve the efficiency of electron collection. Thirdly, the last step of a multi-annealing process was carried out in a reducing ambient. Consequently, many oxygen vacancies were generated on the surface of the ZnO nanorods. The oxygen vacancies acted as recombination centers, which triggered large amounts of recombination of electrons. After coating with TiO_2 film, the recombination of electrons was suppressed. The efficient light harvesting and efficient electron collection contributed to the great improvement in the J_{SC}.

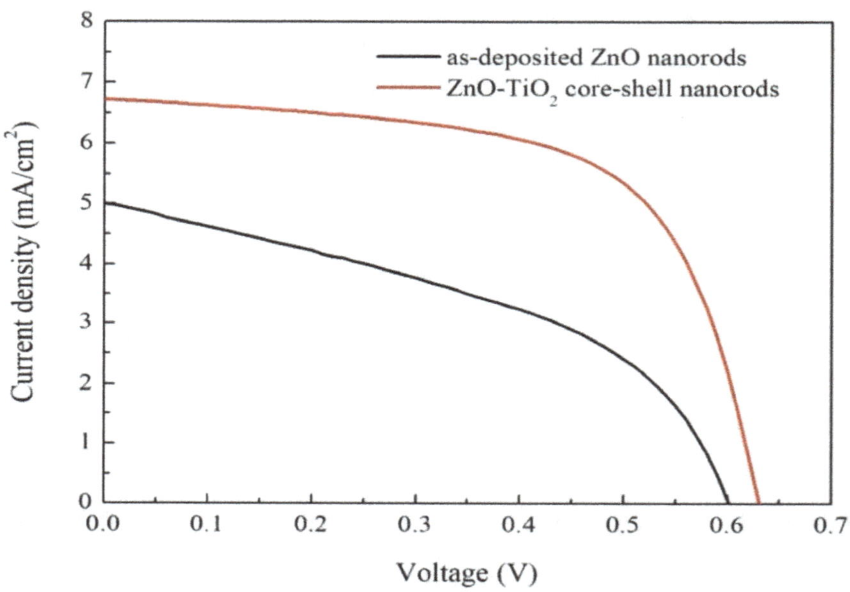

Figure 7. *J-V* characteristics of demonstrated dye-sensitized solar cells (DSSCs) applying as-deposited ZnO nanorods and ZnO–TiO_2 core–shell nanorods as photoanodes.

4. Conclusions

Well-arrayed ZnO–TiO_2 core–shell nanorods were successfully synthesized on AZO substrates by RF magnetron sputtering, multi-annealing, and the mist CVD method. The morphology of the ZnO nanorods was significantly changed by coating with a TiO_2 film. After forming the ZnO–TiO_2 core–shell structures, the diameter and surface-to-volume ratio of the nanorods were greatly increased. The PCE of DSSCs applying ZnO nanorods as photoanodes was increased two-fold from 1.31% to 2.68% by coating with TiO_2.

Author Contributions: Conceptualization, S.H. and C.L.; methodology, S.H. and C.L.; validation, S.H. and C.L.; investigation, S.H.; data curation, Q.Z. and S.H.; writing—original draft preparation, Q.Z.; writing—review and editing, Q.Z. and C.L.; visualization, Q.Z.; supervision, C.L.; project administration, C.L.; funding acquisition, C.L. All authors have read and agreed to the published version of the manuscript.

Acknowledgments: The authors gratefully acknowledge the financial support by Grant-in-Aid for Scientific Research from the Ministry of Education, Culture, Sports, Science and Technology (MEXT), Japan. The authors greatly appreciate the assistance of Associate Professor Noriko Nitta in supporting the TEM measurement.

References

1. O'Regan, B.; Grätzel, M. A low-cost, high-efficiency solar cell based on dye-sensitized. *Nature* **1991**, *353*, 737–740. [CrossRef]
2. Prabavathy, N.; Shalini, S.; Balasundaraprabhu, R.; Velauthapillai, D.; Prasanna, S.; Muthukumarasamy, N. Enhancement in the photostability of natural dyes for dye-sensitized solar cell (DSSC) applications: A review. *Int. J. Energy Res.* **2017**, *41*, 1372–1396. [CrossRef]
3. Ahmad, M.S.; Pandey, A.K.; Rahim, N.A. Advancements in the development of TiO_2 photoanodes and its fabrication methods for dye sensitized solar cell (DSSC) applications. A review. *Renew. Sustain. Energy Rev.* **2017**, *77*, 89–108. [CrossRef]
4. Das, T.K.; Ilaiyaraja, P.; Sudakar, C. Template assisted nanoporous TiO_2 nanoparticles: The effect of oxygen vacancy defects on photovoltaic performance of DSSC and QDSSC. *Sol. Energy* **2018**, *159*, 920–929. [CrossRef]
5. Vaghasiya, J.V.; Sonigara, K.K.; Soni, S.S.; Tan, S.C. Dual functional hetero-anthracene based single component organic ionic conductors as redox mediator cum light harvester for solid state photoelectrochemical cells. *J. Mater. Chem. A* **2018**, *6*, 4868–4877. [CrossRef]
6. Kakiage, K.; Aoyama, Y.; Yano, T.; Oya, K.; Fujisawa, J.-I.; Hanaya, M. Highly-efficient dye-sensitized solar cells with collaborative sensitization by silyl-anchor and carboxy-anchor dyes. *Chem. Commun.* **2015**, *51*, 15894–15897. [CrossRef]
7. Law, M.; Greene, L.E.; Johnson, J.C.; Saykally, R.; Yang, P. Nanowire dye-sensitized solar cells. *Nat. Mater.* **2005**, *4*, 455–459. [CrossRef]
8. Zhang, Q.; Dandeneau, C.S.; Zhou, X.; Cao, G. ZnO nanostructures for dye-sensitized solar cells. *Adv. Mater.* **2009**, *21*, 4087–4108. [CrossRef]
9. Gopalan, S.-A.; Gopalan, A.-I.; Vinu, A.; Lee, K.-P.; Kang, S.-W. Solar Energy Materials and Solar Cells A new optical-electrical integrated bu ff er layer design based on gold nanoparticles tethered thiol containing sulfonated polyaniline towards enhancement of solar cell performance. *Sol. Energy Mater. Sol. Cells* **2018**, *174*, 112–123. [CrossRef]
10. Nandakumar, D.K.; Vaghasiya, J.V.; Yang, L.; Zhang, Y.; Tan, S.C. A solar cell that breathes in moisture for energy generation. *Nano Energy* **2020**, *68*, 104263. [CrossRef]
11. Nandakumar, D.K.; Ravi, S.K.; Zhang, Y.; Guo, N.; Zhang, C.; Tan, S.C. A super hygroscopic hydrogel for harnessing ambient humidity for energy conservation and harvesting. *Energy Environ. Sci.* **2018**, *11*, 2179–2187. [CrossRef]
12. Kolodziejczak-Radzimska, A.; Jesionowski, T. Zinc oxide-from synthesis to application: A review. *Materials (Basel)* **2014**, *7*, 2833–2881. [CrossRef]
13. Lee, J.-C.; Gopalan, A.-I.; Saianand, G.; Lee, K.-P.; Kim, W.-J. Manganese and graphene included titanium dioxide composite nanowires: Fabrication, characterization and enhanced photocatalytic activities. *Nanomaterials* **2020**, *10*, 456. [CrossRef]
14. Tiwana, P.; Docampo, P.; Johnston, M.B.; Snaith, H.J.; Herz, L.M. Electron mobility and injection dynamics in mesoporous ZnO, SnO_2, and TiO_2 films used in dye-sensitized solar cells. *ACS Nano.* **2011**, *5*, 5158–5166. [CrossRef]
15. Gonzalez-Valls, I.; Lira-Cantu, M. Vertically-aligned nanostructures of ZnO for excitonic solar cells: A review. *Energy Environ. Sci.* **2009**, *2*, 19–34. [CrossRef]
16. Vittal, R.; Ho, K.-C. Zinc oxide based dye-sensitized solar cells: A review. *Renew. Sustain. Energy Rev.* **2017**, *70*, 920–935. [CrossRef]
17. Lu, L.; Li, R.; Fan, K.; Peng, T. Effects of annealing conditions on the photoelectrochemical properties of dye-sensitized solar cells made with ZnO nanoparticles. *Sol. Energy.* **2010**, *84*, 844–853. [CrossRef]
18. Ambade, S.B.; Mane, R.S.; Ghule, A.V.; Takwale, M.G.; Abhyankar, A.; Cho, B.W.; Han, S.H. Contact angle measurement: A preliminary diagnostic method for evaluating the performance of ZnO platelet-based dye-sensitized solar cells. *Scr. Mater.* **2009**, *61*, 12–15. [CrossRef]

19. Yan, F.; Huang, L.; Zheng, J.; Huang, J.; Lin, Z.; Huang, F.; Wei, M. Effect of surface etching on the efficiency of ZnO-based dye-sensitized solar cells. *Langmuir.* **2010**, *26*, 7153–7156. [CrossRef]
20. Horiuchi, H.; Katoh, R.; Hara, K.; Yanagida, M.; Murata, S.; Arakawa, H.; Tachiya, M. Electron injection efficiency from excited N3 into nanocrystalline ZnO films: Effect of (N3-Zn^{2+}) aggregate formation. *J. Phys. Chem. B.* **2003**, *107*, 2570–2574. [CrossRef]
21. Law, M.; Greene, L.E.; Radenovic, A.; Kuykendall, T.; Liphardt, J.; Yang, P. ZnO-Al_2O_3 and ZnO-TiO_2 core-shell nanowire dye-sensitized solar cells. *J. Phys. Chem. B.* **2006**, *110*, 22652–22663. [CrossRef] [PubMed]
22. Chandiran, A.K.; Abdi-Jalebi, M.; Nazeeruddin, M.K.; Grätzel, M. Analysis of electron transfer properties of ZnO and TiO_2 photoanodes for dye-sensitized solar cells. *ACS Nano.* **2014**, *8*, 2261–2268. [CrossRef] [PubMed]
23. Atienzar, P.; Ishwara, T.; Illy, B.N.; Ryan, M.P.; O'Regan, B.C.; Durrant, J.R.; Nelson, J. Control of photocurrent generation in polymer/ZnO nanorod solar cells by using a solution-processed TiO_2 overlayer. *J. Phys. Chem. Lett.* **2010**, *1*, 708–713. [CrossRef]
24. Feng, Y.; Ji, X.; Duan, J.; Zhu, J.; Jiang, J.; Ding, H.; Meng, G.; Ding, R.; Liu, J.; Hu, A.; et al. Synthesis of ZnO@TiO_2 core-shell long nanowire arrays and their application on dye-sensitized solar cells. *J. Solid State Chem.* **2012**, *190*, 303–308. [CrossRef]
25. Prabakar, K.; Son, M.; Kim, W.-Y.; Kim, H. TiO_2 thin film encapsulated ZnO nanorod and nanoflower dye sensitized solar cells. *Mater. Chem. Phys.* **2011**, *125*, 12–14. [CrossRef]
26. Zhao, R.; Zhu, L.; Cai, F.; Yang, Z.; Gu, X.; Huang, J.; Cao, L. ZnO/TiO_2 core-shell nanowire arrays for enhanced dye-sensitized solar cell efficiency. *Appl. Phys. A Mater. Sci. Process.* **2013**, *113*, 67–73. [CrossRef]
27. Goh, G.K.L.; Le, H.Q.; Huang, T.J.; Hui, B.T.T. Low temperature grown ZnO@TiO_2 core shell nanorod arrays for dye sensitized solar cell application. *J. Solid State Chem.* **2014**, *214*, 17–23. [CrossRef]
28. Greene, L.E.; Law, M.; Yuhas, B.D.; Yang, P. ZnO-TiO_2 core-shell nanorod/P3HT solar cells. *J. Phys. Chem. C.* **2007**, *111*, 18451–18456. [CrossRef]
29. Li, X.; Li, C.; Kawaharamura, T.; Wang, D.; Nitta, N.; Furuta, M.; Furuta, H.; Hatta, A. Influence of substrates on formation of zinc oxide nanostructures by a novel reducing annealing method. *Nanosci. Nanotechnol. Lett.* **2014**, *6*, 174–180. [CrossRef]
30. Li, X.; Li, C.; Hou, S.; Hatta, A.; Yu, J.; Jiang, N. Thickness of ITO thin film influences on fabricating ZnO nanorods applying for dye-sensitized solar cell. *Compos. Part B Eng.* **2015**, *74*, 147–152. [CrossRef]
31. Hou, S.; Li, C. Aluminum-doped zinc oxide thin film as seeds layer effects on the alignment of zinc oxide nanorods synthesized in the chemical bath deposition. *Thin Solid Films* **2016**, *605*, 37–43. [CrossRef]
32. Li, X.; Li, C.; Kawaharamura, T.; Wang, D.; Nitta, N.; Furuta, M.; Furuta, H.; Hatta, A. Fabrication of zinc oxide nanostructures by mist chemical vapor deposition. *Trans. Mater. Res. Soc. Jpn.* **2014**, *164*, 161–164. [CrossRef]
33. Zhang, Q.; Li, C. TiO_2 coated ZnO nanorods by mist chemical vapor deposition for application as photoanodes for dye-sensitized solar cells. *Nanomaterials* **2019**, *9*, 1339. [CrossRef] [PubMed]
34. Gao, X.; Du, Y.; Meng, X. Cupric oxide film with a record hole mobility of 48.44 cm^2/Vs via direct-current reactive magnetron sputtering for perovskite solar cell application. *Sol. Energy* **2019**, *191*, 205–209. [CrossRef]
35. Fang, X.S.; Bando, Y.; Shen, G.Z.; Ye, C.H.; Gautam, U.K.; Costa, P.M.F.J.; Zhi, C.Y.; Tang, C.C.; Golberg, D. Ultrafine ZnS nanobelts as field emitters. *Adv. Mater.* **2007**, *19*, 2593–2596. [CrossRef]

4

Nano-Photonic Structures for Light Trapping in Ultra-Thin Crystalline Silicon Solar Cells

Prathap Pathi [1,2], Akshit Peer [3] and Rana Biswas [4,*]

[1] Ames Laboratory, Microelectronics Research Center, Iowa State University, Ames, IA 50011, USA; prathap@nplindia.org
[2] Silicon Solar Cell Division, CSIR-National Physical Laboratory, Dr. K.S. Krishnan Road, New Delhi-110012, India
[3] Ames Laboratory, Microelectronics Research Center, Department of Electrical and Computer Engineering, Iowa State University, Ames, IA 50011, USA; apeer@iastate.edu
[4] Ames Laboratory, Microelectronics Research Center, Department of Physics and Astronomy, Department of Electrical and Computer Engineering, Iowa State University, Ames, IA 50011, USA
* Correspondence: biswasr@iastate.edu

Academic Editors: Guanying Chen, Zhijun Ning and Hans Agren

Abstract: Thick wafer-silicon is the dominant solar cell technology. It is of great interest to develop ultra-thin solar cells that can reduce materials usage, but still achieve acceptable performance and high solar absorption. Accordingly, we developed a highly absorbing ultra-thin crystalline Si based solar cell architecture using periodically patterned front and rear dielectric nanocone arrays which provide enhanced light trapping. The rear nanocones are embedded in a silver back reflector. In contrast to previous approaches, we utilize dielectric photonic crystals with a completely flat silicon absorber layer, providing expected high electronic quality and low carrier recombination. This architecture creates a dense mesh of wave-guided modes at near-infrared wavelengths in the absorber layer, generating enhanced absorption. For thin silicon (<2 μm) and 750 nm pitch arrays, scattering matrix simulations predict enhancements exceeding 90%. Absorption approaches the Lambertian limit at small thicknesses (<10 μm) and is slightly lower (by ~5%) at wafer-scale thicknesses. Parasitic losses are ~25% for ultra-thin (2 μm) silicon and just 1%–2% for thicker (>100 μm) cells. There is potential for 20 μm thick cells to provide 30 mA/cm^2 photo-current and >20% efficiency. This architecture has great promise for ultra-thin silicon solar panels with reduced material utilization and enhanced light-trapping.

Keywords: nano-photonics; solar cell; light-trapping; scattering

1. Introduction

Crystalline silicon solar cells are the dominant technology for solar panels, accounting for nearly 90% of the present market share. Crystalline silicon solar cells have achieved ~25% power conversion efficiency (PCE) in small area laboratory cells [1,2] and up to 22% in larger area panels. A very attractive feature of crystalline silicon (c-Si) technology is its high stability over several years, with a degradation rate typically less than 0.5% per year.

The silicon wafers utilized in solar panels are typically 180 μm thick. Hence, the material costs associated with thick silicon wafers are a considerable component of the system costs. It is of great interest to thin the c-Si wafers considerably, and employ recently developed light-trapping techniques [3] to absorb solar photons in thin layers. Moreover, thin silicon solar cells may be flexible and adapted to curved surfaces, increasing tremendously their range of application. Thus, the development of ultra-thin c-Si cells will be an important new technology direction that may have many significant technological impacts.

A major technical hurdle is that the absorption length ($l_d(\lambda) = 4\pi Im(n(\lambda))/\lambda$) of photons (Figure 1) rapidly increases at wavelengths near the band edge (1100 nm or 1.12 eV), where $n(\lambda)$ is the complex refractive index of Si [4]. At the near-infrared wavelength, λ = 900 nm and $l_d(\lambda)$ is 10 μm for Si and grows exponentially for longer wavelengths, exceeding the thickness of the absorbing layer in thin cells. Such photons cannot be effectively absorbed in thin Si layers. It is necessary to employ light trapping techniques to increase the path length of long wavelength red and near-infra-red (IR) photons. A very attractive solar cell architecture for thin amorphous silicon (a-Si:H) and nano-crystalline Si (nc-Si) solar cells [5–12] is to utilize a periodically corrugated back-reflector and grow a conformal solar cell on top of this structure such that all layers have the periodic corrugation. This architecture traps light through (i) strong diffraction leading to a dense mesh of wave-guided modes propagating in the plane of the structure and (ii) propagating surface plasmon modes at the semiconductor-metal interface where the light intensity is considerably enhanced. These effects have resulted in measured enhancement of short circuit current (J_{SC}) exceeding 30% in periodically corrugated nc-Si cells.

A similar diffractive structure has been proposed for thin c-Si solar cells [13–15] consisting of a periodic array of silicon nanocones both on the front and back of the structure combined with a perfect electrical conductor serving as a back reflector. Such thin c-Si cells, with a thickness of just ~2 μm, are predicted to have absorption and photo-current near the Lambertian limit [16,17]. However, preliminary experimental solar cells with this architecture [14] have considerably lower PCE (8%), much lower than predicted [13]. The reason for this is that the photonic crystal is composed of corrugated Si surfaces, where the corrugation leads to a large surface area and surface recombination of photo-excited carriers. The increased recombination losses outweigh the advantages of optical absorption enhancements. A triangular array of dielectric nanospheres on a flat Ag layer has been alternatively proposed [18,19] to be a high performing back reflector, which preserves the flatness of the c-Si layers and is also amenable to fabrication.

Thus, an effective solar light-trapping architecture is to utilize diffractive photonic crystal surfaces that involve flat silicon layers, that minimize carrier recombination losses, and use periodic arrays of insulating materials for diffractive effects. In this paper, we develop and design a practical light-trapping architecture utilizing photonic crystals of insulating materials that preserves electronic quality of the interfaces. There are analogies with the proposed architecture of Ingenito et al. [20] where a front surface was composed of textured "black-silicon" combined with the back surface having a random pyramidal silicon texture and a distributed Bragg reflector (DBR), with predicted absorption near the Lambertian limit. We design an alternative architecture based on periodic nanostructured arrays (rather than random features) and use a simpler metallic back reflector rather than the DBR. The present architecture avoids the alkaline texturization process, which is performed for absorption enhancement of c-Si solar cells. Enhancement of long wavelength absorption in thin Si through rear surface plasmons has been demonstrated for planar Si absorbing layers with Ag nanoparticles on a detached back surface reflector [21,22], an architecture that preserves high electronic quality of interfaces.

Light trapping has also been applied to crystallized silicon on glass (CSG) cells by texturing of the glass substrate [23] leading to J_{SC} of 29 mA/cm^2–29.5 mA/cm^2 for up to 3.5 μm thick poly-Si absorber layers. This is slightly lower than the Lambertian limit (34 mA/cm^2). Front texturing of the glass increases light path through randomization but may not approach the full randomization of light within the absorber layer as suggested by the Lambertian limit. We have theoretically [24] and experimentally [25] studied periodic texturing of the glass in solar cells and found gains that were significantly lower than those possible from texturing the absorber layer itself. As described by Varlamov et al., light trapping in CSG cells has also been implemented via etch-back texturing of the poly-Si itself and does extremely well at long wavelengths (>600 nm) but not in the broad band spectrum, resulting in J_{SC} of 29.5 mA/cm^2 for a 3.6 μm cell.

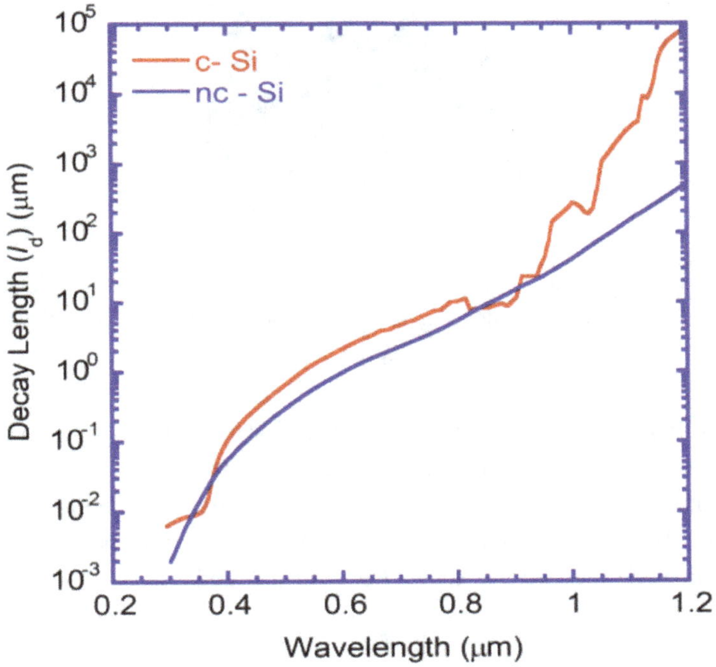

Figure 1. Photon absorption length as a function of wavelength for crystalline silicon (c-Si) and nano-crystalline silicon (nc-Si) (using the complex refractive index (n, k) parameters of Reference [26]).

2. Results

2.1. *Approach and Structure*

We design a practical light trapping architecture where all silicon interfaces are planar and demonstrate a high degree of light trapping, close to the Lambertian limit, that is achieved by photonic crystals of non-absorbing insulating materials. This architecture is concurrently expected to have superior electronic properties, comparable to conventional high-efficiency silicon solar cells. Our solar architecture differs from the previous work of Wang et al. [13] where the silicon was part of the photonic crystal leading to much higher surface recombination of photo-excited carriers.

Our proposed solar cell architecture (Figure 2) consists of (1) an upper photonic crystal array of dielectric titania nano-cones arranged in a triangular lattice, with height d_0 and pitch a; on (2) a passivating layer of titania (thickness d_1); (3) the flat silicon absorber layer (thickness d_2); (4) another passivating layer of titania (thickness d_3); followed by (5) a lower photonic crystal array of titania nanocones with height d_4 and pitch a; that is coated with (6) a metallic (Ag) reflecting layer. The upper photonic crystal has two functions. It diffracts incoming light into the thin absorber layer and realizes a gradual transition from air to the dielectric, thereby reducing impedance mismatch and reflection loss of incoming light back to air. The lower photonic crystal is effective in diffraction of long wavelength light that reaches the back of the cell, but has a much smaller effect on shorter wavelengths that are absorbed within the upper portion of the absorber layer. Without the metallic back-reflector, the long wavelength photons can transmit and escape through the back of the cell. For computational convenience, the front and lower photonic crystal arrays have the same pitch. As required in the high efficiency silicon solar cells, both interfaces of silicon are passivated with thin titanium dioxide (TiO_2) layers. The passivating layers are flat minimizing interfacial recombination. These unique structures have great promise for the fabrication of high efficiency thin c-Si solar cells using PERC (passivated emitter rear contact), PERL (passivated emitter with rear locally diffused), or PERT (passivated emitter, rear totally diffused) configurations.

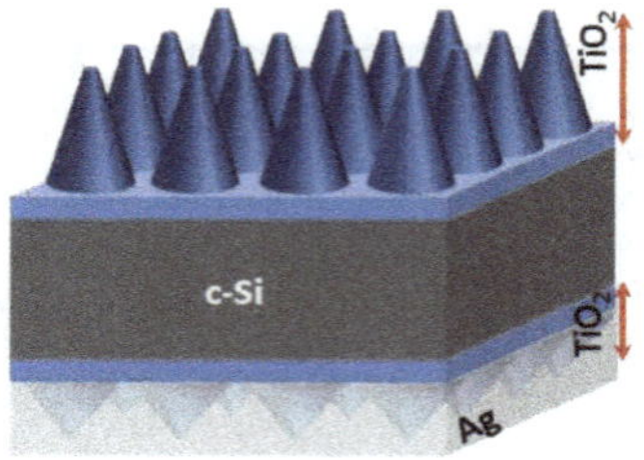

Figure 2. Proposed solar architecture consists of thin flat spacer titanium dioxide (TiO_2) layers on the front and rear surfaces of silicon, nanocone gratings on both sides with optimized pitch and height, and rear cones are surrounded by Ag metal reflector.

It is necessary to systematically design the solar cell structure for optimal performance. We first design optimum photonic crystal based silicon solar architectures with rigorous vectorial simulations using the scattering matrix method, where Maxwell's equations are solved in a plane wave basis, i.e., in Fourier space, for both polarizations of the incident wave. Maxwell's equations in real space are converted to equations for each frequency ω, in Fourier space:

$$\nabla \times E = i\omega H$$

$$\nabla \times H = i\omega\varepsilon(r)E \tag{1}$$

In Fourier space, the solutions of Maxwell's equations are independent for each incoming frequency (ω) or incoming wavelength λ.

Experimentally measured wavelength-dependent complex dielectric functions $\varepsilon(\lambda)$ for Si, Ag, and TiO_2 are utilized. The solar cell is divided into layers in the z direction. Within each layer of the structure (Figure 2), the dielectric function $\varepsilon(x, y)$ is a function of the spatial coordinates (x, y) but not of z. This allows the dielectric function in each layer to be expanded in a two-dimensional basis of reciprocal lattice vectors G, providing the Fourier components of the dielectric function $\varepsilon(G)$. Similarly, the electric fields and magnetic fields ($E(G)$, $H(G)$) are expanded in Bloch waves. Within each layer, Maxwell's equations are solved in Fourier space in an eigenvalue expansion [27,28] to obtain the electric and magnetic fields in each layer. A transfer matrix is used to relate the E and H fields within each layer. Maxwell's equations are integrated with the continuity boundary conditions throughout the unit cell to obtain the scattering matrices of each layer and the entire structure. From the scattering matrix we find the total reflectance R (including diffracted beams) and transmission T (which is 0) at each incident wavelength. The absorption at each wavelength is then $A = 1 - R - T$. Details have been covered in previous publications [29]. We characterize solar cell architectures by their weighted absorption $<A_w>$,

$$< A_w >= \int_{\lambda_1}^{\lambda_2} A(\lambda)\frac{dI}{d\lambda}d\lambda, \tag{2}$$

and short circuit current J_{SC},

$$J_{SC} = \frac{e}{hc}\int_{\lambda_1}^{\lambda_2} \lambda A(\lambda)\frac{dI}{d\lambda}d\lambda \tag{3}$$

Here, $dI/d\lambda$ is the incident solar spectrum. We assume ideal internal quantum efficiency, i.e., absorption of each incoming photon generates an electron-hole pair. We have been very successful in designing optimized thin silicon periodic nano-photonic structures with this rigorous approach [5,29].

2.2. Design of Light Trapping Architecture Methods and Structure

It is of much interest to harvest photons in thin Si-layers/foils over a broad-band of wavelengths below the Si band edge (1.12 μm). The large index mismatch between silicon and air causes significant reflection losses for flat structures. This can be reduced using ergonomic tapered nanostructures which grade the refractive index from Si to the air value. Furthermore, the nanostructures diffract in-coming light and increase the photon path length within the absorber layer.

We performed a systematic set of optimizations for all parameters in this architecture (Figure 3). To conceptually understand the enhancement mechanism, it has been convenient to perform optimization by disassembling some of the components and starting with the simpler structure of two TiO_2 layers passivating silicon. The optimal passivating layer thicknesses was found to be 60 nm and 50 nm on the front and rear Si-surfaces, respectively, close to the expected quarter-wavelength for red light ($\lambda \approx 600$ nm).

After this step the front nano-cones were added and their structure was optimized. Finally, the back cones were added followed by the metallic back reflector. All structural parameters were optimized to achieve maximum anti-reflection and light trapping over the broad-band solar spectrum (300 nm–1100 nm) of interest for a silicon cell. We chose TiO_2 for the dielectric layers due to its dual characteristic of passivating p-type and n-type silicon surfaces and ability to transport electrons.

2.2.1. Front Nanocone Array

Light trapping in the Si absorber layer is achieved through the generation of wave-guided modes propagating parallel to the interface [30,31]. In order to achieve the maximum light absorption, ensuring anti-reflection and light trapping, the structural requirements (pitch, aspect ratio) of the front and rear cone morphology were studied individually, and then combined. The front cone height was first optimized to achieve maximum absorption in silicon. The structure simulated consists of a front texture comprised of flat silicon (1 µm thickness), front spacer layer, front cones, and rear spacer layer without cones (Figure 3a). The optimum cone height was studied for different array pitch values ranging from 0.25 µm to 1.5 µm in steps of 0.25 µm. As shown in Figure 3b,c, the weighted absorption $<A_w>$ and photo-current J_{SC} have a strong inter-dependence of the pitch and cone height. The absorption and J_{SC} are low at small pitch values (below 0.5 µm) irrespective of the cone height. There is a region of high absorption where $<A_w> \approx 0.61$ (in the orange region Figure 3a), where the cone height ranges from 0.6 to 1.2 µm, for pitch values of 0.75 to 1.5 µm, with the optimized cone height increasing with the pitch. J_{SC} shows similar trends with a correspondingly wider region of high $J_{SC} \approx 20.7\ \mathrm{mA/cm^2}$ (Figure 3c), where the optimized cone height increases from 0.6 to 1.4 µm as the pitch increases from 0.75 to 1.5 µm. As expected from mode coupling studies [31], the optimum pitch values are of the same order as the range of wavelengths that need to be absorbed. The simulation reveals it is advantageous to use a pitch larger than 0.75 µm, since optimum height has a tolerance of $>\pm 100$ nm and does not demand high precision processing equipment, an advantage for manufacturing. For example, a pitch of 1.25 µm is coupled with a cone height of 700 nm–900 nm.

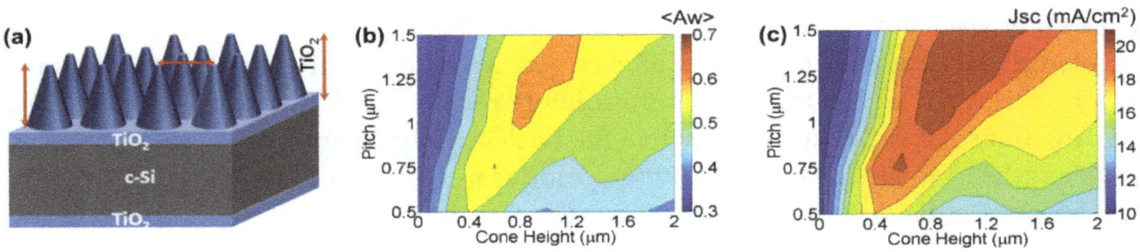

Figure 3. (**a**) Cell structure used for optimization of texture parameter using simulations. The structure consists of thin flat TiO_2 layers on the front and rear surfaces of flat silicon (1 µm). The cones are only on the front surface without rear cones and Ag metal reflector; (**b**) Weighted absorption, $<A_w>$ and (**c**) Short-circuit current density, J_{SC}, as a function of cone height for different pitch values. Figure shows optimum pitch >750 nm and cone height >500 nm with an increasing trend of tolerance of optimum cone height at larger pitch.

In the next optimization sequence, we kept the front cone geometry fixed (with pitch 750 nm and height 600 nm), introduced back cones, and varied the height of the back cones. For computational convenience, we chose the same pitch for the front and back photonic crystals. The height of the back cone was optimized and found to be ~200 nm.

Even with the back cones there is significant transmission through the structure. To completely eliminate this transmission, the back cones were embedded in the silver back reflector; this eliminated transmission and provided a rear mirror to enhance absorption and photo-current.

2.2.2. Absorption Enhancement

A sequence of light-trapping photonic crystal structures (Figure 4), that were systematically built-up on a flat silicon surface, were analyzed to understand the enhancement contributions from the different components. The four sequential configurations (Figure 4) are the (a) Grating-free cell; (b) Front-grating cell; (c) Rear-grating cell, and (d) Dual-grating cell with a metallic back-reflector.

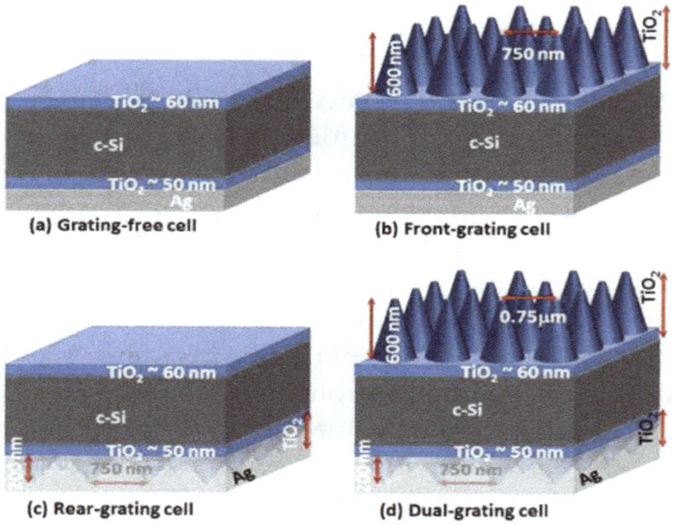

Figure 4. Sequence of light trapping structures on flat silicon. (**a**) Grating-free cell with thin layers of 60 and 50 nm on front and rear surface, respectively, and flat Ag back-reflector; (**b**) Front-grating cell with only front cones of height of 600 nm and a pitch of 750 nm and flat Ag back-reflector; (**c**) *Rear-grating cell* with only rear cones of height of 200 nm and a pitch of 750 nm and corrugated Ag back-reflector; (**d**) *Dual-grating cell* with a combination of front and rear cones with optimized parameters used in 'b' and 'c' with corrugated Ag back-reflector.

Scattering matrix simulations were performed for each solar architecture (Figure 4) with a flat c-Si absorber layer of 2 μm. The optimized value for the photonic crystal arrays were obtained, and the wavelength-dependent absorbance of each architecture (Figure 5a–d) was compared with the Lambertian ($4n^2$) limit, in which the light is completely randomized [32] within the absorber layer.

The starting flat solar cell (Figure 4a) with a flat back reflector (Figure 5a) displays poor absorbance at short wavelengths ($\lambda < 550$ nm), wave-guiding modes in the red and near-IR (between 600 nm and 900 nm), absorbance considerably below the Lambertian limit, and a photo-current of 15 mA/cm^2 (Figure 5e,f). The addition of the front photonic crystal (Figure 4b) reduces the reflectance loss, coupling light more effectively into the cell, and increases the absorbance at shorter wavelengths ($\lambda < 550$ nm) to near the Lambertian limit (Figure 5b). The overall absorbance at longer wavelengths ($\lambda > 800$ nm) is still low, indicating inadequate light trapping.

Alternatively, when the back photonic crystal is added to the flat solar cell (Figure 4c), the near-IR absorbance ($\lambda > 800$ nm) is considerably improved, approaching the Lambertian limit due to the dense mesh of waveguided modes (Figure 5c). However, the absorbance at short wavelengths ($\lambda < 550$ nm) is still low, similar to the flat cell, since these blue-green photons are reflected away from the solar cell. This emphasizes that the front grating is particularly beneficial in reducing reflectance loss and coupling short λ photons into the solar cell. The rear photonic crystal affects only longer-wavelength near-IR photons that reach the back reflector without being absorbed in the absorber layer and generates enhanced absorption at the Lambertian limit (λ > 800 nm).

The optimized front- and rear-photonic crystals were combined (Figure 4d) to generate a high absorbance over the entire wavelength range (Figure 5d). The absorbance and photo-current (Figure 5e,f) are very close to the Lambertian limit. As observed in previous studies [5,13] the absorbance exceeds the Lambertian limit at the specific wavelengths where wave-guiding in the plane of the structure occurs [33,34].

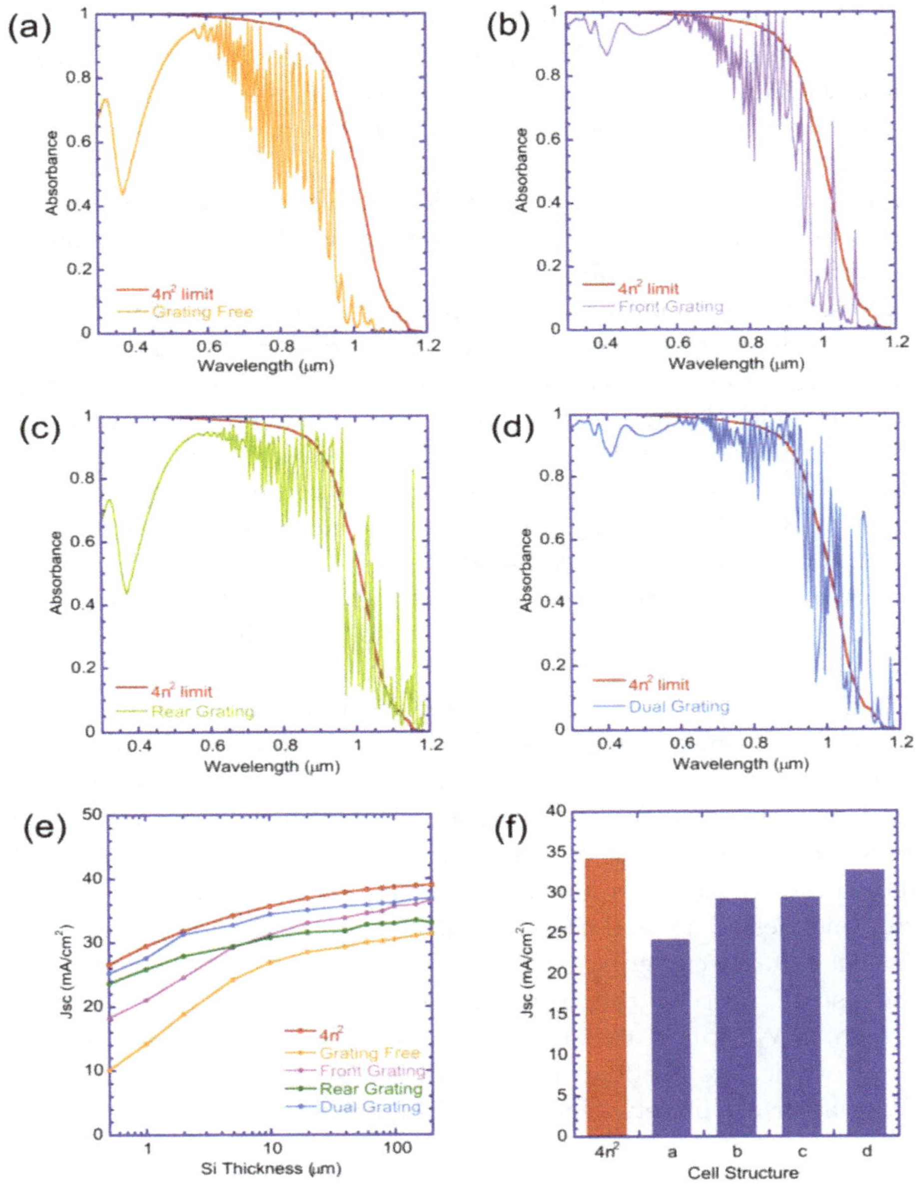

Figure 5. Comparison of absorption spectra of planar silicon with $4n^2$ absorption limit for different light trapping configurations shown in Figure 2 such as (**a**) *grating-free cell*, (**b**) *front-grating cell*, (**c**) *rear-grating cell*, and (**d**) *dual-grating cell* using the optimized grating parameter for 2 μm silicon; (**e**) Comparison of J_{SC} of planar cell for different light trapping configurations shown in Figure 2 with respect to silicon thickness using optimized grating parameters for 2 μm silicon; (**f**) J_{SC} of the cell for a particular thickness of 2 μm for four different configurations.

The dense mesh of wave-guiding modes generates absorption maxima from the diffraction resonances which are observed when the phase difference between modes reflected from the front and rear surfaces are multiples of 2π or $k_z = m\pi/d$, k_z is the z-component of the wave-vector [5]. For a triangular lattice, any reciprocal lattice vector G has components $i(2\pi/a)$ and $(2j - i)(2\pi/a)/\sqrt{3}$, respectively, where i and j are integers. Incident light with wave-vector $k_{||}$ is diffracted according to $k_{||}' = k_{||} + G$, since $k_z = m\pi/d$, $k_z^2 + k_{||}^2 = n(\lambda)^2(\omega/c)^2$ and wave-guiding occurs at resonant wavelengths given by:

$$\lambda(i,j,m) = \frac{2\pi n(\lambda)}{\left[\left(i^2 + \frac{1}{3}(2j-i)^2\right)\left(\frac{2\pi}{a}\right)^2 + \left(\frac{m\pi}{d}\right)^2\right]^{1/2}} \quad (4)$$

where i, j and m are integers. A dense mesh of wave-guided modes occurs in the long λ region (Figure 5), for our choice of parameters, where the wavelength inside silicon $\lambda/n(\lambda)$ is smaller than the pitch (a) of the cones, resulting in large values of i, j and m. The phase coherence of waves is assumed at each interface. Accordingly, the resonant wavelength at which waveguide mode occurs is given by Equation (3).

The diffraction resonances result in propagation of wave-guided modes in the plane of the absorber layer that increases the path length and results in significant enhancement in the absorption at the wavelengths (Figure 5b–d) and enhanced J_{SC}.

It is of interest to study the enhancement and photo-current J_{SC} as a function of silicon thickness (Figure 5e) for the different solar cell architectures. The front cones are more effective at higher c-Si thicknesses (>40 μm), providing a substantial increase from the grating-free flat solar cell. The rear cones are more beneficial at lower thickness (<10 μm) in increasing the computed J_{SC} (Figure 5e) from the flat case, but their benefit levels off at larger thicknesses. At the smaller thicknesses, diffraction induced by the rear cones improved the photon absorption at longer wavelengths. As discussed by Bermel et al. [35], the modes diffracted by the rear cones improves the coupling of electromagnetic field with diffraction modes as c-Si thickness is reduced.

The highest performing architecture is the dual-grating cell (Figure 4d) for all Si-thicknesses. The relative enhancement in J_{SC} of the dual-photonic crystal cell with respect to the flat cell is 94% for the cell thickness of 1 μm as compared to 11% for a 200 μm thick cell. Figure 5f, shows the progression of J_{SC} for a silicon thickness of 2 μm for different dielectric photonic crystal arrays. The flat solar cell has J_{SC} = 18.9 mA/cm^2. The front-grating improved the absorption and J_{SC} to 24.6 mA/cm^2, while the rear-grating alone improved it further to J_{SC} of 27.9 mA/cm^2. The dual-grating cell resulted in the J_{SC} of 31.3 mA/cm^2, which is very close to the $4n^2$ limit of 31.8 mA/cm^2. The enhancement in $<A_w>$ and J_{SC} are found to be 69% and 66%, respectively, in comparison to the grating-free cell. The J_{SC} difference between Lambertian and the dual PC enhanced cell increases with thickness while the enhancement factors decrease with thickness. The dual nanocone arrays offer great promise to trap light in ultra-thin c-Si (<2 μm) and collect suitable currents.

We varied the aspect ratio of the front and rear nano-cones by varying their base radii, R, for a different pitch, a. We found that $R/a \approx 0.4$ provides the best absorption, consistent with previous results for photonic crystal enhanced absorption [24,36]. With this geometry, there is a small flat section of the titania surface in between neighboring nano-cones.

2.2.3. Variation with Angle of Incidence

Strong light absorption over a wide range of incident angles is crucial for the optimum performance of a solar cell during the entire daytime operating hours. Omni-directional absorbance is highly beneficial in capturing diffuse sunlight. The cell response was simulated for the optimized dual photonic crystal (PC) light-trapping architecture as a function of angle of incidence (AoI) (θ) for both incident polarizations. Figure 6a shows the contour plot of angle-dependent absorption of the 2 μm thick c-Si substrate with optimized surface nanostructures for different incident angles (in the azimuthal x–z plane) over the entire wavelength range of the solar spectrum. The absorbance exceeds 0.9 at an angle θ of 10° and is nearly constant above 0.8 until θ ≈ 70°, above which the absorption decreases. Figure 6b shows the predicted photo-current for both p-and s-polarizations over a wide range of θ. J_{SC} is similar for both polarizations at large θ. Interestingly, the photo-current J_{SC} has a maximum at θ ≈ 30° for p-polarization, while it has a maximum for 10° for s-polarization, so that these photonic crystal arrays perform better when the angle of incidence is somewhat away from the normal direction, and the incident light is nearly parallel to the surface of the nano-cones. This is advantageous for solar collection with fixed (non-tracking) solar panels, where the sun sweeps across the sky. The average J_{SC} for both polarizations has a maximum of about 33.7 mA/cm^2 at θ ≈ 10° and is over 30 mA/cm^2 for θ > 70°. A similar trend was observed for the 1D or 2D silicon grating structures [37] and nano-photonic conformal nc-Si solar architectures [5]. For the grating-free cell, J_{SC} for the p-polarization peaks at an θ ≈ 60° as a result of the Brewster angle, which reduces reflectance to zero at the front surface (inset of Figure 6b). Averaging both polarizations results in a constant J_{SC} until θ ≈ 60°, after which the absorption and photo-current suffer sharply. The substantial increase of J_{SC} for the nanocones over wide range of θ is due to the improved in-coupling of light by the PC array as discussed by Heine and Morf [37] for silicon gratings with respect to the incident angle of light.

The wide-angle light trapping and polarization independent characteristics of the nanocone arrays are due to their smooth graded-index profile and their gradual variation of optical density coupling which is better with incident light. The grating structures outperform the biomimetic silicon nanostructures [38] or silicon gratings with low-aspect ratio [39].

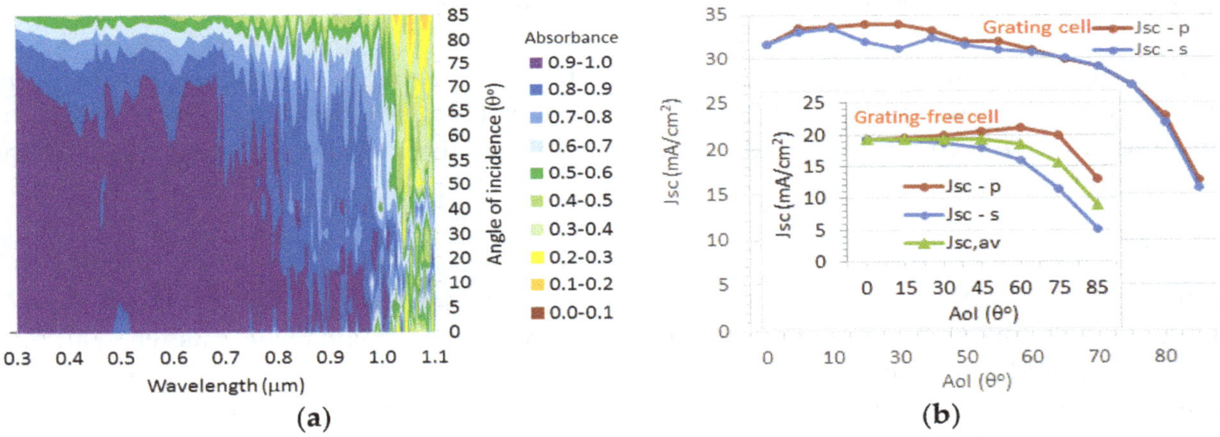

Figure 6. (**a**) Absorption spectra of the dual-grating cell (c-Si thickness = 2 μm) and (**b**) corresponding J_{SC} of the cell as a function of angle of incidence (AoI) in the range 0°–85° in steps of 5° (inset shows the J_{SC} of grating-free cell for p- and s-polarization and its average). The average absorption is more than 80% over a wide wavelength band. J_{SC} is independent of polarization and is less influenced until the AoI reaches 70°, showing the omni-directionality of the nanocone grating structures.

2.2.4. Field Distribution

The photonic crystal structures are effective in exciting wave-guided modes at wavelengths in the order of, or slightly larger than, the pitch of the arrays, according to the wave-guiding condition of the round-trip phase difference from the top and bottom being a multiple of 2π (Equation (3)). The transverse electric (TE) (x-polarized) modes in the c-Si of a thickness of 2 μm are simulated in a cross section of the solar architecture. The field distribution of the mode (Figure 7) shows the electric field intensity $|E|^2$ at an incident wavelength of 500 nm (Figure 7a) and 700 nm (Figure 7b) (typical for the long λ regime) for the dual-photonic crystal cell with the optimized light trapping architecture. At λ = 500 nm, the incident photons are effectively absorbed at the top of the cell (Figure 6a) and do not interact with the rear nanostructure. Since the absorption length (l_d) of light at 750 nm exceeds the thickness of the cell, the incident photons reach the rear photonic crystal array and both front and rear arrays are effective in diffraction. Standing waves are produced in the silicon layer by the nanocone arrays (Figure 7b). The separation between the maxima of the standing waves in the z-direction is the expected value $\lambda/2n$, which is close to 100 nm. This separation between maxima increases as the incident wavelength increases. In the case of nanocone array architecture (Figure 7b), there are regions of high field concentration at the top of the layer, where the field intensity is enhanced by a factor of ~7, which is one order more than that of its grating-free counterpart.

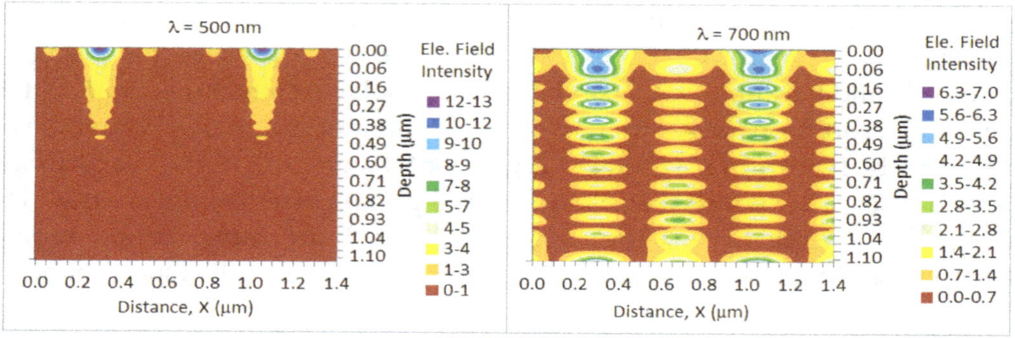

(a) (b)

Figure 7. Electric field intensity distribution across the silicon absorber layer at an incident wavelength of (**a**) 500 nm and (**b**) 700 nm for the dual-grating cell with optimized parameters for the front and rear nano-cone arrays. The incident electric field intensity is normalized to 1.

The guided mode in the dual photonic crystal structure propagates in the x-direction planar to the silicon layer, with a wavelength controlled by the pitch ($a \approx 750$ nm) of the structure. For the flat cell, the guided modes cannot be excited by plane wave illumination from air, since the light line lies above the light line for the dielectric.

2.2.5. Parasitic Losses

One of the critical questions is the magnitude of the parasitic losses within the metal electrode at the back of the cell, especially at long wavelengths, and the predicted enhancements when all parasitic losses are accounted for. Since the Ag electrode has $n_2 = Im(n) > 0$, there are losses in the silver at longer wavelengths. Qualitatively, these losses are expected to be more severe for thinner cells where more light reaches the back-reflector. In the previous sections, the calculated wavelength dependent absorption, $A(\lambda)$ included contributions from the Ag [5,24]. To de-convolute the contribution of the absorption from only the Si layer, it was necessary to extract the electric fields $E(r)$ and magnetic fields $H(r)$ in real space by transformation of the computed fields in Fourier space [24]. The absorption rate R per unit volume at the position r is:

$$R = \frac{1}{8\pi}\omega Im(\varepsilon(\omega))|E(r)|^2 \tag{5}$$

The absorption A is obtained by integrating over the volume V of the absorber layer:

$$A = \frac{1}{V}\int \frac{4\pi n_1 n_2}{\lambda}|E(x,y,z)|^2 dxdydz \tag{6}$$

Such a procedure is computationally laborious and not well suited for the present scattering matrix method where Maxwell's equations are solved in Fourier space rather than real space. In fact, such procedures are more amenable to real space solvers such as the finite difference time domain (FDTD) method. The use of Fourier space has great advantages for parallelization and computational efficiency at the expense of not directly calculating the real space fields and separating the losses.

Nonetheless, we have implemented this procedure of extracting the real space fields and we separately calculate the losses only in the silicon. More than 1000 CPU hours are needed in a multi-processor environment utilizing 64 processors.

We simulated the absorption and J_{SC} (Figure 8) with this real-space method as a function of the Si-absorber layer thickness and compared it with the previous Fourier space simulation that includes the absorption in the metal. The absorption and J_{SC} is lower with the real-space method due to parasitic absorption in Ag. The difference between the real space and Fourier space is largest for the thinner absorber layers, but nearly negligible for thicker Si absorber layers (Figure 8). The difference between the photo-current by the real-space and Fourier space methods is the parasitic absorption, which varies from 1% for 200 μm Si to 15% for 10 μm and 33% for the thinnest 1 μm Si absorber. J_{SC} reaches ~24 mA/cm^2 for the 2 μm Si after accounting for parasitic losses. In comparison, $<A_w>$ is 0.5% to 25% lower as the Si thickness is decreased from 200 μm to 1 μm, respectively. This trend is to be expected since the red and near-IR wavelengths are absorbed in thicker cells and do not reach the back reflector, which consequently has negligible effect on parasitic absorption. A large fraction of the incident light reaches the back reflector in thin cells, increasing the parasitic absorption. Parasitic losses are due to ohmic losses in the structured Ag-dielectric interface. In principle, the parasitic absorption may be reduced by the use of a conductive distributed back reflector (DBR) [20,29] with flat interfaces, but the wavelength bandwidth of high reflectance may be smaller.

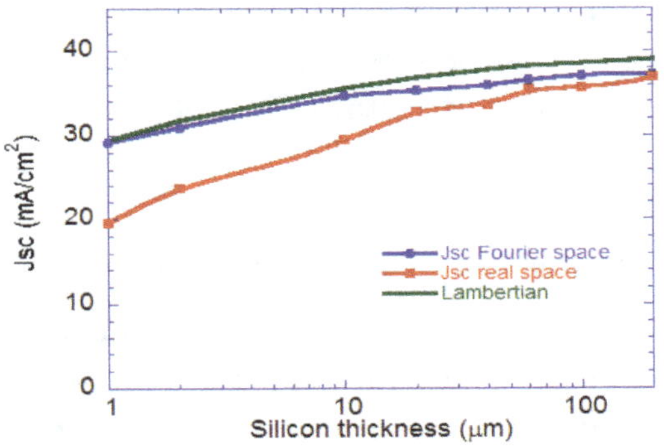

Figure 8. Comparison of the simulated short-circuit photo-current J_{SC} as a function of the thickness of the Si absorber layer in the real-space method utilizing only the absorption in the Si layer, compared to the Fourier space method which includes the absorption in all layers. The difference between the real-space and Fourier space results is the parasitic absorption. The Lambertian limit is shown for comparison.

3. Discussion

Flat interfaces of silicon with passivating titania are expected to have a minimum density of surface defects with a high degree of passivation and electronic quality in comparison to that of the textured or patterned silicon surfaces. The photonic crystal arrays reside within the titania that control light trapping and can be easily fabricated by spraying titania gels on silicon and nano-imprinting the gel in a roll-to-roll method, viable for large scale production. Micro-transfer molding techniques using polydimethylsiloxane (PDMS) templates are suitable for small area laboratory cells. Additionally, the titania layers could be cured during the metallization step, where the metal top contacts are fabricated to the silicon layer. This reduces the additional substrate heating during the deposition step of the antireflection/passivation layer. The spacer layer thickness can be controlled by the processing parameters of pressure, temperature, etc., during imprinting. Our study demonstrates a large tolerance of the optimum design to grating parameters, which does not impose stringent conditions over the processing parameters. The grating structure allows for the use of thinner silicon substrates and avoids the physical or chemical etching or patterning processes that are used for light management in the standard silicon solar cell fabrication. The imprint process is proved to be attractive on plastic substrates for mass production [40].

After accounting for parasitic losses in the thin cells (Figure 8) we obtain predicted short circuit currents J_{SC} of ~30 mA/cm^2 (24 mA/cm^2) for 20 μm (2 μm) thick cells. Utilizing the highest reported cell parameters of V_{oc} = 0.74 V and fill factor (FF) of 0.827 for the Panasonic c-Si cell [2,41], we predict the efficiency to be ~20.2% (14.7%) for a 20 μm (2 μm) thick cell, which reduces Si material utilization by ~15 (compared to a 300 μm cell). This suggests that a ~20 cell μm may yield acceptable efficiency at significantly lower cost.

This work is significantly different from our previous dual photonic crystal solar architecture [24] which used (i) an organic absorber layer; (ii) a polymer lens array on glass for focusing; and (iii) a nanostructured organic layer, none of which are present here.

An alternative to the periodically structured photonic crystals is the use of a random texture on both the front and back surface [20] that randomizes light distribution in the material, a feature that has been extensively utilized in thin amorphous silicon cells.

An issue of much practical importance is that solar panels are subjected to harsh environmental conditions. To protect such panels with nanocone arrays we anticipate using a front glass protective sheet that is commonly employed in most solar panels. Thus, the nanocones are not in direct contact with the outside environment and are expected to have high stability. We have performed simulations with a protective glass sheet that shows the enhancements are similar to those in this paper.

This work mainly targets the state-of-the-art crystalline silicon cell structures (i.e., PERC/PERL/PERT) in the industrial environment irrespective of silicon thickness. These advanced structures require efficient passivation on both surfaces. Rear surface/back surface field (BSF) is contacted through via-holes, which are generally made using screen printer/laser ablation followed by metal filling. Therefore, the charge collection is ensured through via-holes filled with metal, and not through the TiO_2 layer.

Proof of Concept Demonstration of TiO_2 Nanocone Array Fabrication

We have developed a nano-imprinting procedure to demonstrate the feasibility of fabricating nano-cone arrays of titania on silicon substrates, and we illustrate a pathway for experimental fabrication of the simulated light-trapping structures. Briefly, the process started with a polycarbonate master pattern having nanocone gratings in triangular lattice with a period of ~750 nm and height ~150 nm. The master pattern was used to replicate the inverse of the nanocone array on a polydimethylsiloxane (PDMS) mold using soft lithography, using a procedure described in our previous publication [42]. The resulting PDMS mold with the inverse nanocup array was then used to imprint the TiO_2 film. We adopted the process in Chen et al. [43] with slight modifications to fabricate the TiO_2 film. We spin-coated 0.45 M titanium diisopropoxide bis(acetylacetonate) on a clean silicon wafer at 3000 rpm for 30 s. The film was then imprinted under elevated temperature and pressure with the PDMS mold using the process illustrated schematically in Figure 9 with more details in the following Section 4.

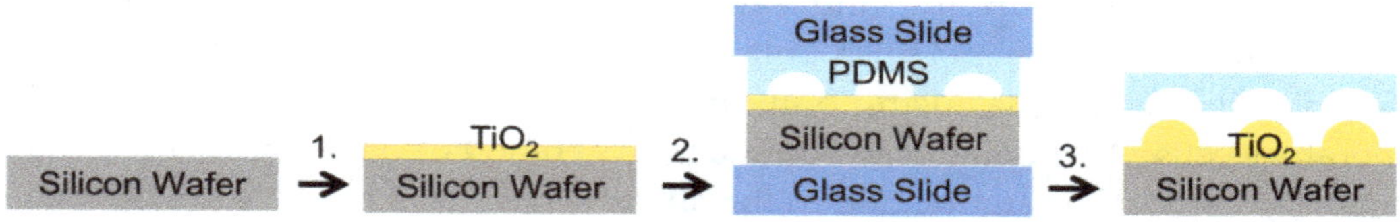

Figure 9. Schematic of the TiO_2 nanoimprinting process. (**1**) TiO_2 film is spin-coated on silicon substrate using a precursor titanium diisopropoxide bis(acetylacetonate); (**2**) The polydimethylsiloxane (PDMS) stamp having nanocups is placed on the spin-coated film with patterned side facing the film. The whole assembly is sandwiched between two glass slides and held together with binder clips (not shown here); (**3**) After keeping at ~170 °C for 15 min, the binder clips are released to reveal the inverse of PDMS nanocups on the TiO_2 film.

Figure 10a shows the atomic force microscopy image of the nanoimprinted TiO_2 film fabricated using soft lithography. The atomic force microscopy (AFM) line scan and three-dimensional view (Figure 10b) shows the well-replicated regular array of nanocones on TiO_2 with a period of ~750 nm and average height of ~35 nm. It is to be noted that the feature height can be varied easily using the initial PDMS stamp of different nanocup depth. Since the TiO_2 material is inherently hard and compact, the shallower stamp has difficulty in penetrating deeper into the film. However, the PDMS nanocups of much higher depths of the order of ~1 micron will yield nanocone gratings with height >500 nm–600 nm which is of interest to solar cell fabrication for efficient light trapping based on simulation results. The similar imprinting procedure has been successfully implemented by our group in nanoimprinting various polymer films such as polystyrene [40] and poly(L-lactic acid) [44].

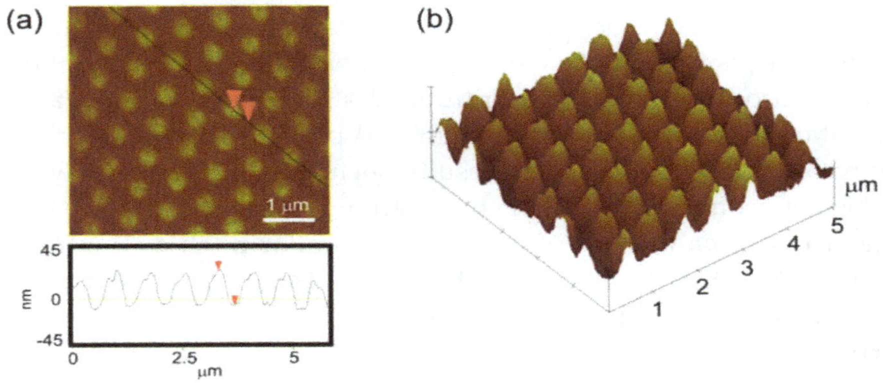

Figure 10. (**a**) Atomic force microscopy (AFM) image of the periodic array of nanocones imprinted on TiO_2 film. The AFM line scan shows the periodicity at ~750 nm and the average height of nanocones at ~35 nm; (**b**) Three-dimensional view of the structure showing titania nanocone arrays.

4. Materials and Methods

The solar cell is divided into layers in the z direction. Within each layer of the structure (Figure 2), the dielectric function $\varepsilon(x, y)$ is a function of the spatial coordinates (x, y) but not of z. This allows the dielectric function in each layer to be expanded in a two-dimensional basis of reciprocal lattice vectors G, providing the Fourier components of the dielectric function $\varepsilon(G)$. Similarly, the electric fields and magnetic fields $(E(G), H(G))$ are expanded in Bloch waves. Within each layer, Maxwell's equations are solved in Fourier space in an eigenvalue expansion. A transfer matrix is used to relate the E and H fields within each layer. Maxwell's equations are integrated with the continuity boundary conditions throughout the unit cell to obtain the scattering matrices of each layer and the entire structure. From the scattering matrix we find the total reflectance R (including diffracted beams) and transmission T (which is 0) at each incident wavelength. The absorption at each wavelength is then $A = 1 - R - T$. This computational algorithm is executed on massively parallel computing clusters (using 192 to 256 processors) with each wavelength/frequency sent to a different processor, which considerably speeds the design time. The advantage of this scattering matrix approach is that a real-space grid is not needed, avoiding complications of large memory requirements for substrates.

Experimentally, the nanocones are fabricated using nanoimprint lithography (NIL) of sol-gel derived TiO_2 layers. The PDMS mold was generated from the polycarbonate master. The precursor solution 0.45 M titanium diisopropoxide bis(acetylacetonate) was spin-coated on a clean silicon wafer at 3000 rpm for 30 s. The PDMS mold was placed such that the patterned side faced the spin-coated titania film and the whole assembly was sandwiched together in two glass slides using binder clips, which served to apply adequate pressure for nanoimprinting. After keeping the assembly at 170 °C for 15 min, the binder clips were released and the PDMS stamp was subsequently lifted to reveal the nanoimprinted film. The sample was then sintered in a furnace at 550 °C for 15 min. The nanoimprinted substrate was then immersed in 50 mM $TiCl_4$ solution for 30 min at 70 °C and washed with deionized

(DI) water, followed by annealing at 550 °C for 30 min. The purpose of annealing is to harden the nanoimprinted TiO_2 film.

The imprinting process is more industrial friendly for roll-to-roll (R2R) or large area fabrication. The evolution of R2R or roll-to-plate (R2P) makes the NIL process cheaper and compatible with the existing fabrication lines in the photovoltaic (PV) industry [45].

5. Conclusions

We have designed a highly absorbing thin film crystalline Si based solar architecture, using periodically patterned front and rear photonic crystals. In contrast to previous approaches, we utilize dielectric photonic crystals, ensuring that the c-Si absorber layer has completely flat interfaces, which is expected to result in high electronic quality and low carrier recombination. The mechanism for enhanced absorption is the generation of a dense mesh of wave-guided modes at near IR wavelengths, propagating in the Si absorbing layer. An optimized structure consists of a triangular lattice of front nanocones of titania with a height of 600 nm and pitch of 750 nm, although other ranges of height/pitch combinations can provide similar results, such as a 1000 nm pitch with a nanocone height of 800 $\pm$ 100 nm. The nanocone height and pitch are strongly inter-related. The back reflector consists of another periodic array of nanocones of titania with similar pitch as the front array, embedded with a silver back reflector. For thin c-Si absorber layers (<2 μm), we achieve an absorption and photo-current enhancement exceeding 100%. For thick c-Si absorber layers, the enhancement is up to 20%. The absorption has approached the Lambertian limit at small thicknesses (<10 μm) and is only slightly lower by (~5%) at larger thickness of 200 μm. Although the absorption at any particular wavelength can exceed the Lambertian limit, the absorption averaged over all wavelengths and the photo-current is below the Lambertian value. The limiting Lambertian value, when averaged over all wavelengths, is an aspect for fundamental studies.

Parasitic losses are significant (>25%) for ultra-thin Si layers below 2 μm thickness, but decrease rapidly as the thickness of the Si absorber layer increases, and are just ~1%–2% for thick (>100 μm) cells. Electric field distributions demonstrate waveguiding modes and high intensity focusing regions within the absorber layer. The strong enhancements predicted by simulation should be tested by fabrication of such nano-photonic structured solar cells. The proposed structure is very promising for high performance flexible, ultra-thin Si solar cells, providing a pathway for low cost high performing silicon solar modules which should be valuable to researchers in this area.

Acknowledgments: This work was supported (in part, Rana Biswas, Akshit Peer) by the U.S. Department of Energy (DOE), Office of Science, Basic Energy Sciences, Materials Science and Engineering Division. The research was performed at Ames Laboratory, which is operated for the U.S. DOE by Iowa State University under contract # DE-AC02-07CH11358. This work was also supported (in part, Prathap Pathi) by the IUSSTF (Indo-US Science and Technology Forum) and DST (Department of Science & Technology), Govt. of India, under the BASE (Bhaskara Advanced Solar Energy) Fellowship program (Award No. 2014/F-3/Prathap Pathi). The research used resources at the National Energy Research Scientific Computing Center (NERSC), which is supported by the Office of Science of the U.S. DOE under Contract No. DE-AC02-05CH11231.

Author Contributions: Rana Biswas and Prathap Pathi conceived the study, performed the simulations, and analyzed the data; Akshit Peer performed experimental studies; Rana Biswas, Prathap Pathi, and Akshit Peer wrote the manuscript.

References

1. National Renewable Energy Laboratory (NREL). National Center for Photovoltaics Solar Efficiency Chart. Available online: http://www.nrel.gov/pv/assets/images/efficiency_chart.jpg (accessed on 30 December 2016).
2. Green, M.A.; Emery, K.; Hishikawa, Y.; Warta, W.; Dunlop, E.D. Solar Efficiency Tables (version 47). *Prog. Photovolt.* **2016**, *24*, 3–11. [CrossRef]
3. Atwater, H.; Polman, A. Plasmonics for improved photonic devices. *Nat. Photonics* **2010**, *9*, 9–16.

4. Palik, E. *Handbook of the Optical constants of Solids*; Academic Press: San Diego, CA, USA, 1998.
5. Biswas, R.; Xu, C. Nano-crystalline silicon solar cell architecture with absorption at the classical $4n^2$ limit. *Opt. Exp.* **2011**, *19*, A672. [CrossRef] [PubMed]
6. Bhattacharya, J.; Chakravarty, N.; Pattnaik, S.; Slafer, W.D.; Biswas, R.; Dalal, V. A Novel photonic-plasmonic Structure for Enhancing Light Absorption in Thin Film Solar Cells. *Appl. Phys. Lett.* **2011**, *99*, 131114. [CrossRef]
7. Ferry, V.E.; Verschuuren, M.A.; Li, H.B.T.; Schropp, R.E.I.; Atwater, H.A.; Polman, A. Improved red-response in thin film a-Si:H solar cells with soft-imprinted plasmonic back reflectors. *Appl. Phys. Lett.* **2009**, *95*, 183503. [CrossRef]
8. Sai, H.; Saito, K.; Hozuki, N.; Kondo, M. Relationship between the cell thickness and the optimum period of textured back reflectors in thin-film microcrystalline silicon solar cells. *Appl. Phys. Lett.* **2013**, *102*, 053509.
9. Han, S.E.; Chen, G. Toward the Lambertian limit of light trapping in thin nanostructured silicon solar cells. *Nano Lett.* **2010**, *10*, 4692–4696. [CrossRef] [PubMed]
10. Zhao, H.; Ozturk, B.; Schiff, E.A.; Sivec, L.; Yan, B.; Yang, J.; Guha, S. Back reflector morphology effects and thermodynamic light-trapping in thin-film silicon solar cells. *Sol. Energy Mater. Sol. Cells* **2014**, *129*, 104–114. [CrossRef]
11. Battaglia, C.; Escarre, J.; Soderstrom, K.; Charriere, M.; Despeisse, M.; Haug, F.-J.; Ballif, C. Nanomoulding of transparent zinc oxide electrodes for efficient light trapping in solar cells. *Nat. Photonics* **2011**, *5*, 535–538. [CrossRef]
12. Soderstrom, T.; Haug, F.J.; Daudrix, V.T.; Ballif, C. Flexible micromorph tandem a-Si/μc-Si solar cells. *J. Appl. Phys.* **2010**, *107*, 014507. [CrossRef]
13. Wang, K.X.; Yu, Z.; Liu, V.; Cui, Y.; Fan, S. Absorption enhancement in ultrathin crystalline silicon solar cells with antireflection and light-trapping nanocone gratings. *Nano Lett.* **2012**, *2012*, 1616–1619. [CrossRef] [PubMed]
14. Wang, S.; Weil, B.D.; Li, Y.; Wang, K.X.; Garnett, E.; Fan, S.; Cui, Y. Large-area free-standing ultrathin single-crystal silicon as processable materials. *Nano Lett.* **2013**, *13*, 4393–4398. [CrossRef] [PubMed]
15. Zhang, R.; Shao, B.; Dong, J.R.; Zhang, J.C.; Yang, H. Absorption enhancement analysis of crystalline Si thin film solar cells based on broadband antireflection nanocone grating. *J. Appl. Phys.* **2011**, *110*, 113105.
16. Yablonovitch, E. Statistical ray optics. *J. Opt. Soc. Am.* **1982**, *72*, 899–907. [CrossRef]
17. Tiedje, T.; Yablonovitch, E.; Cody, G.D.; Brooks, B. Limiting efficiency of silicon solar cells. *IEEE Trans. Electron Devices* **1984**, *31*, 711–716. [CrossRef]
18. Xu, Q.; Johnson, C.; Disney, C.; Pillai, S. Enhanced broadband light trapping in c-Si colar cells using Nanosphere-Embedded Metallic Grating Structure. *IEEE J. Photovolt.* **2016**, *6*, 61–67. [CrossRef]
19. Eisenlohr, J.; Benick, J.; Peters, M.; Bläsi, B.; Goldschmidt, J.C.; Hermle, M. Hexagonal sphere gratings for enhanced light trapping in crystalline silicon solar cells. *Opt. Exp.* **2014**, *22*, A111–A119. [CrossRef] [PubMed]
20. Ingenito, A.; Isabella, O.; Zeman, M. Experimental demonstration of $4n^2$ classical absorption limit in nanotextured ultrathin solar cells with dielectric omnidirectional back reflector. *ACS Photonics* **2014**, *1*, 270–278. [CrossRef]
21. Ouyang, Z.; Pillai, S.; Beck, F.; Kunz, O.; Varalmov, S.; Catchpole, K.R.; Campbell, P.; Green, M.A. Effective light trapping in polycrystalline silicon thin film solar cells with localized surface plasmons. *Appl. Phys. Lett.* **2010**, *96*, 261109. [CrossRef]
22. Pillai, S.; Catchpole, K.R.; Trupke, T.; Green, M.A. Surface plasmon enhanced solar cells. *J. Appl. Phys.* **2007**, *101*, 093107. [CrossRef]
23. Varlamov, S.; Dore, J.; Evans, R.; Ong, D.; Eggleston, B.; Kunz, O.; Schubert, H.; Young, T.; Huang, J.; Soderstron, T.; et al. Polycrystalline silicon on glass thin film solar cells: A transition from solid-phase to liquid-phase crystallized silicon. *Sol. Energy Mater. Sol. Cells* **2013**, *119*, 246–255. [CrossRef]
24. Peer, A.; Biswas, R. Nano-photonic Organic Solar Cell Architecture for Advanced Light Trapping with Dual Photonic Crystals. *ACS Photonics* **2014**, *1*, 840–847. [CrossRef]
25. Chen, Y.; Elshobaki, M.; Gebhardt, R.; Bergeson, S.; Noack, M.; Park, J.-M.; Hillier, A.C.; Ho, K.-M.; Biswas, R.; Chaudhary, S. Reducing optical losses in organic solar cells using microlens arrays: Theoretical-experimental investigation of microlens dimensions. *Phys. Chem. Chem. Phys.* **2015**, *17*, 3723–3730. [CrossRef] [PubMed]
26. Shah, A.V.; Schade, H.; Vanecek, M.; Meier, J.; Vallat-Sauvain, E.; Wyrsch, N.; Kroll, U.; Droz, C.; Bailat, J. Thin-film silicon solar cell technology. *Prog. Photovolt.* **2004**, *12*, 113–142. [CrossRef]

27. Li, Z.Y.; Lin, L.L. Photonic band structures solved by a plane-wave based transfer matrix method. *Phys. Rev. E* **2003**, *67*, 046607. [CrossRef] [PubMed]
28. Biswas, R.; Neginhal, S.; Ding, C.G.; Puscasu, I.; Johnson, E. Mechanisms underlying extraordinary transmission in sub-wavelength hole arrays. *J. Opt. Soc. Am. B* **2007**, *24*, 2489–2495. [CrossRef]
29. Zhou, D.Y.; Biswas, R. Photonic crystal enhanced light-trapping in thin film solar cells. *J. Appl. Phys.* **2008**, *103*, 093102. [CrossRef]
30. Yu, Z.; Raman, A.; Fan, S. Fundamental limit of light trapping in grating structures. *Opt. Exp.* **2010**, *18*, A366–A380. [CrossRef] [PubMed]
31. Yu, Z.; Raman, A.; Fan, S. Fundamental limit of nanophotonic light trapping in solar cells. *Proc. Natl. Acad. Sci. USA* **2010**, *107*, 17491–17496. [CrossRef] [PubMed]
32. Schiff, E.A. Thermodynamic limit to photonic-plasmonic light trapping in thin films on metals. *J. Appl. Phys.* **2011**, *110*, 104501. [CrossRef]
33. Van Lare, M.; Lenzmann, F.; Verschuuren, M.A.; Polman, A. Mode coupling by plasmonic surface scatterers in thin-film silicon solar cell. *Appl. Phys. Lett.* **2012**, *101*, 221110. [CrossRef]
34. Ferry, V.E.; Munday, J.N.; Atwater, H.A. The design considerations for plasmonic photovoltaics. *Adv. Mater.* **2010**, *22*, 4794–4808. [CrossRef] [PubMed]
35. Bermel, P.; Luo, C.; Zeng, L.; Kimerling, L.C.; Joannopoulos, J.D. Improving thin-film crystalline silicon solar cell efficiencies with photonic crystals. *Opt. Exp.* **2007**, *15*, 16986–17000. [CrossRef]
36. Zanotto, S.; Liscidini, M.; Andreani, L.C. Light trapping regimes in thin film silicon cells with a photonic pattern. *Opt. Exp.* **2010**, *18*, 4260–4274. [CrossRef] [PubMed]
37. Heine, C.; Morf, R.H. Sub-micrometer gratings for solar energy applications. *Appl. Opt.* **1995**, *34*, 2476–2482. [CrossRef] [PubMed]
38. Huang, Y.; Chattopadhyay, S.; Jen, Y.; Peng, C.; Liu, T.; Hsu, Y.; Pan, C.; Lo, H.; Hsu, C.; Chang, Y.; et al. Improved broadband and quasi-omnidirectional anti-reflection properties with biomimetic silicon nanostructures. *Nat. Nanotechnol.* **2007**, *2*, 770–774. [CrossRef] [PubMed]
39. Li, J.; Yu, H.; Li, Y.; Wang, F.; Yang, M.; Wong, S.M. Low aspect-ratio hemispherical nanopit surface texturing for enhancing light absorption in crystalline Si thin film-based solar cells. *Appl. Phys. Lett.* **2011**, *98*, 021905. [CrossRef]
40. Ok, J.G.; Youn, H.S.; Kwak, M.K.; Lee, K.; Shin, Y.J.; Guo, L.J.; Greenwald, A.; Liu, Y. Continuous and scalable fabrication of flexible metamaterial films via roll-to-roll nanoimprint process for broadband plasmonic infrared filters. *Appl. Phys. Lett.* **2012**, *101*, 223102. [CrossRef]
41. Masuko, K.; Shigematsu, M.; Hashiguchi, T.; Fujishima, D.; Kai, M.; Yoshimura, N.; Yamaguchi, T.; Ichihashi, Y.; Yamanishi, T.; Takahama, T.; et al. Achievement of more than 25% conversion efficiency with crystalline silicon heterojunction solar cell. *IEEE J. Photovolt.* **2014**, *4*, 1433–1435. [CrossRef]
42. Peer, A.; Biswas, R. Extraordinary optical transmission in nanopatterned ultrathin metal films without holes. *Nanoscale* **2016**, *8*, 4657–4666. [CrossRef] [PubMed]
43. Chen, Q.; Zhou, H.; Hong, Z.; Luo, S.; Duan, H.S.; Wang, H.H.; Liu, Y.; Li, G.; Yang, Y. Planar heterojunction perovskite solar cells via vapor-assisted solution process. *J. Am. Chem. Soc.* **2013**, *136*, 622–625. [CrossRef] [PubMed]
44. Peer, A.; Dhakal, R.; Biswas, R.; Kim, J. Nanoscale patterning of biopolymers for functional biosurfaces and controlled drug release. *Nanoscale* **2016**, *8*, 18654–18664. [CrossRef] [PubMed]
45. Kooy, N.; Mohamme, K.; Pin, L.T.; Guan, O.S. A review of roll-to-roll nanoimprint lithography. *Nanoscale Res. Lett.* **2014**, *9*, 320. [CrossRef] [PubMed]

5

Correlation between CdSe QD Synthesis, Post-Synthetic Treatment and BHJ Hybrid Solar Cell Performance

Michael Eck [1] and Michael Krueger [2,*]

[1] Department of Microsystems Engineering (IMTEK), University of Freiburg, Georges-Köhler-Allee 103, 79110 Freiburg, Germany; eck@imtek.de

[2] Department of Physics, Carl von Ossietzky University of Oldenburg, Carl-von-Ossietzky-Straße 9-11, 26129 Oldenburg, Germany

* Correspondence: michael.krueger@uni-oldenburg.de

Academic Editors: Guanying Chen, Zhijun Ning and Hans Agren

Abstract: In this publication we show that the procedure to synthesize nanocrystals and the post-synthetic nanocrystal ligand sphere treatment have a great influence not only on the immediate performance of hybrid bulk heterojunction solar cells, but also on their thermal, long-term, and air stability. We herein demonstrate this for the particular case of spherical CdSe nanocrystals, post-synthetically treated with a hexanoic acid based treatment. We observe an influence from the duration of this post-synthetic treatment on the nanocrystal ligand sphere size, and also on the solar cell performance. By tuning the post-synthetic treatment to a certain degree, optimal device performance can be achieved. Moreover, we show how to effectively adapt the post-synthetic nanocrystal treatment protocol to different nanocrystal synthesis batches, hence increasing the reproducibility of hybrid nanocrystal:polymer bulk-heterojunction solar cells, which usually suffers due to the fluctuations in nanocrystal quality of different synthesis batches and synthesis procedures.

Keywords: bulk heterojunction solar cells; hybrid solar cells; nanocrystals; quantum dots; conjugated polymers; post-synthetic treatment; thermal annealing; long-term stability

---

1. Introduction

Solar cells possessing a photoactive layer made out of semiconducting nanocrystals (NCs) and a conjugated polymer blend are called hybrid bulk heterojunction (BHJ) solar cells. Their photoactive layer typically possesses a thickness of around 100 nm, and is deposited from a NC/polymer dispersion [1]. A typical device structure is shown in Figure 1a. Within the photoactive layer, the polymer is mostly utilized as donor material for the electrons arising from the photo-induced excitons, while the NCs serve as electron acceptors and extraction material. Already the seminal work of Greenham *et al.* [2] published in 1996 has acknowledged the crucial negative impact of the insulating NC synthesis ligands on the solar cell performance, and a post-synthetic pyridine treatment of NCs was reported to improve overall solar cell performance. Since then, hybrid BHJ solar cells have undergone a remarkable development over the years (see Figure 1b), now exceeding efficiencies of 4% [3–6]; the currently best hybrid solar cells using NCs of different shapes have in common the usage of the low-bandgap polymer PCPDTBT (poly[2,6-(4,4-bis-(2-ethylhexyl)-4H-cyclopenta[2,1-b;3,4-b′]dithiophene)-alt-4,7-(2,1,3-benzothiadiazole)]) and of CdSe as NC material. Moreover, the best solar cells of the respective NC shape (spherical [6], elongated [5], multibranched [4]) all utilize the attachment of a thiol group to the CdSe NC surface. Thereby, high open circuit voltages of 0.71–0.74 V are achieved, leading to power conversion efficiencies (PCE) of 4.05%–4.7%.

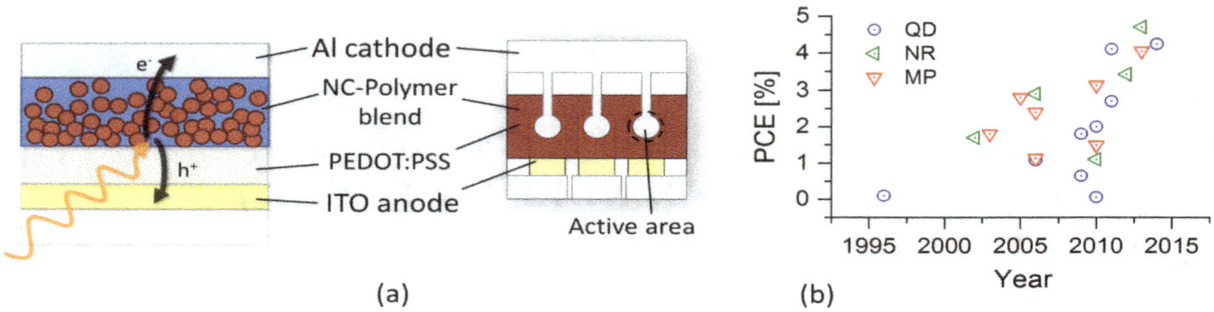

Figure 1. (**a**) Working principle and device structure shown on the cross-sectional and on the top view of the investigated hybrid bulk heterojunction (BHJ) solar cell with the active area size of 0.07 cm^2. (**b**) Power conversion efficiencies of selected hybrid BHJ solar cells published over the years from nanocrystals (NCs) of different shape: quantum dots (QD) [2,3,6–12], nanorods (NR) [5,10,13–15], multipods (MP) [4,10,16–20].

Nevertheless, despite almost reaching the PCEs of (poly[2,6-(4,4-bis-(2-ethylhexyl)-4H-cyclopenta[2,1-b;3,4-b′]dithiophene)-alt-4,7-(2,1,3-benzothiadiazole)]) (PCPDTBT):fullerene-based BHJ solar cells, which are up to 5.24%–5.5% [21,22], the utilization of NCs as electron acceptors has the aforementioned disadvantage of requiring post-synthetic NC treatment to reduce and or exchange the initial synthesis ligands. This treatment, however, may result in the partial destruction of the NC surface [23]—leading to increased charge recombination—and to the formation of NC aggregates, which might introduce shortcuts over the thin active layer. Hence, several publications on hybrid BHJ solar cells report the filtration of the post-synthetically treated NC dispersion, in order to remove large NC aggregates [4,20]. The approach we follow is to avoid NC aggregation within the process of post-synthetic NC treatment prior to incorporation into the polymer:NC blend dispersion. We achieve this by a gradual reduction of the NC ligand sphere [11,15,24] through a protonation step of the hexadecylamine (HDA) capping ligands with hexanoic acid (HA) and their subsequent solvation with the polar solvent methanol. We observed an influence from the post-synthetic treatment duration on the NC ligand sphere size (see Figure 4). Therefore, we believe that we are able to minimize the post-synthetically induced number of NC surface defects, which is supported by the excellent ideality factor of 1.22–1.33 for our CdSe/PCPDTBT solar cells [6], which is relatively close to unity compared to other hybrid BHJ solar cells [25–28], even when compared to organic BHJ photovoltaic cells [29,30].

2. Results and Discussion

2.1. NC Synthesis

CdSe QDs, capped with HDA and trioctylphosphine oxide (TOPO) ligands, were synthesized according to Yuan *et al.* [31]. An exemplary transmission electron microscopy (TEM) picture of the spherical CdSe NCs with a typical diameter of about 6.5 nm as well as the ultraviolet-visible (UV-Vis) absorption and photoluminescence (PL) spectrum are given in Figure 2.

In order to increase the NC size uniformity, the precursor concentration was increased from a 100:1:1 (hexadecylamine/trioctylphosphine oxide):(cadmium-stearic acid):(trioctylphosphine-selenid) (HDA/TOPO:Cd-SA:TOP-Se) molar ratio to a 100:2:2 ratio [24]. Furthermore, to study the influence of unintentionally occurring fluctuations of practically utilized precursor ratios for the NC synthesis, the Cd:Se precursor ratio was varied from 3:2, over 2:2, to 2:3. The parameters of PL intensity, PL peak position, and full width at half maximum (FWHM) were measured and compared to aliquots taken during the NC synthesis at different reaction times and are depicted in Figure 3.

From the parameters of the CdSe quantum dots (QDs)—determined from extracted samples during 2 different syntheses for each precursor ratio—shown in Figure 3, one can observe that indeed a lower minimal FWHM is observed when using higher precursor concentrations. Moreover, a tendency of lower minimum FWHM is also observed for the increased Se precursor concentration; this synthesis furthermore exhibits a higher PL quantum yield in the synthesis brightpoint (point where the PL intensity reaches its maximum during synthesis). These observations confirm the findings firstly described by Qu *et al.* [32]. The properties of CdSe QDs in their brightpoint of different precursor concentrations—utilized for BHJ hybrid solar cell fabrication—are summarized in Table 1.

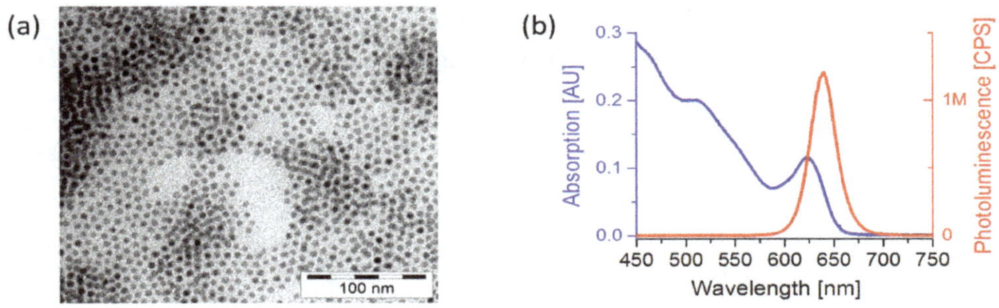

Figure 2. (**a**) Transmission electron microscopy (TEM) image of CdSe nanocrystals (NCs) used for the investigation from a 100:3:2 (hexadecylamine/trioctylphosphine oxide):(cadmium-stearic acid):(trioctylphosphine-selenid) ((HDA/TOPO):(Cd-SA):(TOP-Se)) ratio, synthesized for 30 min at 300 °C by wet-chemical hot injection NC synthesis with an average diameter of 6.5 nm. (**b**) Ultraviolet-visible (UV-Vis) absorption spectrum (blue line) and photoluminescence spectrum of CdSe nanocrystals (NCs) synthesized by a hot injection method recorded at an excitation of 575 nm (red line).

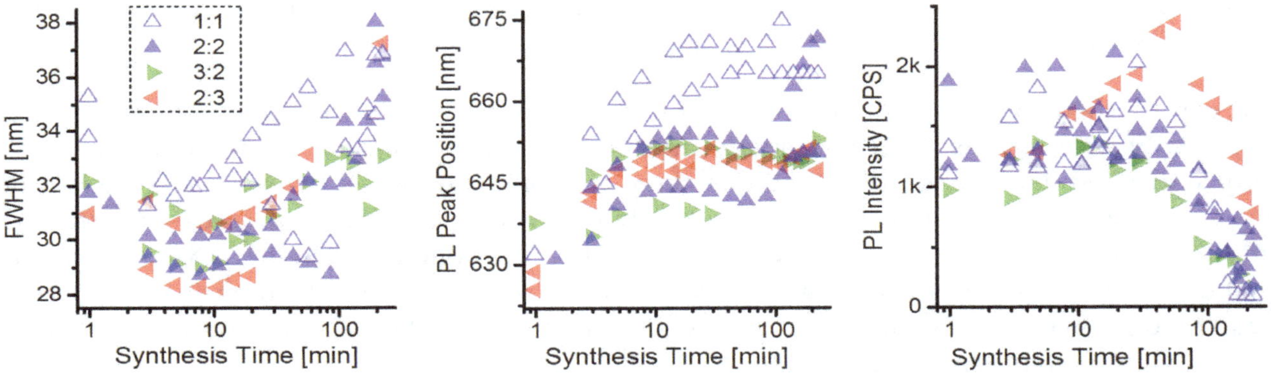

Figure 3. Different Cd:Se precursor ratios and concentrations—full width at half maximum (FWHM), photoluminescence (PL) peak position, and PL intensity evolution during the NC syntheses.

Table 1. Cd:Se precursor ratios, position of the photoluminescence (PL) brightpoint during the nanocrystal (NC) synthesis; average spectral PL peak position, PL full width at half maximum (FWHM), and average PL quantum yield (QY) of utilized quantum dots (QDs); and average power conversion efficiency (PCE) of hybrid bulk heterojunction poly[2,6-(4,4-bis-(2-ethylhexyl)-4H-cyclopenta[2,1-b;3,4-b′]dithiophene)-alt-4,7-(2,1,3-benzothiadiazole)]:CdSe (BHJ PCPDTBT:CdSe) QD solar cells found for the optimal post-synthetic hexanoic acid washing time of CdSe QDs.

Cd:Se Precursor Ratio	Brightpoint (min)	PL Peak Position (nm)	FWHM (nm)	PL QY (%)	Optimal HA Washing Time (min)	PCE with Optimal NC Treatment Time (%)
3:2	17	647 ± 7	30.7 ± 0.8	24 ± 5	21	2.4 ± 0.10
2:2	21	650 ± 3	27.0 ± 0.7	25 ± 5	18	2.8 ± 0.18
2:3	60	646 ± 5	32.2 ± 0.9	37 ± 8	60	1.7 ± 0.04

2.2. Post-Synthetic NC Treatment

As mentioned before, CdSe QDs were taken from the PL brightpoint, since we expect an optimal NC quality in this point, according to Qu *et al.* [32]. A hexanoic acid-based treatment was applied on CdSe QDs synthesized at three different Cd:Se ratios as described in the methods section. As we have presented in previous publications [11,15,33], the NC possesses a sphere of weakly associated ligands, which crystallize after the cool-down of the synthesis reaction around the initial ligand layer chemisorbed to the NC surface. To demonstrate the physical influence of the post-synthetic treatment on the CdSe QDs, the PL intensity and the ligand sphere size were determined for different hexanoic acid (HA) washing times (Figure 4). The peak PL intensity was determined from PL spectroscopy measurements of re-dispersed CdSe QDs, while the ligand sphere size was determined by measuring the hydrodynamic QD diameter (corresponding to the ligand sphere size [34]) by dynamic light scattering (DLS) of the same samples.

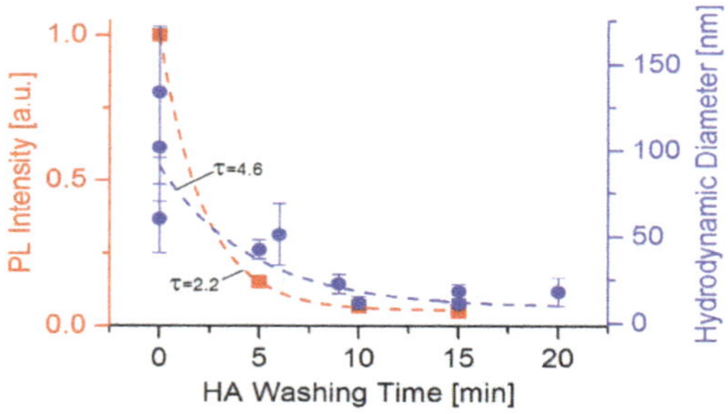

Figure 4. Red squares: Development of PL intensity over the HA washing time for CdSe QDs synthesized in hexadecylamine/trioctylphosphine oxide (HDA/TOPO) (100:3:2, 30 min at 300 °C). Blue dots: Respective development of the hydrodynamic diameter together with the standard deviation for CdSe QDs measured by dynamic light scattering (DLS). For both, the half-life τ in min is given for the respective fitted exponential decay.

It can be seen that both the hydrodynamic diameter and the PL intensity seem to decrease exponentially with increasing HA washing time, corresponding to the facile removal of the outer, weakly associated ligand shell. The DLS measurements of untreated QD samples show a relatively wide value distribution, which might also partially result from QD aggregation to superstructures. Nevertheless, from the DLS measurements we can draw the conclusion that the NC ligand sphere diameter is decreasing with increasing HA washing time, while from the PL intensity measurements one can assume that the original ligand passivation of the NC surface is either reduced or changed (*i.e.*, by a ligand exchange with HA and/or MeOH).

2.3. Optimal Initial Solar Cell Performance

Using CdSe QDs taken from the brightpoint of the respective NC synthesis (*i.e.*, from 21 min synthesis time for the 100:2:2 ratio, 16 min for 100:3:2, and 60 min for 100:2:3), hybrid BHJ Solar cells were fabricated and characterized according to the methods section. In order to obtain the best solar cell performance, the optimal post-synthetic NC treatment time had to be found (see Figure 5).

From these experiments the post-synthetic NC treatment times, resulting in the optimal hybrid solar cell power conversion efficiency (PCE), were determined for 3 differently synthesized QDs. A summary of the optimal NC treatment time based on the experiments presented in Figure 5, the resulting average optimal PCE and the initial properties of the utilized QDs are presented in Table 1.

The parameters summarized in Table 1 show that there is a correlation in between the initial PL QY of CdSe QDs, and the required post-synthetic treatment time for creating the most efficient hybrid solar cell: a high PL QY requires a long post-synthetic NC treatment time. Furthermore, one can see that low PL FWHM leads to high PCEs for BHJ hybrid solar cells. Hence, the PL QY is an indicator of the required post-synthetic treatment time but not for the achievable solar cell performance. This is determined—given that the post-synthetic treatment is optimized—by the FWHM for comparable NC sizes, as they were in this investigation. Lower FWHM results in a higher PCE, and higher FWHM results in a lower PCE. This is reasonable, since a low FWHM presumably corresponds to a high intrinsic uniform NC quality.

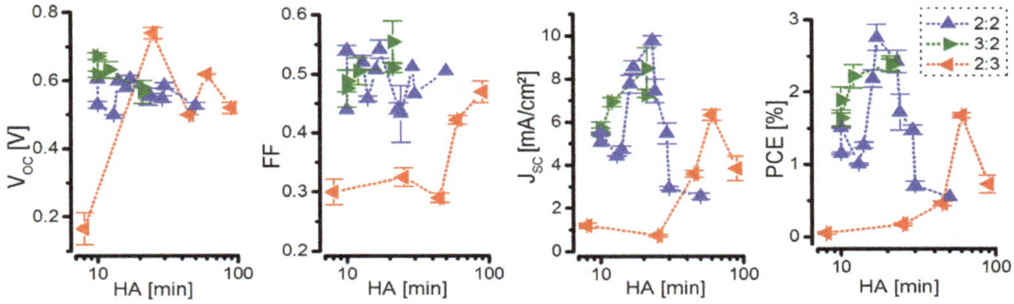

Figure 5. Dependency of open-circuit voltage (V_{OC}), fill factor (FF), short-circuit current density (J_{SC}) and PCE on the HA washing time at 105 °C for hybrid BHJ solar cells containing CdSe NCs taken from the respective PL brightpoint of the NC hot-injection syntheses performed at 300 °C with a 100:3:2, 100:2:2, and 100:2:3 ratio of (HDA/TOPO):(Cd-SA):(TOP-Se). The results are obtained from two synthesis batches for each ratio and from 60 solar cells, which were thermally post-annealed until reaching their optimal performance.

2.4. Influence of the Post-Synthetic NC Treatment on the Device Annealing Time, and Guidance for NC Treatment

To obtain the optimal solar cell performances presented in the prior paragraph, besides the optimal NC treatment time, an optimal time for the thermal solar cell annealing is also necessary. This optimal annealing time is however different for different NC post-synthetic treatment times as shown in Figure 6.

When taking a closer look at this behavior (see Figures 7 and 8), one can even identify indications for finding the optimal NC treatment time with a reduced number of initial experiments.

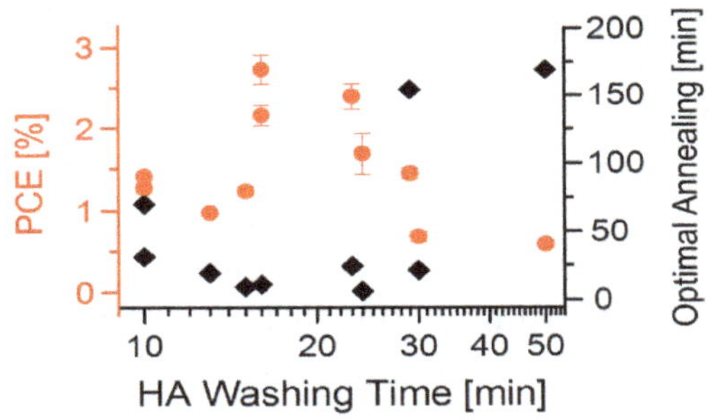

Figure 6. Power conversion efficiencies (PCEs) (red spherical points) obtained from 36 hybrid BHJ solar cells containing CdSe QDs (100:2:2) washed for different times in HA, and the optimal annealing time (black rhombic points) required to reach this efficiency.

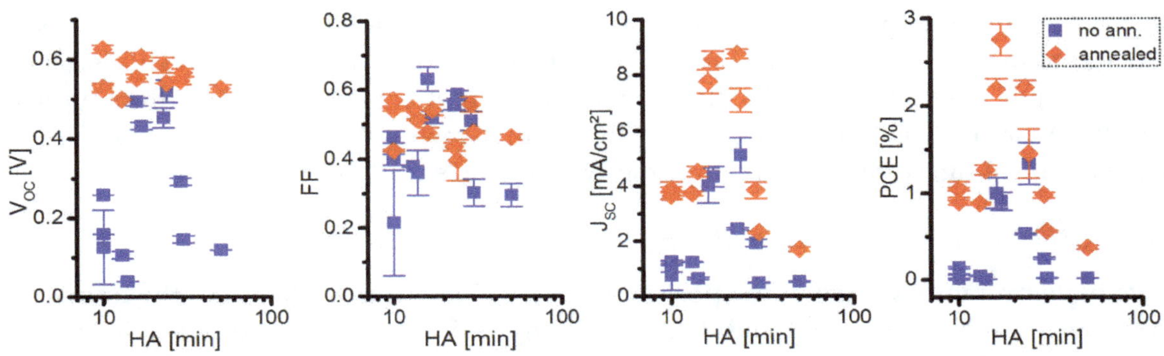

Figure 7. Differently long HA washed CdSe QDs (100:2:2) exhibit different performances when incorporated inside hybrid BHJ solar cells both without annealing and with 10 min of thermal annealing at 145 °C (results obtained from 36 cells).

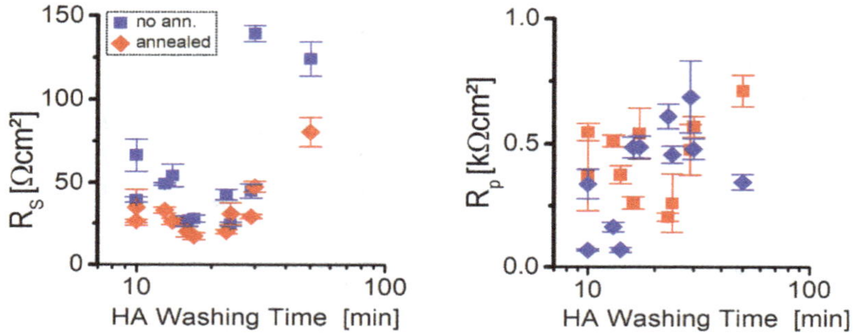

Figure 8. Incorporating differently long HA washed CdSe QDs (100:2:2) into BHJ hybrid solar cells results in different series resistance (R_S) and parallel resistance (R_P) values before and after 10 min of thermal annealing.

As depicted in Figure 7, the short-circuit current density and the open-circuit voltage are already relatively high, near the optimal HA treatment time, even before the thermal annealing step. This is since the series resistance—representing the active layer resistivity and to a lesser extent also the contact resistances [35]—has its lowest values for longer HA washing times (see Figure 8). The reason for this is probably the decreasing NC ligand sphere diameter, measured during the first 15 min of the NC treatment (see Figure 4), leading to an increased conductivity of the QD phase within the active layer and a more intimate NC packing, improving electron hopping processes.

The parallel resistance, which is an indicator for the active layer morphology (e.g., formation of "dead zones" from which no electrons can be extracted, and also to trap-assisted recombination [36]), tends to steadily increase with increased NC washing time, pointing towards an improved active layer morphology (*i.e.*, finer phase segregation and better charge extraction pathways). For post-synthetic NC treatment times exceeding the optimal duration, the parallel resistance tends to remain high in the resulting solar cells, but series resistance greatly increases, supposedly due to destruction of the NC surface by the washing procedure.

After performing a thermal annealing step on a hotplate inside a nitrogen filled glovebox for 10 min at 145 °C, a strong increase in J_{SC} (and accordingly a decrease in R_S) is observed especially for shorter washed QDs, which is probably mostly attributed to the enhanced intermolecular polymer-chain packing according to organic BHJ solar cell literature [36], but here a thermally induced *in situ* reduction of the NC ligand shell might also play a role. The series resistance for the longest washed NCs, however, strongly profits (despite still being of relatively high values) from thermal annealing, which might arise from a partial re-passivation of the NC surface. On the other side, the parallel resistance generally only appears to change after thermal annealing for solar cells containing shortly washed QDs, which we attribute to a higher morphological stability of the active layer of solar cells containing stronger ligand shell-size reduced QDs.

Briefly, after performing a first thermal annealing step on the solar cell, one can conclude whether one has to add to the NC washing time, or further reduce it for the next solar cell. This is, since the J_{SC} increase is stronger after annealing for short-time-washed NCs, accompanied with a typical strong increase of R_P. Moreover, solar cells containing short-time-washed NCs exhibit a higher V_{OC} after the same thermal annealing, compared to long-time-washed NCs containing cells.

2.5. Influence of NC Treatment on Solar Cell Long-Term Performance Stability

As we have previously shown, the post-synthetic NC treatment influences the immediate solar cell performance and the thermal stability of BHJ hybrid solar cells. As we will subsequently show, the post-synthetic NC treatment also influences the long-term performance of solar cells. Figure 9 shows the behavior of BHJ hybrid solar cells containing shorter (10 min and 12 min) and longer washed (21 min) CdSe QDs before being incorporated into the active layer.

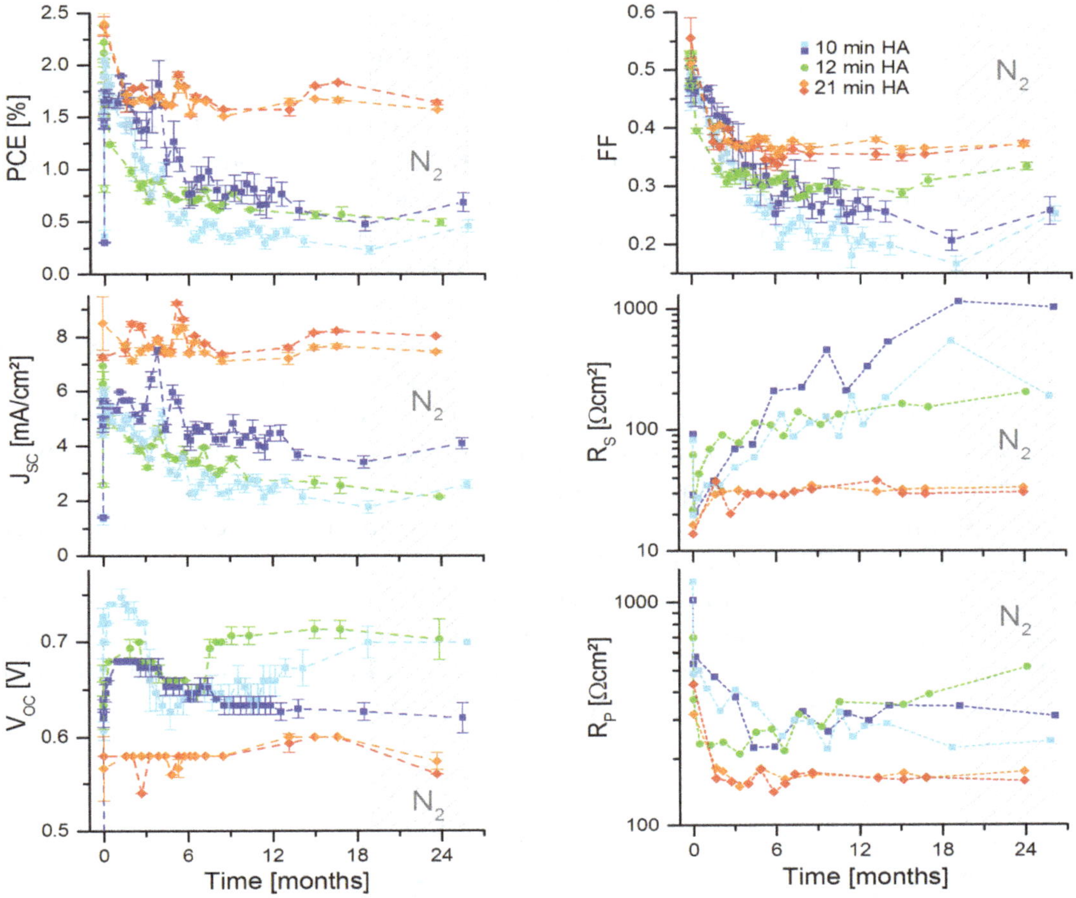

Figure 9. Parameters of PCPDTBT:CdSe QD solar cells from QDs washed by HA for 10 min (turquoise & blue squares), 12 min (green spherical points), and 21 min (orange & red rhombic points), stored in the dark inside a glovebox and periodically illuminated by a sun-simulator (AM 1.5 G spectrum) and characterized inside the same glovebox.

As a result, we observe improved performance stability for solar cells containing longer-time-washed QDs, with a decrease of the PCE from 2.4% to 1.6% (−33%) compared to a decrease from 2.1% to 0.55% (−74%), during 2 years of investigation. The series resistance R_S increases only slightly for long-time-washed NCs, but continuously increases for solar cells with shorter washed NCs. Possible explanations for this behavior would be the facilitated oxygen penetration through a supposedly less dense active layer for solar cells with higher ligand content, leading to increased oxidation of the CdSe QDs and of the polymer. However, the diffusion of ligands towards the electrodes, forming there an insulating layer would also be a possible explanation for the increasing R_S. R_P and V_{OC} are already low for long-time-washed NCs, possibly due to more shunts within the active layer by presumably more aggregated NCs, and induced surface defects introduced by long NC washing, hence increasing charge recombination processes.

It is also worth mentioning that the short-circuit current has shown a clear dependency on the oxygen concentration, which fluctuated during the first 19 months around 25 ppm O_2 with extremes of 0 ppm and 36 ppm, before the glovebox was continuously flushed with nitrogen for the following 5 months (the H_2O concentration was of <5 ppm at all time). We have observed J_{SC} to decrease with increasing oxygen concentration inside the glovebox, and J_{SC} to be recovered after the oxygen concentration has decreased. This observed reversibility does not occur immediately, but rather slowly (within several days). Hence, the origin of J_{SC} recovery might be due to oxygen depletion within the active layer by O_2 diffusion.

2.6. *Influence of NC Treatment on the Solar Cell Performance Stability in Air*

We have also performed a short investigation of the influence the difference in NC treatment time has on the stability of non-encapsulated glass/ITO/PEDOT:PSS/PCPDTBT:CdSe/Al solar cells taken into air and measured under AM 1.5G illumination. Three different solar cells were used herein. The CdSe QDs utilized in this investigation received a HA-based treatment for 15 min (short washed) and 22 min (long washed). To this comparison, also an organic $PC_{61}BM$:PCPDTBT solar cell, with the active layer composed out of a 2.5:1 weight ratio between $PC_{61}BM$ (Phenyl C_{61} butyric acid methyl ester, Sigma-Aldrich, Saint Louis, MS, USA, >99.5%) and PCPDTBT was added to specifically investigate the influence of the NCs.

According to Figure 10, a decreasing J_{SC} is observed for both hybrid and organic solar cells when taking the devices from the glovebox out into air. Possible reasons for the reduced J_{SC} are the (partially) reversible uptake of oxygen within the polymer phase [37], the oxidation of the aluminum cathode [38], and the uptake of water into the hygroscopic Poly(3,4-ethylenedioxythiophene):poly(styrenesulfonic acid) (PEDOT:PSS) layer [39,40]. We also observe that the J_{SC} decrease is more similar to the organic solar cell, when utilizing long-washed NCs; this might be due to the fact that the oxygen diffusion within the higher-density NC ligand depleted active layer is slowed down. The V_{OC} shows a strong increase for hybrid solar cells containing short washed NCs, while a V_{OC} increase is seen much later with long washed NCs, and not at all for organic solar cells—as has already been reported for organic solar cells [40,41]. Hence, we assume that the V_{OC} increase in air is connected to the utilization of NCs, and possibly due to adsorption of H_2O onto CdSe NCs.

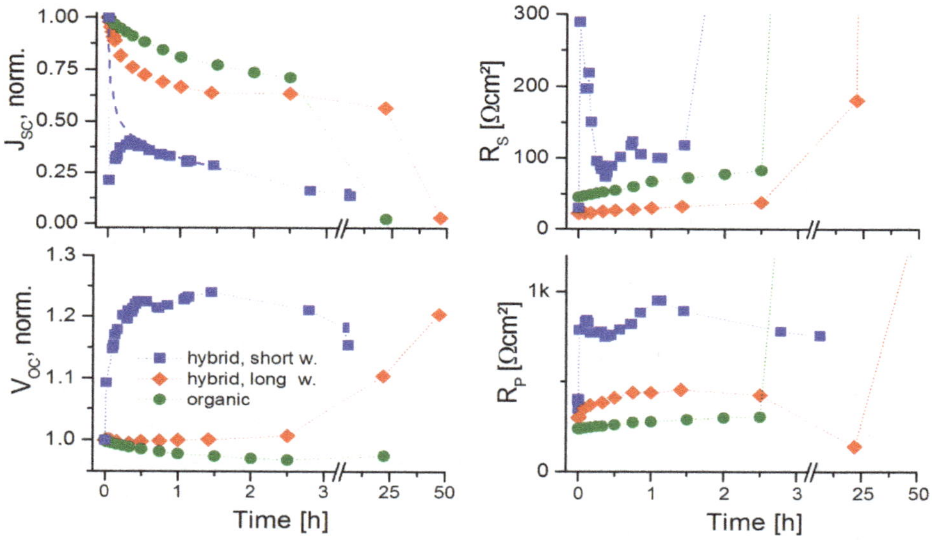

Figure 10. Development by time of short circuit current densities, series resistances, open circuit voltages, and parallel resistances for hybrid BHJ CdSe/PCPDTBT solar cells with short-time HA washed and long-time HA washed NCs, and for an organic BHJ $PC_{61}BM$/PCPDTBT solar cell for comparison when taken into air. Since the J_{SC} of the solar cell containing short washed QDs displayed a fast initial decrease from its original value, an extrapolated graph has been added to guide the eye.

3. Methods

3.1. *Post-Synthetic NC Treatment*

The post-synthetic NC treatment was performed in the following way: a portion of the synthesis product containing 1 mg of CdSe QDs was dissolved in 2.5 mL hexanoic acid (⩾99.5%, Sigma-Aldrich, Saint Louis, MS, USA), stirred for several minutes in a snap-cap glass tube at 110 °C on a hot plate. Subsequently a double volume (compared to HA) of methanol (anhydrous, 99.8%, Sigma-Aldrich, Saint Louis, MS, USA) was added to reduce the QD concentration to 1/3rd of its initial value.

The stirring was continued with the added methanol at 110 °C for half the stirring time compared to that in pure HA. Then, the dispersion was centrifuged by an Eppendorf MiniSpin® plus centrifuge (Eppendorf AG, Hamburg, Germany) for 1 min at 14.5 krpm with the rotor-pre heated in a furnace to 90 °C in order to hinder the re-crystallization of the ligands. Afterwards the QDs were redispersed in chloroform ($CHCl_3$) with a concentration of 2 mg/mL and stirred at 105 °C for 1 min. Consequently, triple the volume of methanol was added, and the NCs were further stirred for 3 min at 105 °C for precipitation. Afterwards, the NCs were collected by centrifugation for 30 s at 14.5 krpm. Chlorobenzene (anhydrous, 99.8%, Sigma-Aldrich, Saint-Louis, MS, USA) was then added to obtain a CdSe QD dispersion of 24 mg/mL. Hence, when mentioning the HA washing times throughout this publication, one must also consider the rest of the NC washing protocol subsequently executed, which additionally includes half the HA washing time with added MeOH, 1 min stirring in $CHCl_3$, and finally 3 min stirring in MeOH. For example, 20 min of HA washing means: 20 min HA plus 10 min MeOH, centrifugation, redisperison, 1 min $CHCl_3$ plus 3 min MeOH, centrifugation, and final redispersion in chlorobenzene (CB).

3.2. *Solar Cell Fabrication*

The obtained CdSe QD/CB dispersion was mixed in a weight ratio of 88:12 (QD:polymer) with a 20 mg/mL solution of PCPDTBT (Mn = 10–20 kDa, 1-Material, Dorval, Canada) in CB. The final ink was spun cast by a WS400-6NPP-Lite spin coater from Laurell Technologies (North Wales, PA, USA with 800 rpm for 30 s followed by a 60 s drying step at 1800 rpm, resulting in an active layer thickness of about 80 nm. The spin coating was done on a structured $\leqslant 10\ \Omega_{sq}$ ITO substrate from Präzisions Glas & Optik GmbH (Iserlohn, Germany), which was treated for 5 min with oxygen plasma and spin coated with Baytron AI4083 PEDOT:PSS from HC Starck GmbH (Goslar, Germany) at 2000 rpm for 30 s and dried for 20 min at 160 °C, to form a 70 nm thick hole blocking layer. After thermal evaporation of an 80 nm aluminum layer as electrode, the cells were thermally annealed for different times at 145 °C on a hot plate. The solar cells were usually annealed for 10 min and subsequently characterized. This procedure was repeated until the solar cell PCE started to decrease.

3.3. *Solar Cell Characterization*

The solar cells were characterized inside a nitrogen filled glovebox by a computer controlled Keithley 2602A source-meter (Keithley Instruments, Solon, OH, USA). The cells were individually illuminated by a LS0400 LOT-Oriel sun simulator (LOT-QuantumDesign GmbH, Darmstadt, Germany), housing a xenon lamp and using an AM 1.5G filter. The light intensity is adjusted by a calibrated silicon reference solar cell to match 100 mW/cm^2.

3.4. *Dynamic Light Scattering (DLS)*

DLS measurements were performed with a Zetasizer Nanosizer ZS of Malvern Instruments Ltd (Malvern, UK). The investigated QDs were dispersed in chloroform to obtain a dispersion with an absorbance of 0.1 A.U. at the first excitonic peak. For the measurements a Hellma® fluorescence cuvette out of Suprasil® quartz glass (Hellma, Müllheim, Germany) was utilized.

4. Conclusions

The optimal NC treatment time before integration into BHJ hybrid solar cells can be determined by systematic analysis of the resulting solar cell device performances. NCs of higher PL QY require a longer post-synthetic NC treatment time. The optimal thermal annealing time for solar cell devices is different for different NC treatment times, and provides crucial information for tuning the NC treatment time in the right direction. Hybrid BHJ solar cells containing stronger ligand sphere reduced QDs exhibit an improved performance stability in the long term and in air. Further investigations would have to be conducted into the influence of the post-synthetic NC treatment on the active layer morphology and on the induced trap density on the NCs and inside the solar cell. We have shown in this publication that the QD synthesis procedure can lead to decisive influences on the optimal post-synthetic NC treatment before integration into hybrid BHJ solar cells and determine the overall solar cell performance. This might be one reason for the widely scattered results of groups working in this field and for QD batch to batch variations, since properties of nanomaterials are often determined by surface properties such as surface defects or ligand attachment and arrangement, which are hard to control and cannot be determined by TEM and UV-Vis spectroscopy, which are the methods of choice for QD characterization. We also believe that our findings for CdSe QDs are also of general importance for other QD systems and for the development of respective post-synthetic treatment protocols before integration into various applications.

Acknowledgments: We thank the German Science Foundation (DFG) Graduate School "Micro Energy Harvesting" (GRK1322) for funding.

Author Contributions: M.E. and M.K. conceived and designed the experiments; M.E. performed the experiments and analyzed the data; M.E. and M.K. wrote the paper; all authors read and approved the final manuscript.

Abbreviations

The following abbreviations are used in this manuscript:

NC	Nanocrystal
BHJ	Bulk heterojunction
PCPDTBT	(poly[2,6-(4,4-bis-(2-ethylhexyl)-4H-cyclopenta[2,1-b;3,4-b′]dithiophene)-alt-4,7-(2,1,3- benzothiadiazole)])
PEDOT:PSS	(Poly(3,4-ethylenedioxythiophene):poly(styrenesulfonic acid))
PCE	Power conversion efficiency
QD	Quantum dot
NR	Nanorod
MP	Multipod
HDA	Hexadecylamine
TOPO	Trioctylphosphine oxide
HA	Hexanoic acid
PL	Photoluminescence
Cd-SA	Cadmium-stearic acid
TOP-Se	Trioctylphosphine-selenid
UV-Vis	Ultraviolet-visible
FWHM	Full width at half maximum
CB	Chlorobenzene
MeOH	Methanol
DLS	Dynamic light scattering
A.U.	Absorbance units
ITO	Indium tin oxide
QY	Quantum yield
V_{OC}	Open-circuit voltage
FF	Fill factor
J_{SC}	Short-circuit current density
R_S	Series resistance
R_P	Parallel resistance

References

1. Zhou, Y.; Eck, M.; Krueger, M. Bulk-heterojunction hybrid solar cells based on colloidal nanocrystals and conjugated polymers. *Energy Environ. Sci.* **2010**, *3*, 1851–1864. [CrossRef]
2. Greenham, N.C.; Peng, X.; Alivisatos, A.P. Charge separation and transport in conjugated-polymer/semiconductor-nanocrystal composites studied by photoluminescence quenching and photoconductivity. *Phys. Rev. B* **1996**, *54*, 17628–17637. [CrossRef]
3. Ren, S.; Chang, L.-Y.; Lim, S.-K.; Zhao, J.; Smith, M.; Zhao, N.; Bulović, V.; Bawendi, M.; Gradečak, S. Inorganic–organic hybrid solar cell: Bridging quantum dots to conjugated polymer nanowires. *Nano Lett.* **2011**, *11*, 3998–4002. [CrossRef]

4. Greaney, M.J.; Araujo, J.; Burkhart, B.; Thompson, B.C.; Brutchey, R.L. Novel semi-random and alternating copolymer hybrid solar cells utilizing CdSe multipods as versatile acceptors. *Chem. Commun.* **2013**, *49*, 8602. [CrossRef] [PubMed]
5. Zhou, R.; Stalder, R.; Xie, D.; Cao, W.; Zheng, Y.; Yang, Y.; Plaisant, M.; Holloway, P.H.; Schanze, K.S.; Reynolds, J.R.; *et al.* Enhancing the efficiency of solution-processed polymer: Colloidal nanocrystal hybrid photovoltaic cells using ethanedithiol treatment. *ACS Nano* **2013**, *7*, 4846–4854. [CrossRef] [PubMed]
6. Eck, M.; van Pham, C.; Züfle, S.; Neukom, M.; Sessler, M.; Scheunemann, D.; Erdem, E.; Weber, S.; Borchert, H.; Ruhstaller, B.; *et al.* Improved efficiency of bulk heterojunction hybrid solar cells by utilizing CdSe quantum dot-graphene nanocomposites. *Phys. Chem. Chem. Phys.* **2014**, *16*, 12251–12260. [CrossRef] [PubMed]
7. Han, L.; Qin, D.; Jiang, X.; Liu, Y.; Wang, L.; Chen, J.; Cao, Y. Synthesis of high quality zinc-blende CdSe nanocrystals and their application in hybrid solar cells. *Nanotechnology* **2006**, *17*, 4736–4742. [CrossRef] [PubMed]
8. Heinemann, M.D.; von Maydell, K.; Zutz, F.; Kolny-Olesiak, J.; Borchert, H.; Riedel, I.; Parisi, J. Photo-induced charge transfer and relaxation of persistent charge carriers in polymer/nanocrystal composites for applications in hybrid solar cells. *Adv. Funct. Mater.* **2009**, *19*, 3788–3795. [CrossRef]
9. Olson, J.D.; Gray, G.P.; Carter, S.A. Optimizing hybrid photovoltaics through annealing and ligand choice. *Sol. Energy Mater. Sol. Cells* **2009**, *93*, 519–523. [CrossRef]
10. Dayal, S.; Reese, M.O.; Ferguson, A.J.; Ginley, D.S.; Rumbles, G.; Kopidakis, N. The effect of nanoparticle shape on the photocarrier dynamics and photovoltaic device performance of poly(3-hexylthiophene):CdSe nanoparticle bulk heterojunction solar cells. *Adv. Funct. Mater.* **2010**, *20*, 2629–2635. [CrossRef]
11. Zhou, Y.; Riehle, F.S.; Yuan, Y.; Schleiermacher, H.-F.; Niggemann, M.; Urban, G.A.; Krueger, M. Improved efficiency of hybrid solar cells based on non-ligand-exchanged CdSe quantum dots and poly(3-hexylthiophene). *Appl. Phys. Lett.* **2010**, *96*. [CrossRef]
12. Zhou, Y.; Eck, M.; Veit, C.; Zimmermann, B.; Rauscher, F.; Niyamakom, P.; Yilmaz, S.; Dumsch, I.; Allard, S.; Scherf, U.; *et al.* Efficiency enhancement for bulk-heterojunction hybrid solar cells based on acid treated CdSe quantum dots and low bandgap polymer PCPDTBT. *Sol. Energy Mater. Sol. Cells* **2011**, *95*, 1232–1237. [CrossRef]
13. Huynh, W.U.; Dittmer, J.J.; Alivisatos, A.P. Hybrid nanorod-polymer solar cells. *Science* **2002**, *295*, 2425–2427. [CrossRef] [PubMed]
14. Sun, B.; Greenham, N.C. Improved efficiency of photovoltaics based on CdSe nanorods and poly(3-hexylthiophene) nanofibers. *Phys. Chem. Chem. Phys.* **2006**, *8*, 3557–3560. [CrossRef] [PubMed]
15. Celik, D.; Krueger, M.; Veit, C.; Schleiermacher, H.F.; Zimmermann, B.; Allard, S.; Dumsch, I.; Scherf, U.; Rauscher, F.; Niyamakom, P. Performance enhancement of CdSe nanorod-polymer based hybrid solar cells utilizing a novel combination of post-synthetic nanoparticle surface treatments. *Sol. Energy Mater. Sol. Cells* **2012**, *98*, 433–440. [CrossRef]
16. Sun, B.Q.; Marx, E.; Greenham, N.C. Photovoltaic devices using blends of branched CdSe nanoparticles and conjugated polymers. *Nano Lett.* **2003**, *3*, 961–963. [CrossRef]
17. Sun, B.; Snaith, H.J.; Dhoot, A.S.; Westenhoff, S.; Greenham, N.C. Vertically segregated hybrid blends for photovoltaic devices with improved efficiency. *J. Appl. Phys.* **2005**, *97*. [CrossRef]
18. Zhou, Y.; Li, Y.; Zhong, H.; Hou, J.; Ding, Y.; Yang, C.; Li, Y. Hybrid nanocrystal/polymer solar cells based on tetrapod-shaped $CdSe_x Te_{1-x}$ nanocrystals. *Nanotechnology* **2006**, *17*, 4041–4047. [CrossRef] [PubMed]
19. Wang, P.; Abrusci, A.; Wong, H.M.P.; Svensson, M.; Andersson, M.R.; Greenham, N.C. Photoinduced charge transfer and efficient solar energy conversion in a blend of a red polyfluorene copolymer with CdSe nanoparticles. *Nano Lett.* **2006**, *6*, 1789–1793. [CrossRef] [PubMed]
20. Dayal, S.; Kopidakis, N.; Olson, D.C.; Ginley, D.S.; Rumbles, G. Photovoltaic devices with a low band gap polymer and CdSe nanostructures exceeding 3% efficiency. *Nano Lett.* **2010**, *10*, 239–242. [CrossRef] [PubMed]
21. Dennler, G.; Scharber, M.C.; Brabec, C.J. Polymer-fullerene bulk-heterojunction solar cells. *Adv. Mater.* **2009**, *21*, 1323–1338. [CrossRef]
22. Peet, J.; Kim, J.Y.; Coates, N.E.; Ma, W.L.; Moses, D.; Heeger, A.J.; Bazan, G.C. Efficiency enhancement in low-bandgap polymer solar cells by processing with alkane dithiols. *Nat. Mater.* **2007**, *6*, 497–500. [CrossRef] [PubMed]
23. Zillner, E.; Fengler, S.; Niyamakom, P.; Rauscher, F.; Köhler, K.; Dittrich, T. Role of ligand exchange at CdSe quantum dot layers for charge separation. *J. Phys. Chem. C* **2012**, *116*, 16747–16754. [CrossRef]

24. Riehle, F.S. The Rational Synthesis Of Defect-Free CdE (E=S,Se) Nanocrystals. From Precursor Reactivity to Surface Stability. Ph.D. Thesis, University of Freiburg, Freiburg im Breisgau, Germany, 2013.
25. Gao, F.; Li, Z.; Wang, J.; Rao, A.; Howard, I.A.; Abrusci, A.; Massip, S.; McNeill, C.R.; Greenham, N.C. Trap-induced losses in hybrid photovoltaics. *ACS Nano* **2014**, *8*, 3213–3221. [CrossRef] [PubMed]
26. Zhu, L.; Richardson, B.J.; Yu, Q. Inverted hybrid CdSe-polymer solar cells adopting PEDOT:PSS/MoO_3 as dual hole transport layers. *Phys. Chem. Chem. Phys.* **2016**, *18*, 3463–3471. [CrossRef] [PubMed]
27. Ramar, M.; Suman, C.K.; Manimozhi, R.; Ahamad, R.; Srivastava, R. Study of Schottky contact in binary and ternary hybrid CdSe quantum dot solar cells. *RSC Adv.* **2014**, *4*, 32651–32657. [CrossRef]
28. Lek, J.Y.; Lam, Y.M.; Niziol, J.; Marzec, M. Understanding polycarbazole-based polymer: CdSe hybrid solar cells. *Nanotechnology* **2012**, *23*. [CrossRef] [PubMed]
29. Kippelen, B.; Brédas, J.-L. Organic photovoltaics. *Energy Environ. Sci.* **2009**, *2*, 251–261. [CrossRef]
30. Steinmann, V.; Kronenberg, N.M.; Lenze, M.R.; Graf, S.M.; Hertel, D.; Meerholz, K.; Bürckstümmer, H.; Tulyakova, E.V.; Würthner, F. Simple, highly efficient vacuum-processed bulk heterojunction solar cells based on merocyanine dyes. *Adv. Energy Mater.* **2011**, *1*, 888–893. [CrossRef]
31. Yuan, Y.; Riehle, F.-S.; Gu, H.; Thomann, R.; Urban, G.; Krueger, M. Critical parameters for the scale-up synthesis of quantum dots. *J. Nanosci. Nanotechnol.* **2010**, *10*, 6041–6045. [CrossRef] [PubMed]
32. Qu, L.; Peng, X. Control of photoluminescence properties of CdSe nanocrystals in growth. *J. Am. Chem. Soc.* **2002**, *124*, 2049–2055. [CrossRef] [PubMed]
33. Krueger, M.; Eck, M.; Zhou, Y.; Riehle, F.-S. Semiconducting nanocrystal/conjugated polymer composites for applications in hybrid polymer solar cells. In *Semiconducting Polymer Composites*; Wiley-VCH Verlag GmbH & Co. KGaA: Weinheim, Germany, 2012; pp. 361–397.
34. Eichhöfer, A.; Hänisch, C.V.; Jacobsohn, M.; Banin, U. Dynamic light scattering at CdSe nanocrystals and CdSe cluster-molecules. In Proceedings of the MRS Fall Meeting 2000, Boston, MA, USA, 27–30 November 2000; Karim, A., Merhari, L., Norris, D.J., Rogers, J.A., Xia, Y., Eds.; Cambridge University Press: Cambridge, UK, 2000; pp. D9.53.1–D9.53.9.
35. Shen, Y.; Li, K.; Majumdar, N.; Campbell, J.C.; Gupta, M.C. Bulk and contact resistance in P3HT:PCBM heterojunction solar cells. *Sol. Energy Mater. Sol. Cells* **2011**, *95*, 2314–2317. [CrossRef]
36. Kim, M.-S.; Kim, B.-G.; Kim, J. Effective variables to control the fill factor of organic photovoltaic cells. *ACS Appl. Mater. Interfaces* **2009**, *1*, 1264–1269. [CrossRef] [PubMed]
37. Schafferhans, J.; Baumann, A.; Wagenpfahl, A.; Deibel, C.; Dyakonov, V. Oxygen doping of P3HT:PCBM blends: Influence on trap states, charge carrier mobility and solar cell performance. *Org. Electron.* **2010**, *11*, 1693–1700. [CrossRef]
38. Tang, J.; Wang, X.; Brzozowski, L.; Barkhouse, D.A.; Debnath, R.; Levina, L.; Sargent, E.H. Schottky quantum dot solar cells stable in air under solar illumination. *Adv. Mater.* **2010**, *22*, 1398–1402. [CrossRef] [PubMed]
39. Kwon, S.; Lim, K.-G.; Shim, M.; Moon, H.C.; Park, J.; Jeon, G.; Shin, J.; Cho, K.; Lee, T.-W.; Kim, J.K. Air-stable inverted structure of hybrid solar cells using a cesium-doped ZnO electron transport layer prepared by a sol–gel process. *J. Mater. Chem. A* **2013**, *1*. [CrossRef]
40. Züfle, S.; Neukom, M.T.; Altazin, S.; Zinggeler, M.; Chrapa, M.; Offermans, T.; Ruhstaller, B. An effective area approach to model lateral degradation in organic solar cells. *Adv. Energy Mater.* **2015**, *5*. [CrossRef]
41. Kawano, K.; Pacios, R.; Poplavskyy, D.; Nelson, J.; Bradley, D.D.; Durrant, J.R. Degradation of organic solar cells due to air exposure. *Sol. Energy Mater. Sol. Cells* **2006**, *90*, 3520–3530. [CrossRef]

6

Transparent Platinum Counter Electrode Prepared by Polyol Reduction for Bifacial, Dye-Sensitized Solar Cells

Alvien Ghifari, Dang Xuan Long, Seonhyoung Kim, Brian Ma and Jongin Hong *

Department of Chemistry, Chung-Ang University, 84 Heukseok-ro, Dongjak-gu, Seoul 06974, Korea; alvien.ghifari@gmail.com (A.G.); dxlong.bk@gmail.com (D.X.L.); kimsh9560@gmail.com (S.K.); kehgeeng@gmail.com (B.M.)

* Correspondence: hongj@cau.ac.kr

Abstract: Pt catalytic nanoparticles on F-doped SnO_2/glass substrates were prepared by polyol reduction below 200 °C. The polyol reduction resulted in better transparency of the counter electrode and high power-conversion efficiency (PCE) of the resultant dye-sensitized solar cells (DSSCs) compared to conventional thermal reduction. The PCEs of the DSSCs with 5 μm-thick TiO_2 photoanodes were 6.55% and 5.01% under front and back illumination conditions, respectively. The back/front efficiency ratio is very promising for efficient bifacial DSSCs.

Keywords: dye-sensitized solar cell; counter electrode; bifacial; platinum; ethylene glycol

1. Introduction

Dye-sensitized solar cells (DSSCs) have shown promise as low-cost photovoltaics compared to commercially available Si solar cells. They hold great potential for building-attached photovoltaics (BAPVs) and building-integrated photovoltaics (BIPVs), because of their adjustable color/transparency and superior performance in dim light [1–4]. Recently, bifacial DSSCs, which can convert incoming sunlight to electricity through both front and back sides, have been an attractive alternative for photovoltaic devices [5,6]. The standard DSSC consists of a substrate coated with transparent conducting oxides (TCOs), a dye-grafted mesoscopic TiO_2 photoanode, a platinized counter electrode (CE), and an electrolyte containing a redox couple. Among the key components, the CE plays a prominent role in maintaining a flow of current by regenerating the redox mediator. Unfortunately, Pt is a highly expensive and scarce metal, and thus alternative materials, including carbon-based materials [7,8], transition metal compounds [9], conducting polymers [10,11], and their composites [12,13] have been explored.

Nevertheless, Pt is still favored because of its superior electrocatalytic activity and good conductivity. The conventional methods for preparing Pt CEs are vacuum sputtering of a Pt target and thermal decomposition of a platinic acid (H_2PtCl_6) precursor [14,15]. However, these high-energy-consuming approaches increase the cost and energy payback time of DSSCs. In addition to logistical issues, the light absorption at the Pt CEs should be minimized for bifacial operation. Therefore, it is crucial to develop new preparation methods for highly transparent Pt CEs at low temperatures.

Polyol-based synthesis is a versatile and straightforward liquid-phase method that uses high-boiling and multivalent alcohols to synthesize nanomaterials without the requirements of high pressure and autoclaves [16,17]. Polyol can simultaneously act as a reducing agent and water-equivalent solvent. Its chelating ability or controlling the nucleation of nanomaterials. For example, Ouyang and coworkers reported nanostructured Pt CEs prepared by polyol reduction of H_2PtCl_6 in ethylene glycol

(EG) [18,19]. They also utilized EG vapor for solventless chemical reduction of the platinum precursor below 200 °C [20]. Song et al. employed the hydrolysis of urea in a two-step EG solution reduction to achieve uniform dispersion and a smaller size of Pt nanoparticles on conducting glass substrates [21]. Unfortunately, the Pt CEs were prepared by drop-casting the EG-based Pt precursor solution. It has proven challenging to prepare a thin layer of Pt nanoparticles and thus provide highly transparent CEs for the bifacial DSSCs. Therefore, we prepared a highly transparent Pt counter electrode by spin-coating a platinic EG solution and further chemical reduction below 200 °C. We also investigate the photovoltaic performance of the DSSCs from the viewpoint of bifacial operation.

2. Materials and Methods

2.1. Fabrication of Pt Counter Electrodes

F-doped SnO_2 (FTO) glass (TEC 8, Pilkington; sheet resistance = 8Ω/□) substrates were ultrasonically cleaned using acetone, isopropyl alcohol, and deionized water. The substrates were baked at 150 °C for 10 min to completely remove residual water. Then 10 mM platinic acid in ethylene glycol was spin-coated on the FTO substrates. Subsequently, the sample was placed in a muffle furnace and heated to a certain temperature (e.g., 130, 150, 170, 190, and 210 °C). The annealing was maintained for 12 h, and then the sample was cooled down to room temperature. For comparison, 40 mM of the platinic acid solution in 2-propanol was prepared and then spin-coated on the same substrates, followed by heating at 425 °C for 1 h in the muffle furnace [22–24]. The schematic diagram of Pt CE preparation is shown in Figure 1.

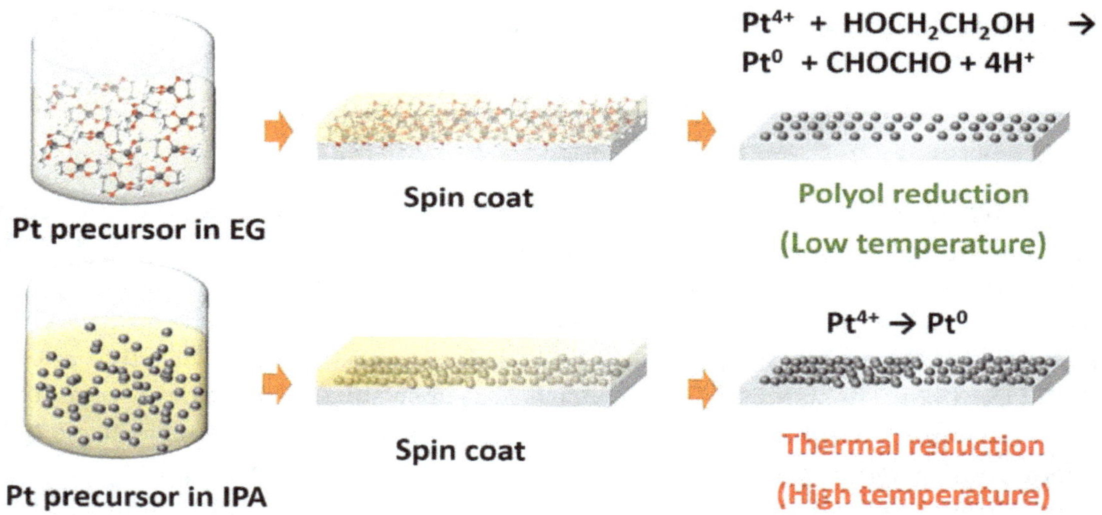

Figure 1. Schematic diagram of counter electrode preparation: polyol reduction (PR) and thermal decomposition (TD).

2.2. Characterization

The surface images of the fabricated Pt CEs were obtained using a field-emission scanning electron microscope (FE-SEM; Sigma, Carl Zeiss AG, Oberkochen, Germany). X-ray photoelectron spectroscopy (XPS) spectra were acquired using a Thermo Scientific K-alpha XPS system (Waltham, MA, United States) with an Al Kα X-ray source monochromator (1486.6 eV). The optical transmittance spectra were recorded using UV/Vis spectroscopy (V-730, JASCO, Tokyo, Japan). An electrochemical measurement station (CompactStat, Ivium Technologies, Eindhoven, The Netherlands) was used to conduct all electrochemical characterization. Cyclic voltammetry (CV) was recorded at a scan rate of 50 mV/s with a three-electrode system that consists of a Pt CE as a working electrode, a Pt wire as a counter electrode, and Ag/AgCl as a reference electrode. A solution of 10.0 mM LiI, 1.0 mM I_2, and 0.1 M $LiClO_4$ in CH_3CN was used as the electrolyte to investigate the electrocatalytic properties of the Pt CEs for redox reactions. Tafel and electrochemical impedance spectroscopy (EIS) measurements were performed on symmetric cells consisting of CE|electrolyte|CE. Tafel polarization measurements were conducted at a scan rate of 10 mV/s. In EIS measurements, a 10 mV amplitude sinusoidal potential perturbation was input over a frequency range from 1 MHz to 0.1 Hz at zero bias potential. The EIS spectra were analyzed using the equivalent circuit fitting routine in the ZView software (AMETEK, Leicester, UK).

2.3. Device Fabrication and Characterization

The FTO glass substrates were cleaned with O_2 plasma for 10 min, dipped in an aqueous solution of 40 mM $TiCl_4$ at 75 °C for 30 min, and then rinsed several times with deionized water. The TiO_2 paste (Transparent TiO_2, ENBKOREA, Gumi, Korea) was screen-printed on the FTO glass, and the printed film was calcinated at 300 °C for 30 min and 575 °C for 1 h in the muffle furnace. The final areas of the TiO_2 photoanodes were 0.2025 cm^2. The photoanodes were treated in the $TiCl_4$ solution and then heated at 500 °C for 30 min on a hot plate. After O_2 plasma treatment, the photoanodes were dipped into a 0.5 mM dye solution of cis-diisothiocyananoto-bis(2,2-bipyridyl-4,4′-dicarboxylate) ruthenium(II) (N719, Merck KGaA, Darmstadt, Germany) in ethanol for 24 h. The dye-grafted photoanode and Pt CE were assembled with 25 μm-thick thermoplastic film (Surlyn, Solaronix, Aubonne, Switzerland) and sealed by heating. An iodide-based redox electrolyte (Iodolyte AN-50, Aubonne, Switzerland) was injected into the pre-drilled holes in the side of the counter electrode and then sealed. The photovoltaic characteristics were investigated under AM 1.5 global one sun illumination (100 mW/cm^2) using a solar cell I–V measurement system (K3000 LAB, McScience, Suwon, Korea). The photocurrent density (J_{sc}), open-circuit voltage (V_{oc}), fill factor (FF), and power conversion efficiency (η) were recorded simultaneously. Monochromatic incident photon-to-current conversion efficiency (IPCE) was collected to evaluate the spectral response of the solar cells (K3100, McScience, Suwon, Korea). EIS measurement on the fabricated devices was carried out by the same protocol mentioned above.

3. Results

Polyol-based synthesis is a versatile technique to prepare Pt nanostructures. In this study, EG was chosen as the solvent and reducing agent of Pt precursors because of its low boiling point and viscosity. Also, byproducts from the EG reduction, such as glycoaldehyde and diacetyl, could be easily removed [25]. Figure 2a–f shows the surface of the FTO glass substrates decorated with Pt nanoparticles formed by polyol reduction at different temperatures (130, 150, 170, 190, and 210 °C) and thermal decomposition at 425 °C (hereafter, "polyol-reduced" is abbreviated as "PR" and "thermally decomposed" is abbreviated as "TD"). Tiny Pt nanoparticles (<10 nm) were formed and dispersed on the FTO surface for all reaction temperatures. The chemical reduction by EG allows for depositing Pt nanoparticles at low temperatures, and is thus feasible for plastic substrates. As the reduction temperature increased, the aggregation of the Pt nanoparticles diminished, and no aggregation could be observed at the temperature of 190 °C. This indicates that the slower evaporation of EG could result in the growth of larger nanoparticles, more agglomeration, and dendrites. However, the aggregated nanoparticles appeared again at a temperature higher than the boiling point of EG (i.e., 197 °C) because of the inability to control particle nucleation and growth at the elevated temperature [17]. Also, the pyrolysis of H_2PtCl_6 at 425 °C directly led to the formation of large Pt nanoparticles prominently populating the FTO. Accordingly, we predict that their size and distribution will affect the transparency and catalytic activity of the resultant electrodes.

XPS was performed to determine the formation of metallic Pt during our chemical reduction. In Figure 3, the spectra of survey XPS indicates that the samples contained Sn, O, Pt, and Cl—no other elements except carbon were detected. Carbon can result from precursors or sample handling. Interestingly, as the reduction temperature increased, the signal for Cl 2p related to ionic Pt species decreased and then disappeared. Narrow XPS scans of the Pt-4f core level region are also provided in Figure 3. The C 1 s peak at 285.0 eV was used to calibrate all the XPS data. The Pt 4f signal is composed of three pairs of deconvoluted doublets. The first doublet (71.4 eV and 74.8 eV) corresponds to the platinum in the zero-valent state (i.e., Pt(0)), while the second doublet (72.3 eV and 75.6 eV) can be assigned to platinum in the 2+ valence state (i.e., Pt(II)) [26]. The third (73.7 eV and 77.0 eV) is caused by the Pt^{4+} species, such as $[PtCl_6]^{2-}$, on the surface (i.e., Pt(IV)) [27]. Table 1 summarizes the binding energies and relative integrated peak areas of the deconvoluted peaks at the Pt-4f core level region. The increase in the reduction temperature resulted in the change in the valence state from Pt^{2+} to Pt^{0}. After the polyol reaction, the Pt species clustered together with different compositions. Accordingly, the reaction temperature is of significant importance in preparing metallic Pt nanoparticles.

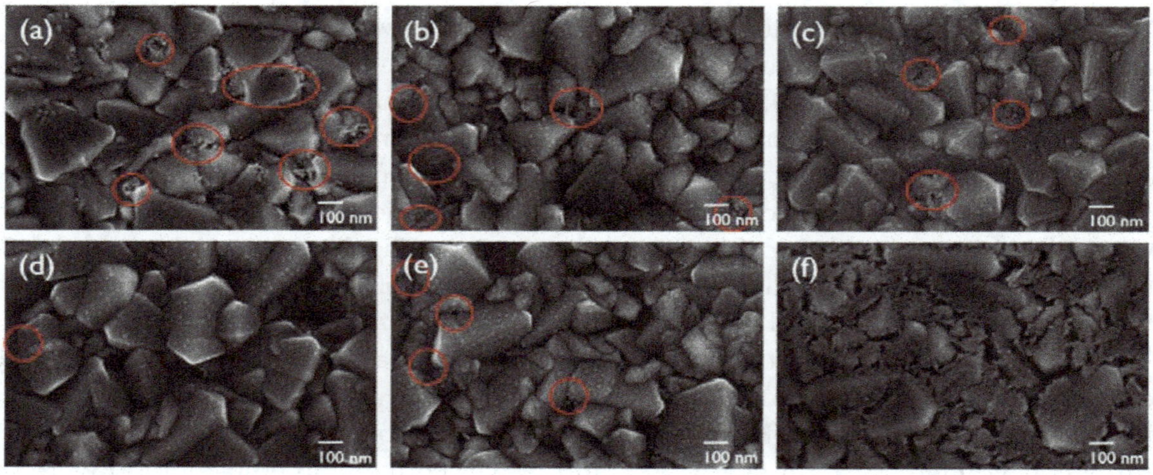

Figure 2. Field-emission scanning electron microscope (FE-SEM) images of Pt films on FTO substrates: (**a**) PR-130, (**b**) PR-150, (**c**) PR-170, (**d**) PR-190, (**e**) PR-210, and (**f**) TD-425.

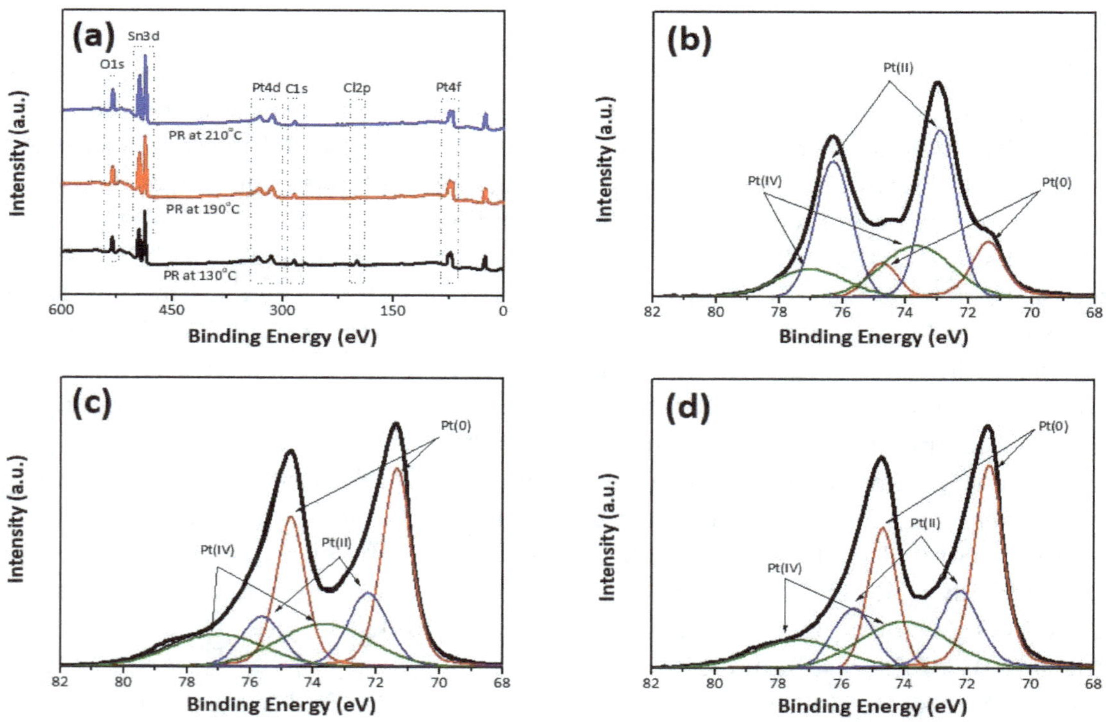

Figure 3. X-ray photoelectron spectroscopy (XPS) scans of polyol-reduced Pt on FTO substrates: (**a**) survey, (**b**) PR-130, (**c**) PR-190, and (**d**) PR-210.

Table 1. Binding energies (eV) and relative proportions (S, in %) of each component measured by XPS.

	Pt(0)			Pt(II)			Pt(IV)		
	$4f_{5/2}$	$4f_{7/2}$	S	$4f_{5/2}$	$4f_{7/2}$	S	$4f_{5/2}$	$4f_{7/2}$	S
PR-130	74.8	71.4	23.7	76.3	72.9	38.8	77.0	73.7	37.5
PR-190	74.7	71.4	48.4	75.6	72.3	22.7	77.0	73.7	28.9
PR-210	74.7	71.4	44.8	75.6	72.3	20.9	77.0	73.7	34.3

In the DSSC operation, an I_3^-/I^- redox couple is commonly utilized as a redox mediator. The CE should collect electrons from the external circuit and effectively catalyze the reduction of I_3^- to I^-. The electrocatalytic activity of the PR CEs was determined using cyclic voltammetry (CV), as shown in Figure 4a. All CV curves show two typical pairs of oxidation and reduction peaks (Ox-1/Re-1 and Ox-2/Re-2), which are described by Equations (1) and (2), respectively [28]:

$$I_3^- + 2e^- \rightleftharpoons 3I^- \quad (1)$$

$$3I_2 + 2e^- \rightleftharpoons 2I_3^- \quad (2)$$

The catalytic reduction activity of the CE can be determined by the first oxidation and reduction peaks (Ox-1 and Re-1). Its electrochemical reversibility can be assessed from the peak-to-peak separation (ΔE_p), which is the difference between the anodic and cathodic peak potentials. A smaller ΔE_p reflects the higher electrocatalytic activity of the CE.

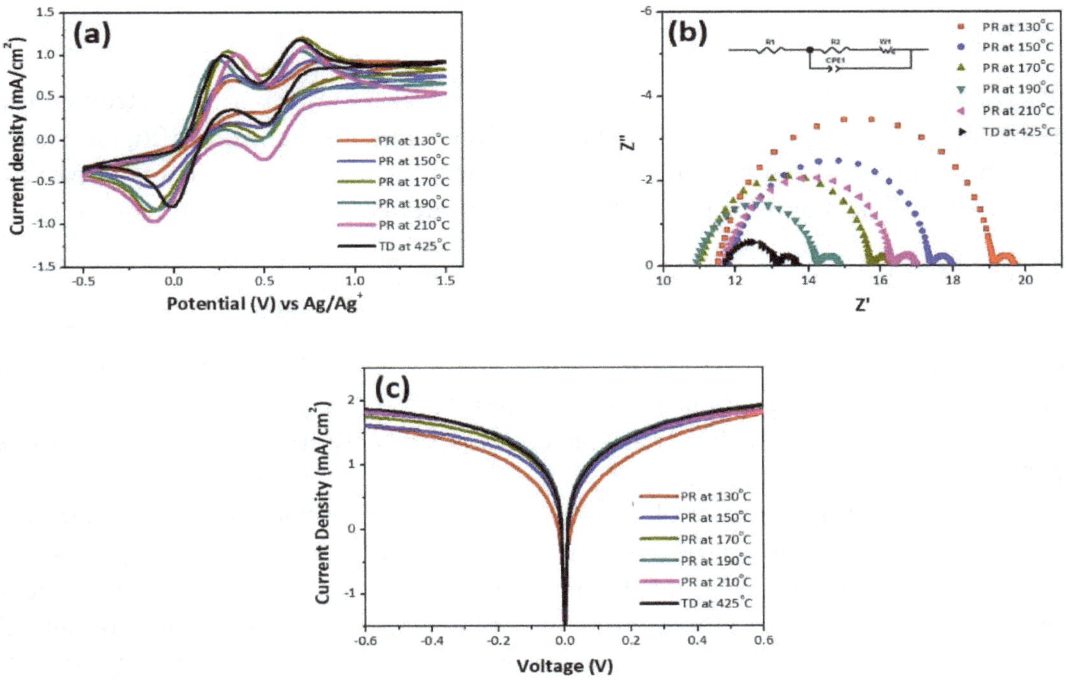

Figure 4. (**a**) Cyclic voltammograms for the redox of I_3^-/I^- species, (**b**) electrochemical impedance spectroscopy (EIS) Nyquist plots for the symmetric cells, and (**c**) Tafel polarization curves.

EIS measurements were also conducted using the symmetrical cells constructed with two identical electrodes. The Nyquist plots (Figure 4b) consist of two apparent semicircles at a high frequency (first semicircle) and a low frequency (second semicircle), respectively. The Randles-type circuit (insert in Figure 4b) was used to simulate the plots. The corresponding parameters are summarized in Table 2. R_s is the series resistance, and can be derived by the intercept of the high-frequency semicircle on the real axis (Z' axis). R_{ct} is the charge transfer resistance at the interface between CE and electrolyte, and can be determined from the radius of the first semicircle on the real axis. All the PR-CEs have nearly the same R_s, and thus its effect on photovoltaic performance can be ignored. A smaller R_{ct} accelerates triiodide reduction, and thus the PR-190 CE has the superior catalytic activity. The low-frequency semicircle results from Nernst diffusion impedance (Z_N) of the redox species in the electrolyte, which is inversely proportional to the diffusion coefficient of I_3^- in the cells [28]. The decrease in Z_N leads to an increase in the electrocatalytic activity of the CE.

Table 2. Electrochemical parameters of polyol-reduction Pt films at different reaction temperatures.

No	ΔE_p (V)	J_0 (mA/cm^2)	J_{lim} (mA/cm^2)	R_s (Ω/cm^2)	R_{ct} (Ω/cm^2)	Z_N (Ω/cm^2)
PR-130	0.47	4.99	0.21	8.05	5.19	0.44
PR-150	0.44	9.16	0.21	8.26	3.79	0.44
PR-170	0.43	11.14	0.24	7.69	3.21	0.41
PR-190	0.35	14.76	0.26	7.63	2.23	0.44
PR-210	0.46	12.44	0.26	8.13	3.17	0.48
TD-425	0.29	11.62	0.27	11.79	1.27	0.59

Figure 4c shows Tafel polarization plots of the symmetrical cells comprising the Pt CEs and I_3^-/I^- electrolytes. The Tafel plot can be separated into three consecutive zones: polarization, Tafel, and limit diffusion zones. The Tafel and limit diffusion zones are valuable for obtaining both the limiting current density (J_{lim}) and current density (J_0), which correlate with the electrocatalytic activity of the CEs [29]. The intersection of the cathodic branch and the equilibrium potential line can be considered J_0, and thus the steep Tafel slope implies a large J_0. Theoretically, J_0 can also be calculated using Equation (3):

$$J_0 = \frac{RT}{nFR_{ct}} \tag{3}$$

where R is the gas constant, T is the absolute temperature, n is the number of electrons participating in the electrochemical reduction of I_3^-, and F is Faraday's constant. In the limit diffusion zone, J_{lim} can be determined by the intersection of the cathodic branch and the y-axis. J_{lim} is directly proportional to a diffusion coefficient of I_3^- (D) at the same potential, and can be expressed as Equation (4):

$$J_{lim} = \frac{2neDCN_A}{l} \tag{4}$$

where e is the elementary charge, C is the concentration of I_3^-, N_A is the Avogadro constant, and l is the distance between two electrodes. Notably, the values of J_0 and J_{lim} followed the same trend observed in both CV and EIS analyses. Accordingly, the electrocatalytic activity of I_3^-/I^- is as follows: PR-190 > PR-170 > PR-210 > PR-150 > PR-130.

The light absorption at the counter electrode should be minimized to improve light harvesting in the bifacial DSSCs. Figure 5a shows the transmittance spectra of the prepared Pt CEs and bare FTO glass. Polyol reduction manifested higher transparency compared to thermal decomposition in the whole visible light regime, which will be beneficial to bifacial applications. Unfortunately, the increase in reduction temperature resulted in the decrease in transmittance of PR CEs. We think that metallic Pt nanoparticles possibly have a negative effect on the light transparency. Figure 5b,c shows, respectively, the current density-voltage (J–V) characteristics and IPCE spectra of the DSSCs (5 μm-thick TiO_2), which were illuminated from the TiO_2 photoanode side (i.e., front illumination). The photovoltaic parameters of the front- and rear-illuminated DSSCs are summarized in Table 3. The improved electrocatalytic activity of the PR–Pt CE resulted in better photovoltaic performance: power conversion efficiency (η) increased from 5.50% (PR-130) to 6.55% (PR-190). Also, PR-190 exhibited better photovoltaic performance than TD-425 because of the improvement of R_s and Z_N related to the electrocatalytic activity of the CE. It should be noted that all the DSSCs employing PR–Pt CEs maintained more than 76% of their front-illumination efficiency when lit from the Pt CE side (i.e., back illumination). The decrease of η in the back-illumination condition is related to the transmission losses due to the

Pt-based electrocatalyst and the I_3^-/I^- electrolyte [6]. The ratio of the back-illumination efficiency to the front-illumination efficiency ($\eta(R)$) followed the trend of the transmittance of the PR–Pt CEs observed above.

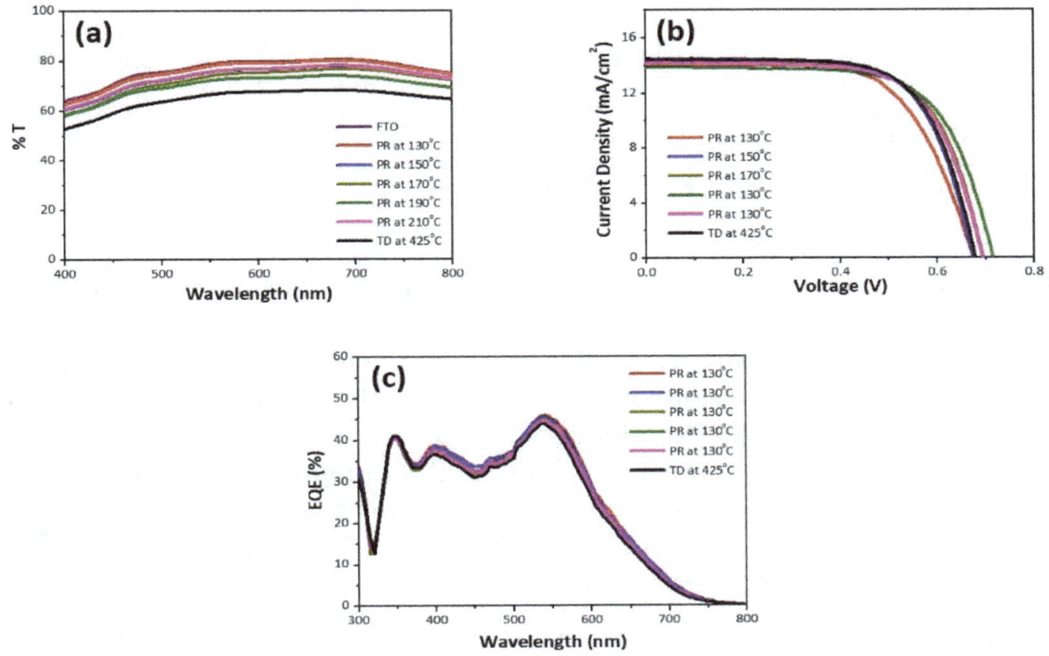

Figure 5. (**a**) Transmittance spectra of the Pt counter electrodes, (**b**) J–V characteristic curves of the dye-sensitized solar cells (DSSCs) (5 μm-thick TiO_2) under one sun front illumination, and (**c**) their incident photon-to-current conversion efficiency (IPCE) spectra.

Table 3. Photovoltaic parameters of DSSCs (5 μm-thick TiO_2) with different Pt counter electrodes [a].

No	Illumination	V_{oc} (V)	J_{sc} (mA/cm^2)	*FF* (%)	η (%)	η (*R*) (%)
PR-130	Front	0.666 ± 0.007	13.82 ± 0.23	59.70 ± 5.48	5.50 ± 0.66	81.03 ± 1.68
	Back	0.664 ± 0.008	10.67 ± 0.32	62.74 ± 4.85	4.45 ± 0.50	
PR-150	Front	0.673 ± 0.003	13.99 ± 0.51	67.45 ± 1.43	6.35 ± 0.33	77.57 ± 1.57
	Back	0.670 ± 0.004	10.65 ± 0.62	69.04 ± 1.22	4.93 ± 0.35	
PR-170	Front	0.694 ± 0.006	13.96 ± 0.31	67.20 ± 1.33	6.51 ± 0.30	76.73 ± 1.33
	Back	0.689 ± 0.003	10.62 ± 0.52	68.28 ± 0.90	5.00 ± 0.30	
PR-190	Front	0.702 ± 0.011	13.77 ± 0.12	67.66 ± 1.55	6.55 ± 0.28	76.61 ± 3.24
	Back	0.699 ± 0.00	10.44 ± 0.23	68.71 ± 1.76	5.01 ± 0.24	
PR-210	Front	0.695 ± 0.004	13.76 ± 0.45	67.02 ± 0.84	6.41 ± 0.29	76.70 ± 2.14
	Back	0.692 ± 0.008	10.42 ± 0.68	68.22 ± 0.94	4.92 ± 0.3	
TD-425	Front	0.682 ± 0.00	13.97 ± 0.46	67.36 ± 0.72	6.42 ± 0.28	67.92 ± 2.66
	Back	0.672 ± 0.005	9.40 ± 0.16	68.97 ± 0.66	4.35 ± 0.05	

[a] The parameters were obtained from at least five cells fabricated with each CE condition; each cell was measured five times.

To date, various transparent CEs have been developed for bifacial applications, such as photovoltaic windows and façades. Table 4 shows the photovoltaic performance of the bifacial DSSCs with platinum and non-platinum CE materials compared to our work. Although DSSC fabrication conditions (e.g., TiO_2 thickness, dyes, electrolytes) are not the same, our polyol reduction technique proved sufficient for fabricating bifacial DSSCs.

Table 4. Photovoltaic parameters of bifacial DSSCs with different CE materials.

Material	Method	T (%)	V_{oc} (V)	J_{sc} (mAcm^{-2})	*FF* (%)	η (front) (%)	η (back) (%)	η (R) (%)	Refs.
PEDOT	Electropolymerization	~90 [a]	0.751	14.60	67	7.40	5.23	70.7	[6]
Carbon	Carbonization	~75	0.721	10.52	60	6.07	5.04	83.0	[7]
Ni_3S_4	Hydrothermal	~70	0.700	13.58	65	6.56	4.86	73.9	[9]
PANI/MoS_2	Composite	~63	0.799	17.93	63	7.99	3.40	42.6	[12]
Pt	Photo-reduction	~95	0.810	13.53	66.6	7.29	5.85	80.3	[30]
Pt	Thermal decomposition	~80 [a]	0.757	15.00	70.7	8.02	4.43	55.2	[30]
Pt	Polyol reduction [b]	~80	0.702	13.77	67.6	6.55	5.01	76.6	-

[a] Transmittance was calculated from absorbance. [b] Our work.

4. Conclusions

In this work, highly transparent Pt CEs were prepared using the polyol reduction technique at low temperatures (<200 °C). Our facile and versatile technique provided better electrocatalytic activity and transparency than conventional thermal decomposition methods, and thus brought significant improvement to the photovoltaic performance of the bifacial DSSCs. In particular, the bifacial DSSC with PR-190 attained 6.55% for front illumination and 5.01% for back illumination, while that with TD had 6.42% for front illumination and 4.35% for back illumination. Because our process does not require an elevated temperature, we are currently exploring the fabrication of flexible bifacial DSSCs that employ polymeric substrates. In addition, further optimization (e.g., $[Co(bpy)_3]^{3+/2+}$ electrolytes) should result in better bifacial DSSC performance.

Author Contributions: Conceptualization, A.G. and D.X.L.; methodology, A.G., S.K., and D.X.L.; formal analysis, A.G., S.K., B.M., and D.X.L.; investigation, A.G., D.X.L., and J.H., data curation, A.G. and D.X.L.; writing—original draft preparation, A.G., B.M., and J.H.; writing—review and editing, B.M. and J.H.; visualization, A.G. and J.H.; supervision, J.H.; project administration, J.H.; funding acquisition, J.H. All authors have read and agreed to the published version of the manuscript.

Acknowledgments: We acknowledge the Chung-Ang University Research Scholarship in 2019.

References

1. Fakharuddin, A.; Jose, R.; Brown, T.M.; Fabregat-Santiago, F.; Bisquert, J. A perspective on the production of dye-sensitized solar modules. *Energ. Environ. Sci.* **2014**, *7*, 3952–3981. [CrossRef]
2. Wu, W.; Wang, J.; Zheng, Z.; Hu, Y.; Jin, J.; Zhang, Q.; Hua, J. A strategy to design novel structure photochromic sensitizers for dye-sensitized solar cells. *Sci. Rep.* **2015**, *5*, 8592. [CrossRef] [PubMed]
3. Freitag, M.; Teuscher, J.; Saygili, Y.; Zhang, X.; Giordano, F.; Liska, P.; Hua, J.; Zakeeruddin, S.M.; Moser, J.E.; Grätzel, M.; et al. Dye-sensitized solar cells for efficient power generation under ambient lightning. *Nat. Photonics* **2017**, *11*, 372–378. [CrossRef]
4. Yuan, H.; Wang, W.; Xu, D.; Xu, Q.; Xie, J.; Chen, X.; Zhang, T.; Xiong, C.; He, Y.; Zhang, Y.; et al. Outdoor testing and ageing of dye-sensitized solar cells for building integrated photovoltaics. *Solar Energy* **2018**, *165*, 233–239. [CrossRef]
5. Li, P.; Tang, Q. Highly transparent metal selenide counter electrodes for bifacial dye-sensitized solar cells. *J. Power Sources* **2016**, *317*, 43–48. [CrossRef]
6. Kang, J.S.; Kim, J.; Kim, J.-Y.; Lee, M.J.; Kang, J.; Son, Y.J.; Jeong, J.; Park, S.H.; Ko, M.J.; Sung, Y.-E. Highly efficient bifacial dye-sensitized solar cells employing polymeric counter electrodes. *ACS Appl. Mater. Interfaces* **2018**, *10*, 8611–8620. [CrossRef]
7. Bu, C.; Liu, Y.; Yu, Z.; You, S.; Huang, N.; Liang, L.; Zhao, X.-Z. Highly transparent carbon counter electrode prepared via an in situ carbonization method for bifacial dye-sensitized solar cells. *ACS Appl. Mater. Interfaces* **2013**, *5*, 7432–7438. [CrossRef]
8. Zhang, Y.; Zhao, Y.; Duan, J.; Tang, Q. S-doped CQDs tailored transparent counter electrodes for high-efficiency bifacial dye-sensitized solar cells. *Electrochim. Acta.* **2018**, *216*, 588–595. [CrossRef]
9. Hu, Z.; Xia, K.; Zhang, J.; Hu, Z.; Zhu, Y. Highly transparent ultrathin metal sulfide films as efficient counter electrodes for bifacial dye-sensitized solar cells. *Electrochim. Acta* **2015**, *170*, 39–47. [CrossRef]

10. Bu, C.; Tai, Q.; Liu, Y.; Guo, S.; Zhao, X. A transparent and stable polypyrrole counter electrode for dye-sensitized solar cell. *J. Power Sources* **2013**, *221*, 78–83. [CrossRef]
11. Li, H.; Xiao, X.; Han, G.; Hou, H. Honeycomb-like poly(3,4-ethylenedioxythiophene) as an effective and transparent counter electrode in bifacial dye-sensitized solar cells. *J. Power Sources* **2017**, *342*, 709–716. [CrossRef]
12. He, B.; Zhang, X.; Zhang, H.; Li, J.; Meng, Q.; Tang, Q. Transparent molybdenum sulfide decorated polyaniline complex counter electrodes for efficient bifacial dye-sensitized solar cells. *Solar Energy* **2017**, *147*, 470–478. [CrossRef]
13. Ahmed, A.S.A.; Xiang, W.; Hu, X.; Qi, C.; Amiinu, I.S.; Zhao, X. Si_3N_4/MoS_2-PEDOT:PSS composite counter electrode for bifacial dye-sensitized solar cells. *Solar Energy* **2018**, *173*, 1135–1143. [CrossRef]
14. Moraes, R.S.; Saito, E.; Leite, D.M.G.; Massi, M.; da Silva Sobrinho, A.S. Optical, electrical and electrochemical evaluation of sputtered platinum counter electrodes for dye sensitized solar cells. *Appl. Surf. Sci.* **2016**, *364*, 229–234. [CrossRef]
15. Zhou, R.; Guo, W.; Yu, R.; Pan, C. Highly flexible, conductive and catalytic Pt networks as transparent counter electrodes for wearable dye-sensitized solar cells. *J. Mater. Chem. A* **2015**, *3*, 23028–23034. [CrossRef]
16. Dong, H.; Chen, Y.C.; Feldmann, C. Polyol synthesis of nanoparticles: Status and options regarding metals, oxides, chalcogenides, and non-metal elements. *Green Chem.* **2015**, *17*, 4107–4132. [CrossRef]
17. Fievet, F.; Ammar-Merah, S.; Brayner, R.; Chau, F.; Giraud, M.; Mammeri, F.; Peron, J.; Piquemal, J.Y.; Sicard, L.; Viau, G. The polyol process: A unique method for easy access to metal nanoparticles with tailored sizes, shapes and compositions. *Chem. Soc. Rev.* **2018**, *47*, 5187–5233. [CrossRef]
18. Cho, S.J.; Ouyang, J. Attachment of platinum nanoparticles to substrates by coating and polyol reduction of a platinum precursor. *J. Phys. Chem. C* **2011**, *115*, 8519–8526. [CrossRef]
19. Sun, K.; Fan, B.; Ouyang, J. Nanostructured platinum films deposited by polyol reduction of a platinum precursor and their application as counter electrode dye-sensitized solar cells. *J. Phys. Chem. C* **2010**, *114*, 4237–4244. [CrossRef]
20. Cho, S.J.; Neo, C.Y.; Mei, X.G.; Ouyang, J.Y. Platinum nanoparticles deposited on substrates by solventless chemical reduction of a platinum precursor with ethylene glycol vapor and its application as highly effective electrocatalyst in dye-sensitized solar cells. *Electrochim. Acta* **2012**, *85*, 16–24. [CrossRef]
21. Song, M.Y.; Chaudhari, K.N.; Park, J.; Yang, D.-S.; Kim, J.H.; Kim, M.-S.; Lim, K.; Ko, J.; Yu, J.-S. High efficient Pt counter electrode prepared by homogeneous deposition method for dye-sensitized solar cell. *Appl. Energy* **2012**, *100*, 132–137. [CrossRef]
22. Decoppet, J.-D.; Moehl, T.; Babkair, S.S.; Alzubaydi, R.A.; Ansari, A.A.; Habib, S.S.; Zakeeruddin, S.M.; Schmidt, H.-W.; Grätzel, M. Molecular gelation of ionic liquid–sulfolane mixtures, a solid electrolyte for high performance dye-sensitized solar cells. *J. Mater. Chem. A* **2014**, *2*, 15972–15977. [CrossRef]
23. Leonardi, E.; Penna, S.; Brown, T.M.; Di Carlo, A.; Reale, A. Stability of dye-sensitized solar cells under light soaking test. *J. Non-Cryst. Solids* **2010**, *356*, 2049–2052. [CrossRef]
24. Veerender, P.; Saxena, V.; Jha, P.; Koiry, S.P.; Gusain, A.; Samanta, S.; Chauhan, A.K.; Aswal, D.K.; Gupta, S.K. Free-standing polypyrrole films as substrate-free and Pt-free counter electrodes for quasi-solid dye-sensitized solar cells. *Org. Electron.* **2012**, *13*, 3032–3039.
25. Skrabalak, S.E.; Wiley, B.J.; Kim, M.; Formo, E.V.; Xia, Y. On the polyol synthesis of silver nanostructures: Glycolaldehyde as a reducing agent. *Nano Lett.* **2008**, *8*, 2077–2081. [CrossRef] [PubMed]
26. Dablemont, C.; Lang, P.; Mangeney, C.; Piquemal, J.-Y.; Petkov, V.; Herbst, F.; Viau, G. FTIR and XPS study of Pt nanoparticle functionalization and interaction with alumina. *Langmuir* **2008**, *24*, 5832–5841. [CrossRef]
27. Liu, Z.; Lee, J.Y.; Chen, W.; Han, M.; Gan, L.M. Physical and electrochemical characterizations of microwave-assisted polyol preparation of carbon-supported PtRu nanoparticles. *Langmuir* **2004**, *20*, 181–187. [CrossRef]
28. Wan, J.; Fang, G.; Yin, H.; Liu, X.; Liu, D.; Zhao, M.; Ke, W.; Tao, H.; Tang, Z. Pt-Ni alloy nanoparticles as superior counter electrodes for dye-sensitized solar cells: Experimental and theoretical understanding. *Adv. Mater.* **2014**, *26*, 8101–8106. [CrossRef]
29. Chang, P.-J.; Cheng, K.-Y.; Chou, S.-W.; Shyne, J.-J.; Yang, Y.-Y.; Hung, C.-Y.; Lin, C.-Y.; Chen, H.-L.; Chou, H.-L.; Chou, P.-T. Tri-iodide reduction activity of shape- and composition-controlled PtFe nanostructures as counter electrodes in dye-sensitized solar cells. *Chem. Mater.* **2016**, *28*, 2110–2119. [CrossRef]
30. Popoola, I.K.; Gondal, M.A.; AlGhamdi, J.M.; Qahtan, T.F. Photofabrication of highly transparent platinum counter electrodes at ambient temperature for bifacial dye sensitized solar cells. *Sci. Rep.* **2018**, *8*, 12864.

7

Influence of Nitrogen Doping on Device Operation for TiO_2-Based Solid-State Dye-Sensitized Solar Cells: Photo-Physics from Materials to Devices

Jin Wang [1], Kosti Tapio [2], Aurélie Habert [1], Sebastien Sorgues [3], Christophe Colbeau-Justin [3], Bernard Ratier [4], Monica Scarisoreanu [5], Jussi Toppari [2], Nathalie Herlin-Boime [1,*] and Johann Bouclé [4,*]

1 IRAMIS/NIMBE/LEDNA, UMR 3685, CEA Saclay, 91191 Gif sur Yvette, France; 516208050@qq.com (J.W.); aurelie.habert@cea.fr (A.H.)
2 Nanoscience Center, Department of Physics, University of Jyväskylä, P.O. Box 35, 40014 Jyväskylä, Finland; kosti.t.o.tapio@jyu.fi (K.T.); j.jussi.toppari@jyu.fi (J.T.)
3 Laboratoire de Chimie Physique, UMR8000, Université Paris-Sud, 91405 Orsay, France; sebastien.sorgues@u-psud.fr (S.S.); christophe.colbeau-justin@u-psud.fr (C.C.-J.)
4 XLIM UMR 7252, Université de Limoges/CNRS, 87060 Limoges Cedex, France; bernard.ratier@unilim.fr
5 National Institute for Lasers Plasma and Radiation Physics, P.O. Box MG 36, R-077125 Bucharest, Romania; monica.scarisoreanu@inflpr.ro
* Correspondence: nathalie.herlin@cea.fr (N.H.); johann.boucle@unilim.fr (J.B.)

Academic Editors: Guanying Chen, Zhijun Ning and Hans Agren

Abstract: Solid-state dye-sensitized solar cells (ssDSSC) constitute a major approach to photovoltaic energy conversion with efficiencies over 8% reported thanks to the rational design of efficient porous metal oxide electrodes, organic chromophores, and hole transporters. Among the various strategies used to push the performance ahead, doping of the nanocrystalline titanium dioxide (TiO_2) electrode is regularly proposed to extend the photo-activity of the materials into the visible range. However, although various beneficial effects for device performance have been observed in the literature, they remain strongly dependent on the method used for the production of the metal oxide, and the influence of nitrogen atoms on charge kinetics remains unclear. To shed light on this open question, we synthesized a set of N-doped TiO_2 nanopowders with various nitrogen contents, and exploited them for the fabrication of ssDSSC. Particularly, we carefully analyzed the localization of the dopants using X-ray photo-electron spectroscopy (XPS) and monitored their influence on the photo-induced charge kinetics probed both at the material and device levels. We demonstrate a strong correlation between the kinetics of photo-induced charge carriers probed both at the level of the nanopowders and at the level of working solar cells, illustrating a direct transposition of the photo-physic properties from materials to devices.

Keywords: solid-state dye-sensitized solar cells; TiO_2; nitrogen doping; photo-physics; photo-response; spiro-OMeTAD

---

1. Introduction

Since the pioneering work of Bach *et al.* in 1998 [1], solid-state dye-sensitized solar cells (ssDSSC) based on the organic molecular glass 2′,7,7′-Tetrakis-(*N*,*N*-di-4-methoxy phenylamino)-9,9′-spirobifluorene (spiro-OMeTAD) as p-type solid-state electrolyte have demonstrated constant performance improvement, thanks to the rational engineering of organic sensitizers and doping strategies. In particular, doping of hole-transporting materials (HTM) by cobalt complexes [2] or organic compounds such as 1,1,2,2-tetrachloroethan (TeCA) [3] have led to power conversion efficiencies over 7% under standard illumination conditions, illustrating the relevance of hybrid solid-state approaches for solar energy conversion. Improved spectral coverage of the solar spectrum as well as enhanced light harvesting efficiencies are now achieved by exploiting various metal-free organic dyes [4], such as porphyrin [5] or arylamine derivatives [6]. More recently, and apart from perovskite solar cells [7,8], which are not the primary topic of this article, ssDSSC based on TiO_2 porous electrodes have received additional benefits from the intensive developments made on HTMs [9–11]. Consequently, efficiencies over 8% were demonstrated using p-type perovskite materials [12,13] or Copper phenanthroline complexes [14] such as HTM.

In a typical device, the TiO_2 porous electrode acts simultaneously as a high specific area substrate for dye adsorption, and as the electron transporting material [15,16]. The metal oxide photo-electrode is therefore crucial for both light absorption and current generation. Several approaches have been proposed to improve the charge generation and collection efficiencies of nanostructured metal oxide electrodes using alternative electrode morphologies, such as nanorods or hierarchical structures [17]; substitution of TiO_2 by other metal oxides such as ZnO or SnO_2 [18]; deposition of insulating oxide shell on TiO_2 [19,20]; insertion of metallic (Au, Ag, . . .) [21,22] or non-metallic (S, N, *etc.*) [23–26] elements. Among all of these strategies, the doping of TiO_2 materials by nitrogen can exploit several effects, which can positively affect device performance. Asahi *et al.* reported in 2001 that, compared to pure TiO_2, N-doped TiO_2 exhibits a broader absorption range that extends into the visible up to 500 nm [27]. Since then, some attention has been paid to the origin of this additional absorption feature [28]. In a theoretical approach [29,30], Di Valentin *et al.* pointed out that if N atoms are in substitutional positions, N2p states can lie 0.13 eV above the top of the valence band of TiO_2. Moreover, if N is found in interstitial positions or at the surface of the nanoparticles, NO species can be formed and introduce π^*-NO-states into the band gap (0.73 eV above the top of the valence band). Asahi *et al.* investigated the effects of different nitrification conditions, and found out that substitutional N can be chemically more stable in the presence of oxygen vacancies [31]. Therefore, benefits induced by the N-doping rapidly raised the interest of researchers in the fields of photo-catalysis and photovoltaics.

In the field of hybrid solar cells, Ma *et al.* demonstrated an N-TiO_2 based liquid DSSC presenting an improved efficiency and stability than that prepared from a pristine TiO_2 electrode [24]. An enhancement of the incident photon to charge carrier efficiency (IPCE) was observed within the 380–520 nm range, related to the contribution of nitrogen to the absorption of N-TiO_2 powders. Tian *et al.* analyzed the electron lifetime in the sol-gel synthesized N-TiO_2 solar cells and found that the formation of O-Ti-N in the TiO_2 lattice could retard charge recombination reactions at the TiO_2 electrode/electrolyte interface [32]. In addition, the insertion of N atoms can alter the Fermi level of electrons in the oxide [33], potentially leading to a slight increase of the open-circuit voltage of the cell, as well as of the overall device efficiency. However, most of the studies reported on N-doping were carried out on liquid DSSC. In our previous study [34], we reported on solid-state DSSCs based on N-doped TiO_2 electrodes processed from nanocrystals synthesized by laser pyrolysis. We provided evidence of a significant contribution from the N-doped electrode to the generation of charge carriers, as a secondary current generation pathway. Although recent studies based on quasi-solid state DSSC devices demonstrated similar trends [35], no clear correlation between material properties and device performance were drawn.

In this work, we systematically report on the influence of nitrogen doping on ssDSSC device operation and photo-physics properties. To this end, we synthesized a set of TiO_2 nano-particles doped with different levels of nitrogen. Using complementary characterization techniques, both at the material and device levels, we focus on the impact of the presence of N atoms on the photo-generated charges in the metal oxide, and discuss its implications on the photovoltaic performance of the cells. X-ray photoelectron spectroscopy (XPS) is exploited to monitor the exact location of N atoms in the TiO_2 crystalline sites, while time-resolved microwave conductivity (TRMC) is used to assess its impact on photo-generated charge dynamics. Photo-conductivity measurements performed on test devices based on un-doped and N-doped porous TiO_2 electrodes are finally discussed with regard to charge recombination kinetics measured on full devices using transient photo-voltage. Considering the different treatments applied to the powders for the preparation of porous electrodes suitable for device testing, such methodology is particularly relevant to reveal the influence of the nitrogen doping both in the starting powders and in the final solar cells.

2. Results

2.1. Properties of the Starting Powders

Laser pyrolysis was used to synthesize a set of N-doped TiO_2 nanopowders [36]. This technique is an efficient method for the production of well-controlled nanocrystals with tunable properties, well-adapted for the photovoltaic application [37,38]. Here, the precursor mixture included ammonia as the source of N atoms. More details on the experimental conditions used for the synthesis of the powders can be found in Supplementary materials (Table S1), which also summarizes the chemical composition of the samples. In all cases, the as-synthesized nanopowders contain free carbon phases easily removed by thermal annealing under air. This treatment, which does not alter the main physical properties of the samples [36], was therefore applied to all powders considered in the following parts of this article. Figure 1 presents the transmission electron microscopy (TEM) image of a typical TiO_2 powders (doped with nitrogen in this case).

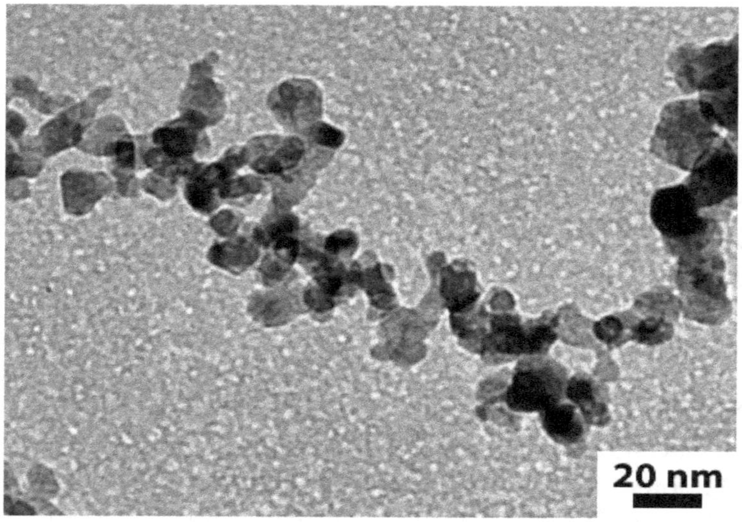

Figure 1. Transmission electron microscopy (TEM) image of a typical N-doped TiO_2 powder (after annealing treatment in air). The nitrogen content is 0.2 wt % in this case.

We observed a typical "chain-like" morphology of nano-scaled grains (mean diameter within 10 to 20 nm), typical of gas phase synthesis methods.

Table 1 summarizes the main physical parameters of the powders synthesized in this work, including nitrogen content, crystalline phase and crystallite diameter extracted from X-ray diffraction (XRD), as well as specific area and mean particle diameter estimated from the BET (Brunauer, Emmett, and Teller) method. The Spurr and Scherrer equations were used on the XRD patterns (Supplementary materials, Figure S1) to extract the Anatase to rutile crystalline fraction and the mean crystal diameter [39,40].

Table 1. Main physico-chemical properties of the TiO_2 and N-doped TiO_2 powders including N content determined by elemental analysis, anatase crystalline fraction and mean crystal diameter obtained by X-ray diffraction (XRD), as well as Brunauer, Emmett, and Teller (BET) specific area and mean grain diameter.

Sample	N content (wt %)	Data extracted from XRD		BET analysis	
		Fraction of anatase (%)	Mean crystal diameter (nm)	Specific area ($m^2 \cdot g^{-1}$)	Mean grain diameter (nm)
TiO_2	<<0.1	94	15.6	77	20
N-TiO_2-0.1	0.1	80	12.0	86	18
N-TiO_2-0.2	0.2	90	11.4	86	18
N-TiO_2-0.3	0.3	94	12.4	90	17
N-TiO_2-0.6	0.6	94	15.0	96	16

The N content in the final powders is directly driven by the level of ammonia in the precursor mixture (see Supplementary materials Table S1). The sizes estimated by BET are slightly larger than those extracted from the XRD analysis, revealing the presence of amorphous regions in the powders, or a slight particle agglomeration [36]. Both the crystalline phase and particle diameter are only slightly dependent on the doping level, leading to five powders with rather comparable morphologic features.

Regarding their optical properties, N-doped TiO_2 powders exhibit a dominant yellow color (see inset of Figure 2), indicating a shift in their absorption threshold towards the visible range. Figure 2 shows the Kubelka–Munk function of the powders calculated from diffused reflectance measurements. Comparing to pure titania, N-doped TiO_2 powders exhibit an additional absorption band between 370 and 550 nm. This optical feature, which was discussed in previous reports [24,34], can be assigned to the influence of nitrogen on the energetic level of TiO_2, through a mixing of N and O 2p states [27]. We observe that the intensity of this additional absorption band increases with the N doping level.

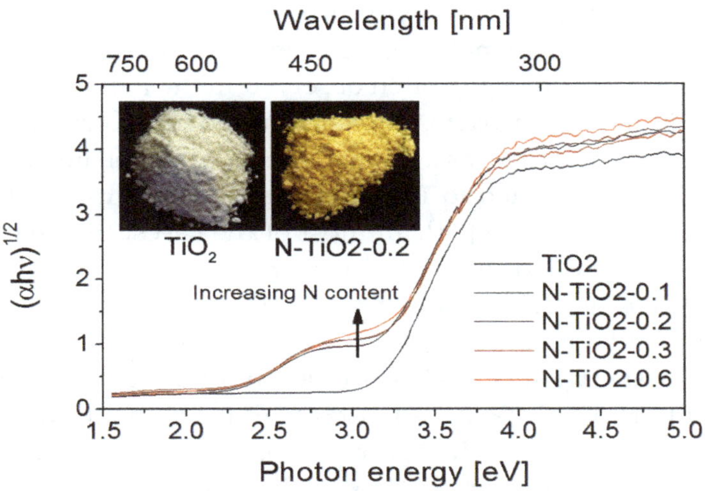

Figure 2. Optical data extracted from diffuse reflectance measurements on the TiO_2 and N-doped TiO_2 powders. The inset presents pictures of the N-TiO_2-0.2 sample (N content of 0.2 wt %) compared to the TiO_2 reference sample.

A more detailed picture of the local environment of nitrogen atoms in the metal oxide structure is drawn from X-ray photoelectron spectroscopy (XPS) applied on the powders. Data associated with annealed powders are presented in Figure 3, together with their deconvolution, while data associated with as-prepared samples are presented in Supplementary materials (Figures S2 and S3, respectively).

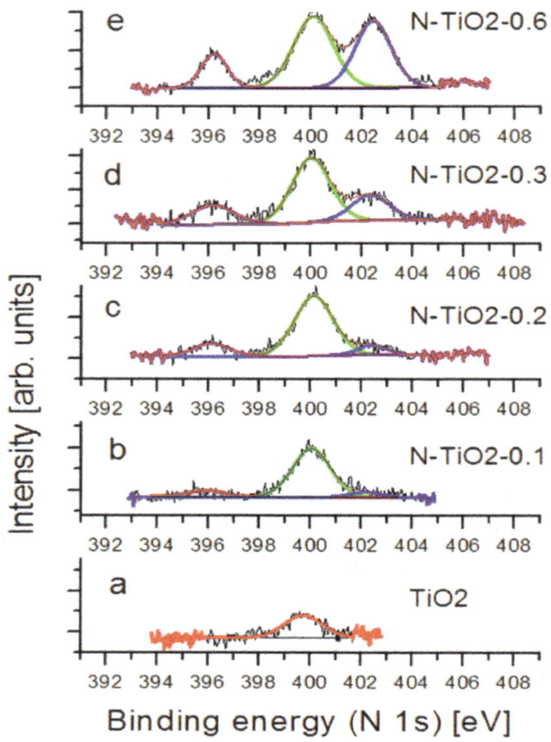

Figure 3. XPS spectra (N 1s) of the undoped and N-doped TiO_2 powders as a function of N content (from 0 to 0.6 wt %).

Concerning the pure TiO_2 sample, the only peak appearing at 399.7 eV in the spectra, both before and after annealing (see Supplementary materials), can be attributed to N_2 species chemically adsorbed on the TiO_2 surface, which is a feature being often observed in the literature [41]. The situation is more complex in the case of the N-doped powders. Three main contributions centered at 396.2 eV, 399.9 eV, and 402.2 eV are observed, and their exact assignment is still under debate in the literature [27,32,41,42]. The peak at 396.2 eV is usually attributed to Ti-N bonds and it thus implies that N atoms are situated at substitutional sites in the TiO_2 lattice [24]. In most of the cases, the peak at 399.9 eV is assigned to O-Ti-N bonds—in other words, to interstitial nitrogen atoms [29,32,43,44]. This assignment is consistent with the high electronegativity of oxygen, which reduces the electron density on nitrogen compared to Ti-N. As a result, the binding energy of O-Ti-N is slightly larger than that of Ti-N. However, several groups associate this peak with NO in interstitial sites or NO_2 in substitutional sites [31]. We reckon that part of this peak can also result from chemically adsorbed N_2 species on the TiO_2 surface, like in undoped TiO_2. However, comparing the absolute intensities of the signals, we can safely conclude that the peak at 399.9 eV in our N-TiO_2 samples can be assigned to interstitial nitrogen. The last peak at 402.2 eV is often observed in the literature [24]. Asahi *et al.* [27], as well as Tian *et al.* [32], assigned it to atomically adsorbed N species. Therefore, we associate this peak as the signature of nitrogen present at the surface of the TiO_2 particles. These surface N atoms should however be quite different than the adsorbed N_2 species observed on the surface of the pure TiO_2 samples (Figure 3a), as the associated binding energies are quite dispersed on all N-doped spectra (Figure 3b–e). Considering the nature of our samples, this third feature is likely to be associated to NO or NO_2 groups on the particle surface. Table 2 summarizes the relative contributions of the three features associated to nitrogen in the doped samples (see the Supplementary materials for data associated with non-annealed samples, Table S2).

Table 2. Relative contributions of the X-ray photoelectron spectroscopy (XPS) peaks observed for samples doped by nitrogen at various contents. The table is presenting data for annealed powders (see Supplementary materials for data associated with as-prepared samples).

Sample	Relative contributions of XPS features		
	Substitutional N (peak at 396 eV)	**Interstitial N (peak at 400 eV)**	**Surface N (peak at 402 eV)**
N-TiO_2-0.1	11%	89%	Not measurable
N-TiO_2-0.2	16%	74%	9%
N-TiO_2-0.3	19%	58%	23%
N-TiO_2-0.6	15%	46%	39%

Comparing the relative contributions of the different peaks in the as-prepared and annealed powders, we observe a significant decrease of the intensity of the 396.2 eV peak with annealing (Table S2 and Table 2), which indicates a significant oxidation of substitutional Nitrogen. DFT calculations performed by Di Valentin *et al.* [40] show that the transition from substitutional to interstitial N is an exothermic process. Under oxygen-poor conditions, substitutional N position is favored. This is typically our case during particle growth, as powders are synthesized under inert or reducing conditions. In the opposite case, the annealing treatment performed at 400 °C in the presence of oxygen results in the rapid oxidation of substitutional nitrogen atoms. Thus, for low nitrogen content, most of the N atoms are located in interstitial positions for annealed powders (about 89% of all nitrogen atoms for sample doped at 0.1 wt %). However, for higher doping levels, not only the percentage of substitutional N increases, but a strong increase of the contribution of surface nitrogen is also observed. For comparison, Wang *et al.* reported that most of the N atoms were located only in interstitial sites for TiO_2 doped with 1.53 atom % of nitrogen [44]. In our case, because substitutional N atoms are always present in as-prepared samples, these types of N atoms are already present at low doping levels. In particular, our analysis suggests that the thermal treatment leads to an overall decrease of the substitutional N present in as-prepared samples (due to the oxygen-free synthesis conditions) and to a migration of the N atoms to the surface of the particles, especially at high doping levels. This feature is an important drawback of nitrogen doping, as free charge generation and current collection is drastically limited by surface states in DSSC and ssDSSC.

2.2. *Photovoltaic Performance of ssDSSC Based on N-Doped TiO_2 Electrodes*

Several independent sets of solid-state dye-sensitized solar cells (ssDSSC) were prepared from the un-doped and doped metal oxide nanopowders following procedures already described [34,37]. Briefly, a metal oxide paste is initially formulated and used to deposit the electrode on FTO/compact TiO_2 substrates. After sintering, porous electrodes of around two microns thick are obtained, which are further sensitized by the D102 indoline dye and infiltrated by the reference spiro-OMeTAD molecular glass acting as HTM. The conventional dopants (lithium salt and *tert*-butylpyridine) are used in our case. SEM cross sections of the infiltrated D102-sensitized electrodes are presented in Supplementary materials for the undoped and N-doped electrodes (Figure S4). In all cases, the solid-state electrolyte is clearly visible down to the bottom of the electrode. This observation is consistent with the morphologies of the starting nanopowders. Rather similar particle morphologies result in similar porous electrode morphology and, keeping all other parameters equal (nature of the dye, HTM concentration, and infiltration parameters), lead to comparable pore filling fractions. The absorption coefficient of the dye-sensitized electrodes, before HTM infiltration (Supplementary materials, Figure S5), does not clearly reveal any additional band associated to N-doping. However, a slight increase in absorption with increasing N content is observed over the entire wavelength range, with a more pronounced effect in the the 400–550 nm region. This region covers both the absorption band related to nitrogen doping in TiO_2 and the absorption band of the D102 dye, usually centered at 480 nm. It is thus reasonable to suggest that nitrogen doping is likely to slightly contribute to this increase of absorption.

However, we believe that the main effect of doping is related to a better sensitization of the electrodes, as previously reported [24,34]. A better dye grafting may also result from a change in surface potential induced by doping [45]. The electrical characteristics of the solar cells under illumination (AM1.5G, 100 mW· cm^{-2}) are presented in Figure 4, while Table 3 summarizes the corresponding photovoltaic parameters (the electrical characteristics in the dark are presented in Figure S6).

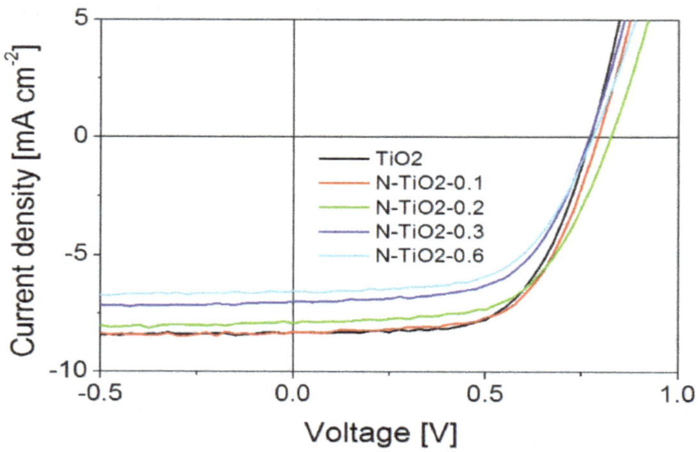

Figure 4. Current density/voltage characteristics of ssDSSC solar cells under standard illumination conditions (AM1.5G, 100 mW· cm^{-2}) for the different N contents.

Table 3. Photovoltaic parameters of solar cells based on N-doped TiO_2 electrodes as a function of Nitrogen content.

Nature of porous electrode	V_{OC} (V)	J_{SC} (mA· cm^{-2})	FF	η (%)
TiO_2	0.77	8.31	0.62	4.0
N-TiO_2-0.1	0.79	8.31	0.62	4.1
N-TiO_2-0.2	0.82	7.86	0.60	3.9
N-TiO_2-0.3	0.77	7.00	0.61	3.3
N-TiO_2-0.6	0.78	6.55	0.60	3.0

Let us note here that our reference device based on pure TiO_2 shows state-of-the-art performance considering the material used [37,46]. A slight improvement of solar cell efficiency is evident for low nitrogen content (up to 0.1 wt %) compared to the pure TiO_2 electrode, before a drastic decrease in performance at higher doping levels. Although the beneficial influence of nitrogen at low doping levels on the overall power conversion efficiency seems not so clear, the trend is unambiguously confirmed through several independent set of devices (not shown here), and is consistent with our preliminary report [34]. The incident photon to charge carrier efficiency (IPCE, or external quantum efficiency EQE) spectra of the cells (Figure S7, Supplementary materials) still especially exhibit an increase of photocurrent generation in the 400–500 nm region that can be related to nitrogen doping. We also note a significant decrease of IPCE as a function of doping level in the 550–650 nm region. This drop in photocurrent generation efficiency is related to the existence of surface-related electronic features, as revealed by photoluminescence spectroscopy on the nanopowders (Figure S8, Supplementary materials). A typical emission at 2.03 eV, associated with radiative recombination of self-trapped excitons at the particle surface, is observed for all powders [38]. This emission is found more pronounced in the presence of N atoms (especially when they are located at the particle surface), and results in a rapid recombination of photo-generated excitons following excitation in this spectral range. This decrease of IPCE in the 550–600 nm region is in fact counterbalancing the relative increase around 450 nm, leading to lower photocurrent as the doping level increases.

Going back to device performance, when the doping level reaches 0.2 to 0.3 wt %, device performance starts to significantly drop. Xie *et al.* also reported this trend, although no threshold was clearly pointed out [47]. For liquid cells, Guo *et al.* observed an optimal doping level of about 0.4 atom % [45], which is consistent with our data if we consider that our best performing device is associated with a nitrogen content of 0.1 wt %.

If we carefully check the photovoltaic parameters of the cells, the open circuit voltage of the N-doped devices is slightly improved compared to pure TiO_2, as usually observed in the literature [24,48]. This increase is likely to be due to a slight shift of the quasi-Fermi level of electrons in N-TiO_2 [29], especially if we consider that nitrogen is more concentrated at the particle surface. The short-circuit current density is at the maximum for both the un-doped device and for the one based on the lowest Nitrogen content (N-TiO_2-0.1), before decreasing for a higher doping level. In most of the cases, larger currents are evident for N-doped devices when the doping level is near this optimum concentration. This improved photocurrent, also observed in our preliminary study [34], was attributed to a beneficial contribution of nitrogen on the optical absorption of the electrode in the visible range [28,49]. Moreover, in the studies reporting significant increases in photocurrents and efficiency with doping, N atoms were mainly detected in substitutional or interstitial positions [48,50]. Smaller currents for N-doped electrodes compared to un-doped reference cells have also been reported in some cases [32]. The next sections of this article will focus on the elucidation of the relation between charge kinetics and nitrogen location going from materials to devices using techniques adapted to each scale.

2.3. Charge Kinetics Probed by Transient Photo-Voltage at the Device Level

To get a better insight into the exact influence of nitrogen on charge kinetics in ssDSSC devices, we performed transient photo-voltage measurements under working conditions, as a function of the incident light intensities. Under open circuit conditions, the transient photo-generated charges can decay only through recombination at the TiO_2-dye-HTM interface, as no carrier can be extracted in the external circuit. Figure 5 presents the corresponding recombination time for ssDSSC based on pure and N-doped TiO_2 electrodes.

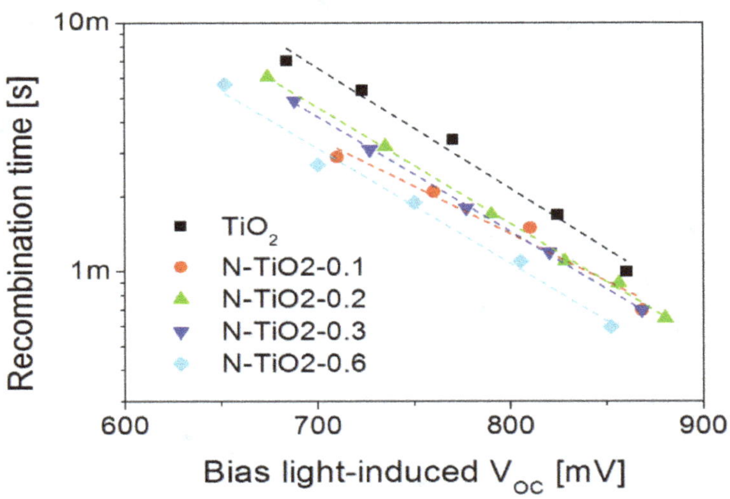

Figure 5. Recombination kinetics of ssDSSC probed by transient photo-voltage decay measurements, as a function of doping level.

Clearly, the charge lifetime decreases in all N-doped devices. Furthermore, recombination seems to become faster as the nitrogen content increases, especially under high light intensities (when the bias-light induced open-circuit voltage of the cells exceeds 800 mV). This initial observation indicates that N-doping is responsible for accelerated charge recombination in the devices, especially under standard illumination conditions. Our observation is opposite to that of Tian *et al.* who reported that N-doped TiO_2 can retard charge recombination due to the presence of O-Ti-N bonds [32]. However, faster recombination has been observed for N-doped electrodes elaborated from particles with large diameters (>20 nm) [49,51,52]. Such a phenomenon being unlikely in our case (see Table 1), this decrease in charge lifetime seems to be a direct consequence of the presence of N dopants in our TiO_2 materials. In particular, the preferential location of nitrogen atoms at the particle surface for high doping levels is consistent with shorter charge carrier lifetimes and lower performance. However, charge kinetics probed at the level of working devices is potentially strongly affected by the various processing steps used for cell fabrication, including sintering steps at high temperature which may significantly alter the nitrogen distribution. Therefore, the next section focuses on the characterization of charge dynamics at the material level.

2.4. Charge Kinetics Probed by Transient Techniques at the Material Level

In order to understand the role played by nitrogen atoms on charge generation, time-resolved microwave conductivity (TRMC) measurements were performed on the starting TiO_2 and N-TiO_2 powders. This technique, briefly described in Supplementary materials, is not a common tool of photo-physicists and photo-chemists. It was however successfully used to analyze charge kinetics of metal oxide materials for photo-catalysis [53–55] and photovoltaics [56–58]. It appears particularly well suited to investigate the influence of doping on the photo-conductivity properties of TiO_2 in the context of this study [59]. Figure 6 presents the absolute and normalized TRMC signals for the pure and N-doped TiO_2 before and after light excitation at 355 nm. In order to remove the effect of the number of photons, all the TRMC signals presented in this work are divided by the number of photons in nanoEinstein (nein), corresponding to the number of nanomoles of photons.

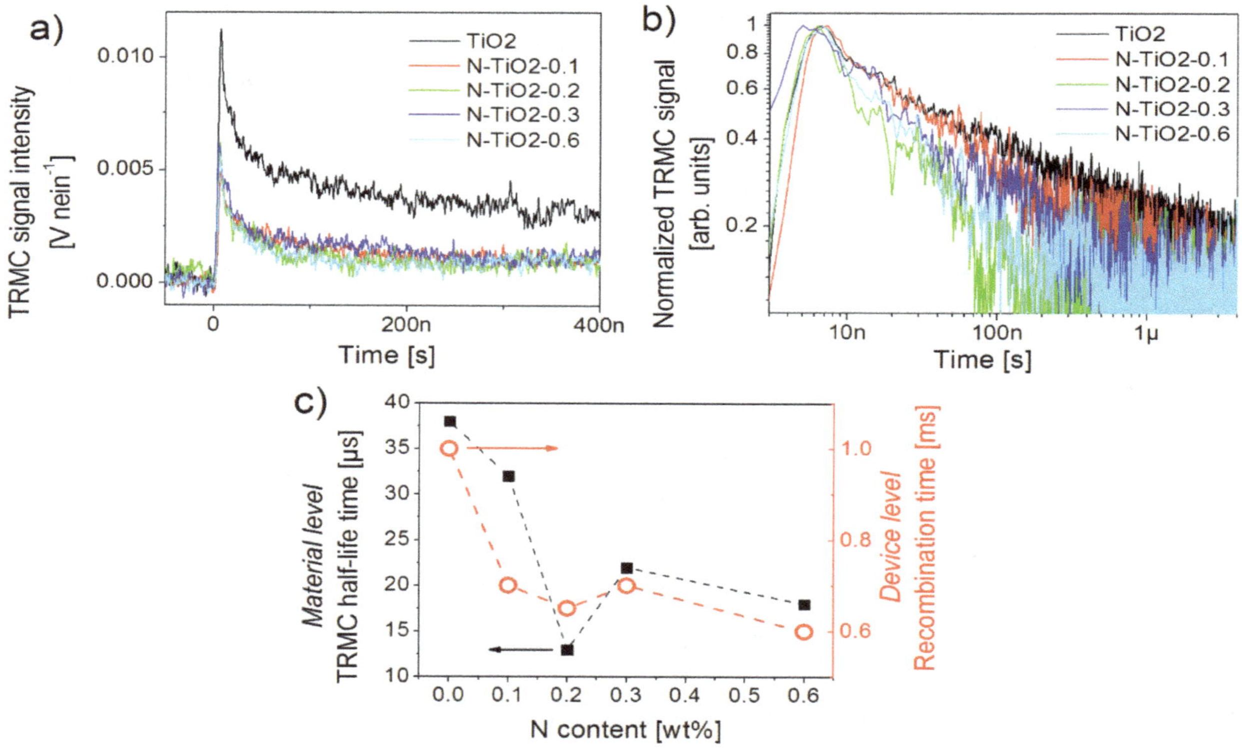

Figure 6. (**a**) absolute and (**b**) normalized time-resolved microwave conductivity (TRMC) signals of pure and N-doped TiO_2 powders with light excitation at 355 nm for the different N-doping levels; (**c**) TRMC decay half-times (black squares) measured at the material level on the TiO_2 and N-TiO_2 powders, and compared to recombination times extracted from transient photo-voltage at the device level.

The maximum intensity of the TRMC signal depends on three factors: the absorption coefficient of the material at the excitation wavelength; the interaction between the microwave electronic field and the materials, *i.e.*, the dielectric constant of the material; the recombination rate of electrons and holes during the laser pulse. Considering that the absorption coefficient at 355 nm of both un-doped and doped TiO_2 is mainly driven by transitions from O 2p to Ti 3d orbitals, we safely assume a similar absorption coefficient for all samples. We also assume a rather similar dielectric constant for the powders, as they present comparable crystalline structure and morphology (see Table 1), assuming low doping levels (always below 1 wt %). In these conditions, the strong decrease of TRMC signal with doping is consistent with additional recombination of free charge carriers during the laser pulse. Such observation is consistent with a previous report on N-doped titania materials [59], as well as with our transient photo-voltage analysis performed at the level of working devices (Figure 5). Accordingly, faster TRMC decay rates are observed with increasing N content in the nanopowders compared to the reference (Figure 6b). In addition, and except for sample N-TiO_2-0.2, the decay rate is faster for large nitrogen content (Figure 6c).

This is quite expected as when N atoms are inserted into the TiO_2 lattice, additional defects, such as oxygen vacancies, are spontaneously generated to ensure the global electric neutrality of the system. The photo-generated electrons are therefore more likely to be trapped in such defects, with a probability following in principle the nitrogen content. In our previous work, DFT calculations confirmed the occurrence of oxygen vacancies induced by the presence of nitrogen dopants [34]. Such processes have also been reported by other groups [59]. Finally, the extracted TRMC half-life time shows a strong correlation with the recombination kinetics measured on devices by transient photo-voltage decays (Figure 6c). This correlation, which was also confirmed through charge kinetics extracted from impedance spectroscopy applied on the same devices (not shown here), shows, in this particular case, the relevance for a multi-scale photo-physical approach from materials to devices. Our observations suggest that the features introduced by nitrogen doping are preserved during device fabrication, and that the dynamics of photo-generated charge carriers probed in the TiO_2 nanopowders still mainly drives device performance under simulated sunlight.

Considering the photo-activity of the N-doped samples in the visible region, we now give a better look at the influence of nitrogen on charge kinetics probed by TRMC using an excitation in the visible range. First, no TRMC signal is evidenced for the un-doped TiO_2 powder, in accordance with a flat absorption in this region. However, significant TRMC signals are recorded for the N-doped nanopowders, which is consistent with their optical absorption (Figure 2). No significant differences in the TRMC decay profiles of doped samples are, however, observed by exciting them either at 355, 420, 450, or 480 nm (see Supplementary materials, Figure S9 corresponding to sample N-TiO_2-0.6). Similar reports were made in the literature [59], which are associated with the fact that both the O 2p to Ti 3d and the N 2p to Ti 3d transitions, induced through UV and visible excitation, respectively, show a similar final state. By fixing the excitation to 450 nm, the amplitudes of the TRMC signals increase with the nitrogen content up to 0.6 wt % in the powder (Figure 7), indicating that the photo-conductivity of the samples increases with the nitrogen content during the first 100 ns. This observation is consistent with the creation of a larger amount of electron-hole pairs for high doping levels. Considering our previous conclusions, this result suggests that a competition between charge transport and recombination is occurring in doped samples. In order to better interpret these effects, we finally analyze more carefully the electrical photo-conductivity of porous TiO_2 electrodes processed from the doped and un-doped nanopowders, *i.e.*, an intermediate elaboration level between the powder and the device.

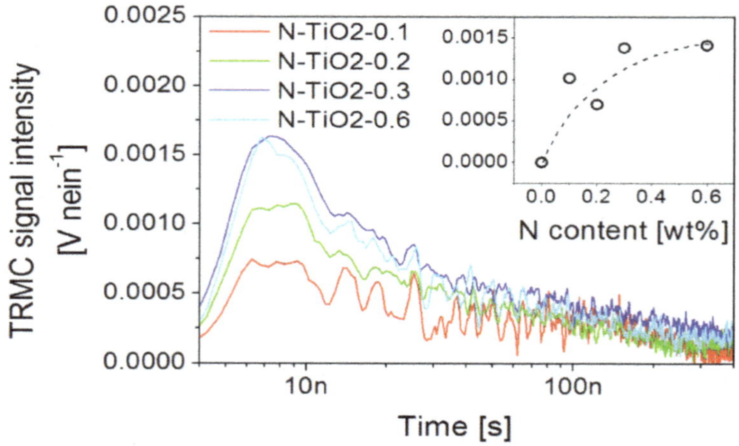

Figure 7. TRMC signals associated with an excitation at 450 nm, for N-doped powder samples. The inset shows the maximum TRMC signal amplitude (recorded after 9 ns in all cases) as a function of N content in the nanopowders.

2.5. Photo-Conductivity Measurements on Porous Electrodes

As the conductivity of semiconductors depends directly on the charge carrier density, it can be temporarily influenced by photo-generated electron-hole pairs into the material (*i.e.*, transitions of electrons from the valence band or donor levels to the conduction band). The technique is briefly presented in Supplementary materials. Considering the bandgap of anatase TiO_2 [60,61], photo-excitation by wavelengths above 400 nm mainly involves electrons from the donor band while UV excitation mainly involves electrons initially in the valence band. Figure 8 presents the current change (ΔI), defined as the difference between the current level in the dark and the saturation value under illumination, as a function of UV light intensity (excitation in the 330–380 nm range, see the Experimental section) for the un-doped and N-doped porous electrodes deposited from the nanopowders. It is worth noting that a sintering procedure similar to that used for device fabrication was also carried out in this case.

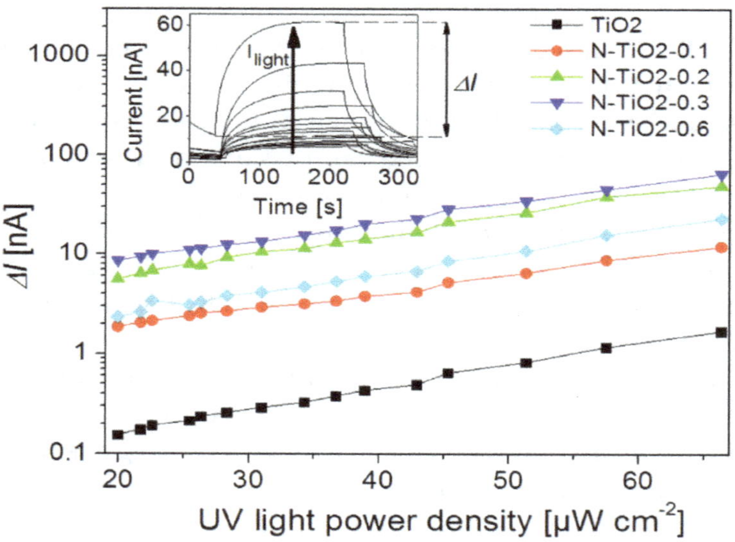

Figure 8. Photo-current response $\Delta I(t)$ under UV excitation of porous TiO_2 electrodes as a function of incident light intensity and for various nitrogen doping levels. The inset presents characteristic photo-response curves for various illumination intensities from 20 to 66 $\mu W \cdot cm^{-2}$, for the sample N-TiO_2-0.2. The current jump ΔI can be calculated from this data by subtracting the dark current from the saturation current as illustrated in the inset.

In all cases, the current—and hence the conductivity—increases under illumination, and saturates after about 180 s (inset of Figure 8). All the nitrogen doped samples show similar photo-response curves. An increased nitrogen content in the film is associated with an increased current change. This behavior indicates that nitrogen doping increases the amount of charge carriers being promoted into the conduction band of TiO_2, as also suggested by the TRMC analysis of the nanopowders and the intensities observed for the different doping levels (Figure 7). Unlike in our TRMC analysis, which showed faster charge recombination with doping (Figure 6a), the high voltage biasing of the devices used during the photo-conductivity measurements (bias voltage of 8 V) assists immediate charge transport to the electrodes preventing rapid charge recombination, so that a transient current signal can indeed be recorded. Thus, in this measurement, the limiting factor for the electrical response of the devices is not the charge recombination time but rather the density of recombination centers, e.g., impurities related to surface N atoms. The differences between the N-doped samples are small, as also noticed in the TRMC measurements made under UV excitation. A decrease in the photo-response is, however, observed for sample N-TiO2-0.6, which suggests that high nitrogen content is detrimental to current collection. This observation is consistent with the high density of surface nitrogen atoms revealed in this case, and with the high recombination rate evidenced for this sample in the previous section.

Figure 9 presents the current change (ΔI) of N-doped films for an excitation in the visible range provided by a long-pass filtered halogen lamp (λ > 410 nm, see Experimental section), as a function of incident light intensity.

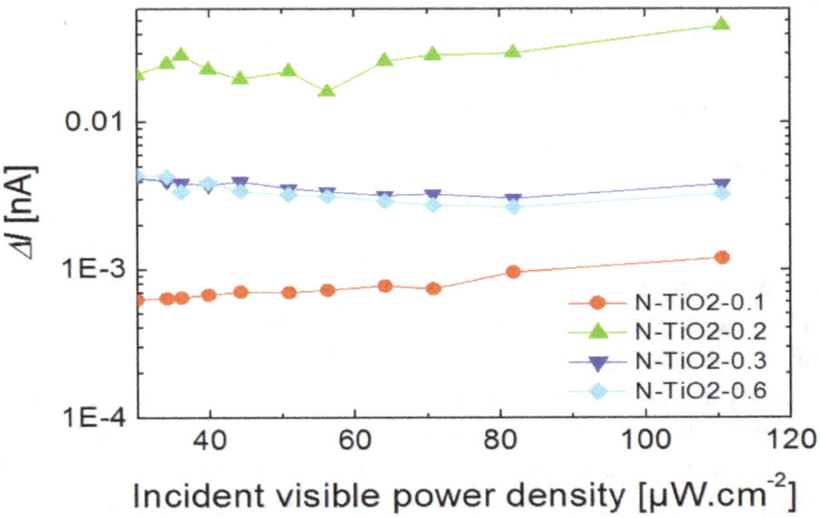

Figure 9. Current change associated to N-doped TiO_2 porous films under visible light illumination as a function of incident light intensity.

The un-doped film is not shown in Figure 9, as it is associated with a very low current level (before the light is switched on) and current change (<1 pA, see Supplementary materials, Figure S10) under illumination by visible light. This behavior is expected as the sample does not significantly absorb the incident light, so that no photo-conductivity is observable. On the opposite side, doped samples rapidly show a significant current change and photo-conductive behavior. We observe that after the initial jump corresponding to the beginning of the illumination, the current experiences a linear increase and no saturation can be observed. This linear increase, which was not observed under UV illumination, can be attributed to thermal effects induced by the infrared wavelengths of the light source. Filtering the infrared part of the lamp is indeed found to completely remove this linear dependency of the photo-response (see Supplementary materials, Figure S11). This effect is an illustration of the important temperature-dependency of n_d, and hence of ΔI.

Keeping this thermal effect in mind, we observe in Figure 9 a larger current level for porous films based on N-doped TiO_2 compared to the un-doped reference. Both the dark current and current change ΔI under visible light are enhanced through doping (by about one order of magnitude). Similarly to previous sections, our data suggest that the photo-induced charge transport properties of the porous films are improved through nitrogen doping, especially within the visible range.

3. Discussion

Nitrogen-doped porous electrodes can potentially improve the performance of dye-sensitized solar cells. However, although several reports discuss the positive effect of doping on device efficiency [24,32–34], it is also clear that true improvements remain subject to our ability to properly control both the amount and the location of the dopants. In the present study, as well as in our previous report [34], several sets of independent ssDSSC show that improvements in short-circuit current densities, open-circuit voltage, and in overall power conversion efficiencies can be achieved at low doping levels (Table 3). Our fabrication strategy exploits the laser pyrolysis process to grow TiO_2 and N-doped TiO_2 nanopowders of comparable morphologies and crystallinities (Table 1); however, nitrogen location, revealed by XPS analysis, is found to be strongly dependent on the doping level: although interstitial and substitutional N atoms are evident, the fraction of surface nitrogen increases with the doping level. While atoms inserted inside the TiO_2 structure give the metal oxide a visible photo-activity, surface nitrogen can act as electron traps or recombination centers. The resulting surface states are found to be limiting the charge lifetime, as revealed by transient photo-voltage measurements performed at the device level (Figure 5), and by TRMC at the material level (Figure 6). Larger recombination rates are indeed observed for doped electrodes, as soon as nitrogen is incorporated in the material. In parallel, photo-conductivity measurements on porous films prepared from the synthesized TiO_2 nanopowders (similar to device electrodes) show that better transport can be achieved in the doped electrodes, as long as recombination can be reduced through an external applied voltage for example (Figure 8). More specifically, our analysis gives direct evidence for the positive role played by the visible range photo-activity of the doped samples (both using TRMC and photo-response measurements). During illumination within the visible range, a larger fraction of charges can contribute to current collection, which is consistent with the signature observed in the IPCE spectra of the cells around 450 nm (Figure S7). However, in full working devices, such benefit of nitrogen doping remain subject to the competition with the intense recombination processes, which are found to mainly drive device operation. We strongly believe that the main limitation to current generation for the N-doped cells is associated with the preferential location of N atoms on the TiO_2 particle surface. The XPS signatures of such type of nitrogen (Figure 3 and Table 2) start to be clearly observed for sample N-TiO_2-0.2, when photocurrent also begins to drop. This drop is consistent with the spectral signature observed around 600 nm in the IPCE spectrum of the N-doped devices, which is associated with the influence of surface-related defects induced by the presence of N atoms. Such assumption was also proposed by Lindgren *et al.* for liquid DSSC, for example [28]. For higher doping level up to 0.6 wt %, when recombination is the main limiting factor, we observe a drastic drop of 25% in both photocurrent and device efficiency.

4. Materials and Methods

4.1. Synthesis of N-Doped TiO_2 Nanocrystals

Nanoparticles doped with different levels of nitrogen were synthesized through laser pyrolysis by varying the NH_3 flow in the precursor mixture composed of titanium tetra-isopropoxide (TTIP) and C_2H_4. The details were reported in our previous work [36,38,62]. For the photovoltaic application, the as-prepared nanopowders were annealed at 400 °C for 3h to remove the free carbon phase that remains due to the decomposition of TTIP and/or C_2H_4.

4.2. Device Fabrication

TiO_2 and N-doped TiO_2 porous electrodes were deposited on pre-cleaned FTO glass substrates initially covered with a TiO_2 blocking layer deposited by chemical spray pyrolysis, using spin-coating from formulations based on ethanol, α-terpineol and ethyl-cellulose (EC), as previously described [38]. These films were progressively sintered up to 430 °C during 40 min in air. A $TiCl_4$ treatment (Aldrich, 0.04 M in deionized water) was performed on these films before a final sintering step at 430 °C in air during 45 min. The 1.8 μm thick electrodes were then immersed in D102 dye (Mitsubishi Paper Mills, Tsukuba, Japan) dissolved in an acetonitrile:*tert*-butanol mixture (1:1 in volume) at 80 °C overnight. The sensitized electrodes were rinsed and infiltrated by the hole transporting material (HTM) spiro-OMeTAD (Merck KGaA, Darmstadt, Germany) by spin-coating in ambient conditions, following recipes previously reported [37,38]. Gold counter electrodes were evaporated through a shadow mask at 10^{-6} mbar, leading to two independent active areas of 0.18 cm^2 per cell.

4.3. Characterization Techniques

TEM images were recorded on a Philips CM12 microscope to examine the morphology of both the obtained nanoparticles and device cross sections. The crystalline phases were determined by XRD with a Siemens D5000 instrument using the Cu-Kα radiation. The specific surface of the nanopowders was determined by the BET method using a Micromeritics FlowsorbII 2300 instrument. The chemical environment of nitrogen atoms in TiO_2 and N-TiO_2 was characterized by X-ray photoelectron spectroscopy (XPS) using a Kratos Analytical Axis Ultra DLD spectrometer (Kα X-ray) on the powders. The optical properties of the powders were analyzed using a UV-visible-NIR spectrophotometer (Jasco V-570) in reflectance mode. According to the Kubelka–Munk equation, the optical gap can be estimated from the (F(*R*).hυ) plot as a function of photon energy, where $F(R) = (1 - R)^2/2R$ and R is the reflection coefficient [63]. The current density-voltage characteristics of the devices were recorded using a Keithley 2400 source-measure unit in the dark and under simulated solar emission (NEWPORT class A solar simulator) at 100 $mW \cdot cm^{-2}$ in AM1.5G conditions after spectral mismatch correction.

4.4. Transient Photo-Voltage Measurements

Transient photo-voltage (TPV) decay measurements were measured under open-circuit conditions using a set-up previously described [37]. Two continuous white LEDs (OSRAM) were used to provide the constant illumination of the device up to approximately 100 $mW \cdot cm^{-2}$. An additional pulsed LED (λ = 550 nm, Luxeon STAR, 5W), controlled by a solid state switch, generated a small light pulse on the cell. The transient charge population generated by this LED was always below a few percent of the continuous steady-state charge density in the device, ensuring a small-perturbation regime. The photo-voltage of the device was monitored and recorded using a digital oscilloscope (Tektronix DPO 4032) interfaced using a home-made Labview routine. The photo-voltage decays were adjusted using mono-exponential decay functions.

4.5. Time-Resolved Microwave Conductivity Measurements (TRMC)

The charge-carrier lifetimes in the un-doped and N-doped TiO_2 nanopowders after an illumination were determined by microwave absorption experiments using TRMC. The incident microwaves were generated by a Gunn diode of the Ka band at 30 GHz. The pulsed light source was an OPO laser from EKSPLA where the accord ability extends from 200 to 2000 nm. The full width at half-maximum of one pulse was 7 ns and the repetition rate of the experiments was 10 Hz. The light energy density received by the sample depends on the wavelength. To avoid the excitation energy effect on the signals, all the data are divided by the number of photons. Typically, the energy density at 450 nm is 5.2×10^{-3} $J \cdot cm^{-2}$, corresponding to 1.2×10^{16} photons$\cdot cm^{-2}$.

4.6. Photo-Conductivity Measurements

Photo-conductivity characteristics of the samples were studied by illuminating un-doped and N-doped TiO_2 thin films with UV (double peak between 330 and 380 nm, see Supplementary materials, Figure S12) and visible light (long-pass filtered halogen lamp, $\lambda > 410$ nm, see Supplementary materials, Figure S13) and measuring the photo-response of the conductivity of the film, *i.e.*, the change in the sample conductivity when turning on the light under a constant applied bias voltage. For these experiments, thin porous films were fabricated from the same TiO_2 nanopowders used for the solar cell preparation, except that it was sonicated for 1h using a Hielscher sonicator (UP400s ultrasonicator, Hielscher Ultrasonics GmbH, Teltow, Germany) while cooled in a water path. After sonication, the powder was spin-coated on a glass chip (~1 cm^2) at 1000 rpm for 1 min. To avoid cracks in the films, they were left to dry for 5 to 10 min before performing a first soft annealing step at 50 °C for 5 min in air. Then, the films were annealed at 430 °C for 35 min. Film deposition, as well as the drying and annealing steps, were performed in a laminar flow hood. Gold contact electrodes (100 nm thick) were evaporated under high vacuum using a thin copper wire (diameter $\leqslant$ 120 μm) placed as a mask across the film. For the photoconductivity measurements, samples were placed in a windowed vacuum chamber at 4–6 mbar to remove most of the ambient moisture. The response of photoconductivity as a function of illumination intensity was measured using Stanford Research Systems voltage and current low-noise preamplifiers (models SR560 and SR570). A MinUVIS 30 W mercury lamp (Desaga #751311 , Heidelberg, Germany) and a 50 W halogen lamp (Solux C5 12 V, Rochester, USA) were used for illumination, and light intensity was adjusted by changing the distance between the sample and the light source. The light intensities were measured using PD100M optical power meter (Thorlabs Sweden AB, Mölndal, Sweden) with S302C thermal power head (Thorlabs Sweden AB, Mölndal, Sweden).

5. Conclusions

TiO_2 and N-doped TiO_2 nanoparticles have been synthesized using laser pyrolysis, and used to deposit porous electrodes suitable for solid-state dye-sensitized solar cells. A systematic analysis of the physical properties of the samples as a function of doping level was performed in order to discuss the exact influence of the dopant on material and device photo-physics. At low doping levels, N atoms have been efficiently incorporated in interstitial positions into the metal oxide structure for annealed powders. In this case, device performance is found sensibly improved compared to pure TiO_2. However, a large fraction of surface nitrogen is also observed at higher doping levels, which was found to be responsible for faster recombination kinetics that clearly reduces device efficiency. At the material scale, charge kinetics is found to be in good correlation with kinetics observed on devices. Both suggest faster electron-hole pair recombination being induced in the presence of N atoms compared to the un-doped powder. Our analysis confirms that the rapid drop in photocurrent observed in our ssDSSC at high doping levels can be associated with the preferential location of nitrogen atoms at the particle surface, which favor interfacial recombination. The visible photo-sensitivity of the samples is confirmed by TRMC measurements, which showed an increasing photo-conductivity of the material in visible range as the nitrogen content increases. This trend was further confirmed through photo-response measurement made on porous TiO_2 and N-TiO_2 films. Once again, the photo-conductivity of the material was found to be largely improved under visible excitation, compared to the pure TiO_2 reference.

Although nitrogen doping remains a relevant strategy to improve the efficiencies of dye-sensitized solar cells, our data show that achieving concrete benefits at the device level from the visible photo-activity of N-doped TiO_2 requires a fine control of the doping level and of the location of N atoms in the metal oxide structure. We also point out an interesting correlation between the photo-physical properties of samples probed at various scales from materials to devices. In light of our investigations, the specific features of the starting nanopowder materials are found to drive the device photo-physics, even when surface treatments and sintering steps are used during solar cell fabrication.

Supplementary Materials
Complementary experimental details, Table S1: Experimental conditions of laser pyrolysis for the synthesis of N-doped TiO_2 powders, Figure S1: XRD patterns of all powders, Figure S2: XPS spectra of TiO_2 nanopowder, Figure S3: XPS spectra as-prepared N-TiO_2-0.2 and N-TiO_2-0.6 samples, Table S2: Relative contributions of XPS features for as-prepared N-doped samples, Figure S4: SEM cross-section micrographs of ssDSSC, Figure S5: Absorption coefficient of the dye-sensitized electrodes, Figure S6: J(V) curves of the solar cells in the dark, Figure S7: Normalized IPCE, Figure S8: Photoluminescence spectra of undoped and doped nanopowders, Figure S9: Normalized TRMC signals for different excitation wavelengths, Figure S10: Photoconductivity curves of porous films under visible light excitation, Figure S11: Photoconductivity curves of the porous film under visible light illumination, with and without infrared filter, Figure S12 : Emission spectrum of the UV-lamp used in the photoconductvity measurements, Figure S13: Optical spectrum of the long-pass filtered halogen lamp used in the photoconductivity measurements.

Acknowledgments: The authors acknowledge the INSIS CNRS Energy program (Noxomix project), Academy of Finland (Projects 263262, 283011, 263526) and CEA-IFA Nanophob project for funding, as well as the support of the "Région Limousin" (thematic project "EVASION") and NewIndigo ERA-NET NPP2 ("Aquatest" project). This work was performed in the framework of the Energy and Environment thematic of the SIGMA-LIM Laboratory of Excellence. We thank Heikki Häkkänen and Pasi Myllyperkiö for useful discussions.

Author Contributions: Jin Wang and Nathalie Herlin-Boime conceived and designed the experiments and supervised the analysis of the data, as well as paper writing; Jin Wang performed the main experiments and participated in the writing of the paper; Monica Scarisoreanu participated in the synthesis by laser pyrolysis of some nanopowders used in this study; Aurélie Habert performed the SEM measurements; Bernard Ratier participated in the interpretation of data and paper writing; Kosti Tapio and Jussi Toppari performed the photoconductivity measurements and contributed to the interpretations of the data; Sebastien Sorgues and Christophe Colbeau-Justin performed the TRMC analyses and contributed to the analysis of the data.

References

1. Bach, U.; Lupo, D.; Comte, P.; Moser, J.E.; Weissörtel, F.; Salbeck, J.; Spreitzer, H.; Grätzel, M. Solid-state dye-sensitized mesoporous TiO_2 solar cells with high photon-to-electron conversion efficiencies. *Nature* **1998**, *395*, 583–585.
2. Burschka, J.; Dualeh, A.; Kessler, F.; Baranoff, E.; Cevey-Ha, N.-L.; Yi, C.; Nazeeruddin, M.K.; Grätzel, M. Tris(2-(1H-pyrazol-1-yl)pyridine)cobalt(III) as p-Type Dopant for Organic Semiconductors and Its Application in Highly Efficient Solid-State Dye-Sensitized Solar Cells. *J. Am. Chem. Soc.* **2011**, *133*, 18042–18045. [CrossRef] [PubMed]
3. Xu, B.; Gabrielsson, E.; Safdari, M.; Cheng, M.; Hua, Y.; Tian, H.; Gardner, J.M.; Kloo, L.; Sun, L. 1,1,2,2-Tetrachloroethane (TeCA) as a Solvent Additive for Organic Hole Transport Materials and Its Application in Highly Efficient Solid-State Dye-Sensitized Solar Cells. *Adv. Energy Mater.* **2015**, *5*. [CrossRef]
4. Ahmad, S.; Guillén, E.; Kavan, L.; Grätzel, M.; Nazeeruddin, M.K. Metal free sensitizer and catalyst for dye sensitized solar cells. *Energy Environ. Sci.* **2013**, *6*, 3439–3466. [CrossRef]
5. Li, L.L.; Diau, E.W.G. Porphyrin-sensitized solar cells. *Chem. Soc. Rev.* **2013**, *42*, 291–304. [CrossRef] [PubMed]
6. Liang, M.; Chen, J. Arylamine organic dyes for dye-sensitized solar cells. *Chem. Soc. Rev.* **2013**, *42*, 3453–3488. [CrossRef] [PubMed]
7. Stranks, S.D.; Snaith, H.J. Metal-halide perovskites for photovoltaic and light-emitting devices. *Nat. Nano* **2015**, *10*, 391–402. [CrossRef] [PubMed]
8. Chen, Q.; de Marco, N.; Yang, Y.; Song, T.-B.; Chen, C.-C.; Zhao, H.; Hong, Z.; Zhou, H. Under the spotlight: The organic–inorganic hybrid halide perovskite for optoelectronic applications. *Nano Today* **2015**, *10*, 355–396. [CrossRef]
9. Bui, T.T.; Goubard, F. Small organic molecule hole transporting materials for solid-state dye-sensitized solar cells. *Mater. Tech.* **2013**, *101*. [CrossRef]
10. Aulakh, R.K.; Sandhu, S.; Tanvi; Kumar, S.; Mahajan, A.; Bedi, R.K. Designing and synthesis of imidazole based hole transporting material for solid state dye sensitized solar cells. *Synth. Met.* **2015**, *205*, 92–97. [CrossRef]

11. Hsu, C.Y.; Chen, Y.C.; Lin, R.Y.Y.; Ho, K.C.; Lin, J.T. Solid-state dye-sensitized solar cells based on spirofluorene (spiro-OMeTAD) and arylamines as hole transporting materials. *Phys. Chem. Chem. Phys.* **2012**, *14*, 14099–14109. [CrossRef] [PubMed]
12. Chung, I.; Lee, B.; He, J.; Chang, R.P.H.; Kanatzidis, M.G. All-solid-state dye-sensitized solar cells with high efficiency. *Nature* **2012**, *485*, 486–489. [CrossRef] [PubMed]
13. Lee, B.; Stoumpos, C.C.; Zhou, N.; Hao, F.; Malliakas, C.; Yeh, C.-Y.; Marks, T.J.; Kanatzidis, M.G.; Chang, R.P. Air-Stable Molecular Semiconducting Iodosalts for Solar Cell Applications: Cs_2SnI_6 as a Hole Conductor. *J. Am. Chem. Soc.* **2014**, *136*, 15379–15385. [CrossRef] [PubMed]
14. Freitag, M.; Daniel, Q.; Pazoki, M.; Sveinbjornsson, K.; Zhang, J.; Sun, L.; Hagfeldt, A.; Boschloo, G. High-efficiency dye-sensitized solar cells with molecular copper phenanthroline as solid hole conductor. *Energy Environ. Sci.* **2015**, *8*, 2634–2637. [CrossRef]
15. Bouclé, J.; Ackermann, J. Solid-state dye-sensitized and bulk heterojunction solar cells using TiO_2 and ZnO nanostructures: Recent progress and new concepts at the borderline. *Polym. Int.* **2012**, *61*, 355–373. [CrossRef]
16. Docampo, P.; Guldin, S.; Leijtens, T.; Noel, N.K.; Steiner, U.; Snaith, H.J. Lessons Learned: From Dye-Sensitized Solar Cells to All-Solid-State Hybrid Devices. *Adv. Mater.* **2014**, *26*, 4013–4030. [CrossRef] [PubMed]
17. Chen, H.Y.; Xu, Y.F.; Kuang, D.B.; Su, C.Y. Recent advances in hierarchical macroporous composite structures for photoelectric conversion. *Energy Environ. Sci.* **2014**, *7*, 3887–3901. [CrossRef]
18. Concina, I.; Vomiero, A. Metal oxide semiconductors for dye- and quantum-dot-sensitized solar cells. *Small* **2015**, *11*, 1744–1774. [CrossRef] [PubMed]
19. Gao, C.; Li, X.; Lu, B.; Chen, L.; Wang, Y.; Teng, F.; Wang, J.; Zhang, Z.; Pan, X.; Xie, E. A facile method to prepare SnO_2 nanotubes for use in efficient SnO_2–TiO_2 core-shell dye-sensitized solar cells. *Nanoscale* **2012**, *4*, 3475–3481. [CrossRef] [PubMed]
20. Antila, L.J.; Heikkilä, M.J.; Mäkinen, V.; Humalamäki, N.; Laitinen, M.; Linko, V.; Jalkanen, P.; Toppari, J.; Aumanen, V.; Kemell, M. ALD grown aluminum oxide submonolayers in dye-sensitized solar cells: The effect on interfacial electron transfer and performance. *J. Phys. Chem. C* **2011**, *115*, 16720–16729. [CrossRef]
21. Liu, Y.; Zhai, H.; Guo, F.; Huang, N.; Sun, W.; Bu, C.; Peng, T.; Yuan, J.; Zhao, X. Synergistic effect of surface plasmon resonance and constructed hierarchical TiO_2 spheres for dye-sensitized solar cells. *Nanoscale* **2012**, *4*, 6863–6869. [CrossRef] [PubMed]
22. Tian, Z.; Wang, L.; Jia, L.; Li, Q.; Song, Q.; Su, S.; Yang, H. A novel biomass coated Ag–TiO_2 composite as a photoanode for enhanced photocurrent in dye-sensitized solar cells. *RSC Adv.* **2013**, *3*, 6369–6376. [CrossRef]
23. Sun, Q.; Zhang, J.; Wang, P.; Zheng, J.; Zhang, X.; Cui, Y.; Feng, J.; Zhu, Y. Sulfur-doped TiO_2 nanocrystalline photoanodes for dye-sensitized solar cells. *J. Renew. Sustain. Energy* **2012**, *4*. [CrossRef]
24. Ma, T.; Akiyama, M.; Abe, E.; Imai, I. High-efficiency dye-sensitized solar cell based on a nitrogen-doped nanostructured titania electrode. *Nano Lett.* **2005**, *5*, 2543–2547. [CrossRef] [PubMed]
25. Wang, H.; Li, H.; Wang, J.; Wu, J.; Li, D.; Liu, M.; Su, P. Nitrogen-doped TiO_2 nanoparticles better TiO_2 nanotube array photo-anodes for dye sensitized solar cells. *Electrochim. Acta* **2014**, *137*, 744–750. [CrossRef]
26. Zhao, B.; Wang, J.; Li, H.; Wang, H.; Jia, X.; Su, P. The influence of yttrium dopant on the properties of anatase nanoparticles and the performance of dye-sensitized solar cells. *Phys. Chem. Chem.Phys.* **2015**, *17*, 14836–14842. [CrossRef] [PubMed]
27. Asahi, R.; Morikawa, T.; Ohwaki, T.; Aoki, K.; Taga, Y. Visible-light photocatalysis in nitrogen-doped titanium oxides. *Science* **2001**, *293*, 269–271. [CrossRef] [PubMed]
28. Lindgren, T.; Mwabora, J.M.; Avandaño, E.; Jonsson, J.; Hoel, A.; Granqvist, C.G.; Lindquist, S.E. Photoelectrochemical and optical properties of nitrogen doped titanium dioxide films prepared by reactive DC magnetron sputtering. *J. Phys. Chem. B* **2003**, *107*, 5709–5716. [CrossRef]
29. Di Valentin, C.; Finazzi, E.; Pacchioni, G.; Selloni, A.; Livraghi, S.; Paganini, M.C.; Giamello, E. N-doped TiO_2: Theory and experiment. *Chem. Phys.* **2007**, *339*, 44–56. [CrossRef]
30. Di Valentin, C.; Pacchioni, G. Trends in non-metal doping of anatase TiO_2: B, C, N and F. *Catal. Today* **2013**, *206*, 12–18. [CrossRef]
31. Asahi, R.; Morikawa, T. Nitrogen complex species and its chemical nature in TiO_2 for visible-light sensitized photocatalysis. *Chem. Phys.* **2007**, *339*, 57–63. [CrossRef]

32. Tian, H.; Hu, L.; Zhang, C.; Liu, W.; Huang, Y.; Mo, L.; Guo, L.; Sheng, J.; Dai, S. Retarded charge recombination in dye-sensitized nitrogen-doped TiO_2 solar cells. *J. Phys. Chem. C* **2010**, *114*, 1627–1632. [CrossRef]
33. Di Valentin, C.; Pacchioni, G.; Selloni, A. Origin of the different photoactivity of N-doped anatase and rutile TiO_2. *Phys. Rev. B* **2004**, *70*. [CrossRef]
34. Melhem, H.; Simon, P.; Wang, J.; di Bin, C.; Ratier, B.; Leconte, Y.; Herlin-Boime, N.; Makowska-Janusik, M.; Kassiba, A.; Bouclé, J. Direct photocurrent generation from nitrogen doped TiO_2 electrodes in solid-state dye-sensitized solar cells: Towards optically-active metal oxides for photovoltaic applications. *Sol. Energy Mater. Sol. Cells* **2013**, *117*, 624–631. [CrossRef]
35. Diker, H.; Varlikli, C.; Stathatos, E. N-doped titania powders prepared by different nitrogen sources and their application in quasi-solid state dye-sensitized solar cells. *Int. J. Energy Res.* **2014**, *38*, 908–917. [CrossRef]
36. Pignon, B.; Maskrot, H.; Ferreol, V.G.; Leconte, Y.; Coste, S.; Gervais, M.; Pouget, T.; Reynaud, C.; Tranchant, J.F.; Herlin-Boime, N. Versatility of laser pyrolysis applied to the synthesis of TiO_2 nanoparticles—Application to UV attenuation. *Eur. J. Inorg. Chem.* **2008**, *2008*, 883–889. [CrossRef]
37. Melhem, H.; Simon, P.; Beouch, L.; Goubard, F.; Boucharef, M.; Di Bin, C.; Leconte, Y.; Ratier, B.; Herlin-Boime, N.; Bouclé, J. TiO_2 Nanocrystals Synthesized by Laser Pyrolysis for the Up-Scaling of Efficient Solid-Stage Dye-Sensitized Solar Cells. *Adv. Energy Mater.* **2011**, *1*, 908–916. [CrossRef]
38. Wang, J.; Lin, Y.; Pinault, M.; Filoramo, A.; Fabert, M.; Ratier, B.; Bouclé, J.; Herlin-Boime, N. Single-Step Preparation of TiO_2/MWCNT Nanohybrid Materials by Laser Pyrolysis and Application to Efficient Photovoltaic Energy Conversion. *ACS Appl. Mater. Interfaces* **2015**, *7*, 51–56. [CrossRef] [PubMed]
39. Spurr, R.A.; Myers, H. Quantitative analysis of anatase-rutile mixtures with an X-ray diffractometer. *Anal. Chem.* **1957**, *29*, 760–762. [CrossRef]
40. Klug, H.P.; Alexander, L.E. *X-Ray Diffraction Procedures*; Wiley: New York, NY, USA, 1954.
41. Kang, S.H.; Kim, H.S.; Kim, J.Y.; Sung, Y.E. Enhanced photocurrent of nitrogen-doped TiO_2 film for dye-sensitized solar cells. *Mater. Chem. Phys.* **2010**, *124*, 422–426. [CrossRef]
42. Zhang, M.; Lin, G.; Dong, C.; Kim, K.H. Mechanical and optical properties of composite TiO_xN_y films prepared by pulsed bias arc ion plating. *Curr. Appl. Phys.* **2009**, *9*, S174–S178. [CrossRef]
43. Amadelli, R.; Samiolo, L.; Borsa, M.; Bellardita, M.; Palmisano, L. N-TiO_2 Photocatalysts highly active under visible irradiation for NO_X abatement and 2-propanol oxidation. *Catal. Today* **2013**, *206*, 19–25. [CrossRef]
44. Wang, J.; Tafen, D.N.; Lewis, J.P.; Hong, Z.; Manivannan, A.; Zhi, M.; Li, M.; Wu, N. Origin of Photocatalytic Activity of Nitrogen-Doped TiO_2 Nanobelts. *J. Am. Chem. Soc.* **2009**, *131*, 12290–12297. [CrossRef] [PubMed]
45. Guo, W.; Shen, Y.; Wu, L.; Gao, Y.; Ma, T. Effect of N dopant amount on the performance of dye-sensitized solar cells based on N-Doped TiO_2 electrodes. *J. Phys. Chem. C* **2011**, *115*, 21494–21499. [CrossRef]
46. Schmidt-Mende, L.; Bach, U.; Humphry-Baker, R.; Horiuchi, T.; Miura, H.; Ito, S.; Uchida, S.; Grätzel, M. Organic Dye for Highly Efficient Solid-State Dye-Sensitized Solar Cells. *Adv. Mater.* **2005**, *17*, 813–815. [CrossRef]
47. Xie, Y.; Huang, N.; Liu, Y.; Sun, W.; Mehnane, H.F.; You, S.; Wang, L.; Liu, W.; Guo, S.; Zhao, X.-Z. Photoelectrodes modification by N doping for dye-sensitized solar cells. *Electrochim. Acta* **2013**, *93*, 202–206. [CrossRef]
48. Tian, H.; Hu, L.; Zhang, C.; Mo, L.; Li, W.; Sheng, J.; Dai, S. Superior energy band structure and retarded charge recombination for Anatase N, B codoped nano-crystalline TiO_2 anodes in dye-sensitized solar cells. *J. Mater. Chem.* **2012**, *22*, 9123–9130. [CrossRef]
49. Guo, W.; Shen, Y.; Boschloo, G.; Hagfeldt, A.; Ma, T. Influence of nitrogen dopants on N-doped TiO_2 electrodes and their applications in dye-sensitized solar cells. *Electrochim. Acta* **2011**, *56*, 4611–4617. [CrossRef]
50. Guo, W.; Miao, Q.Q.; Xin, G.; Wu, L.Q.; Ma, T.L. Dye-sensitized solar cells based on nitrogen-doped titania electrodes. *Key Eng. Mater.* **2011**, *451*, 21–27. [CrossRef]
51. Guo, W.; Wu, L.; Chen, Z.; Boschloo, G.; Hagfeldt, A.; Ma, T. Highly efficient dye-sensitized solar cells based on nitrogen-doped titania with excellent stability. *J. Photochem. Photobiol. A* **2011**, *219*, 180–187. [CrossRef]
52. Guo, W.; Shen, Y.; Wu, L.; Gao, Y.; Ma, T. Performance of Dye-Sensitized Solar Cells Based on MWCNT/$TiO_{2-x}N_x$ Nanocomposite Electrodes. *Eur. J. Inorg. Chem.* **2011**, *2011*, 1776–1783. [CrossRef]
53. Kolen'ko, Y.V.; Churagulov, B.R.; Kunst, M.; Mazerolles, L.; Colbeau-Justin, C. Photocatalytic properties of titania powders prepared by hydrothermal method. *Appl. Catal. B Environ.* **2004**, *54*, 51–58. [CrossRef]

54. Carneiro, J.T.; Savenije, T.J.; Moulijn, J.A.; Mul, G. How phase composition influences optoelectronic and photocatalytic properties of TiO_2. *J. Phys. Chem. C* **2011**, *115*, 2211–2217. [CrossRef]
55. Meichtry, J.M.; Colbeau-Justin, C.; Custo, G.; Litter, M.I. Preservation of the photocatalytic activity of TiO_2 by EDTA in the reductive transformation of Cr(VI). Studies by Time Resolved Microwave Conductivity. *Catal. Today* **2014**, *224*, 236–243. [CrossRef]
56. Segal-Peretz, T.; Leman, O.; Nardes, A.M.; Frey, G.L. On the origin of charge generation in hybrid TiO_x/conjugated polymer photovoltaic devices. *J. Phys. Chem. C* **2012**, *116*, 2024–2032. [CrossRef]
57. Abdi, F.F.; Savenije, T.J.; May, M.M.; Dam, B.; van de Krol, R. The origin of slow carrier transport in $BiVO_4$ thin film photoanodes: A time-resolved microwave conductivity study. *J. Phys. Chem. Lett.* **2013**, *4*, 2752–2757. [CrossRef]
58. Saeki, A.; Yasutani, Y.; Oga, H.; Seki, S. Frequency-modulated gigahertz complex conductivity of TiO_2 nanoparticles: Interplay of free and shallowly trapped electrons. *J. Phys. Chem. C* **2014**, *118*, 22561–22572. [CrossRef]
59. Katoh, R.; Furube, A.; Yamanaka, K.I.; Morikawa, T. Charge separation and trapping in N-doped TiO_2 photocatalysts: A time-resolved microwave conductivity study. *J. Phys. Chem. Lett.* **2010**, *1*, 3261–3265. [CrossRef]
60. Tang, H.; Prasad, K.; Sanjines, R.; Schmid, P.E.; Levy, F. Electrical and optical properties of TiO_2 anatase thin films. *J. Appl. Phys.* **1994**, *75*, 2042–2047. [CrossRef]
61. Asahi, R.; Taga, Y.; Mannstadt, W.; Freeman, A.J. Electronic and optical properties of anatase TiO_2. *Phys. Rev. B* **2000**, *61*, 7459–7465. [CrossRef]
62. Simon, P.; Pignon, B.; Miao, B.; Coste-Leconte, S.; Leconte, Y.; Marguet, S.; Jegou, P.; Bouchet-Fabre, B.; Reynaud, C.; Herlin-Boime, N. N-doped titanium monoxide nanoparticles with TiO rock-salt structure, low energy band gap, and visible light activity. *Chem. Mater.* **2010**, *22*, 3704–3711. [CrossRef]
63. Lee, S.; Jeon, C.; Park, Y. Fabrication of TiO_2 tubules by template synthesis and hydrolysis with water vapor. *Chem. Mater.* **2004**, *16*, 4292–4295. [CrossRef]

8

Enhancing the Photocurrent of Top-Cell by Ellipsoidal Silver Nanoparticles: Towards Current-Matched GaInP/GaInAs/Ge Triple-Junction Solar Cells

Yiming Bai [1,*], Lingling Yan [1], Jun Wang [2], Lin Su [1], Zhigang Yin [3], Nuofu Chen [1,*] and Yuanyuan Liu [3]

1 State Key Laboratory of Alternate Electrical Power System with Renewable Energy Sources, North China Electric Power University, Beijing 102206, China; dgunwung@163.com (L.Y.); 18229799348@163.com (L.S.)
2 Institute of Information Photonics and Optical Communications, Beijing University of Posts and Telecommunications, Beijing 100876, China; wangjun@semi.ac.cn
3 Key Laboratory of Semiconductor Materials Science, Institute of Semiconductors, Chinese Academy of Sciences, P.O. Box 912, Beijing 100083, China; yzhg@semi.ac.cn (Z.Y.); liuyy@semi.ac.cn (Y.L.)
* Correspondence: ymbai@ncepu.edu.cn (Y.B.); nfchen@ncepu.edu.cn (N.C.)

Academic Editors: Guanying Chen, Zhijun Ning and Hans Agren

Abstract: A way to increase the photocurrent of top-cell is crucial for current-matched and highly-efficient GaInP/GaInAs/Ge triple-junction solar cells. Herein, we demonstrate that ellipsoidal silver nanoparticles (Ag NPs) with better extinction performance and lower fabrication temperature can enhance the light harvest of GaInP/GaInAs/Ge solar cells compared with that of spherical Ag NPs. In this method, appropriate thermal treatment parameters for Ag NPs without inducing the dopant diffusion of the tunnel-junction plays a decisive role. Our experimental and theoretical results confirm the ellipsoidal Ag NPs annealed at 350 °C show a better extinction performance than the spherical Ag NPs annealed at 400 °C. The photovoltaic conversion efficiency of the device with ellipsoidal Ag NPs reaches 31.02%, with a nearly 5% relative improvement in comparison with the device without Ag NPs (29.54%). This function of plasmonic NPs has the potential to solve the conflict of sufficient light absorption and efficient carrier collection in GaInP top-cell devices.

Keywords: GaInP/GaInAs/Ge triple-junction solar cells; Ag ellipsoidal nanoparticles; thermal treatment parameters; current-match

---

1. Introduction

Currently, GaInP/GaInAs/Ge triple-junction solar cells (TJSCs) for space and terrestrial concentrator applications have attracted increasing attention for their very high conversion efficiencies [1–3] and dramatic reduction in cost [4]. Such TJSCs with different subcell bandgaps divide the broad solar spectrum into three narrower sections, each of which can be converted to electricity more efficiently [5,6], while the 50% theoretical efficiency have not been obtained as expected due to the current mismatch among the subcells [7,8]. There are two aspects of this problem that have to be addressed. The first involves the Ge subcell, which absorbs approximately two times that photons than that needed for current matching with the GaInP and GaAs subcells [6]. The second problem relates to the lowest photocurrent of GaInP top-cells, which limits the efficiency of TJSCs greatly [9]. To the former, some reports suggest that the Ge bottom-cell would be replaced by a material with a bandgap of 1.0 eV, such as GaInNAs [10], but it is confined by the requirements of lattice matching with the other junctions and a higher epitaxy technique. Hence, improving the photocurrent of GaInP top-cell to approach other subcells is an effective way for enhancing the conversion efficiency of TJSCs.

For the commercialized case of the lattice-matched $Ga_{0.49}In_{0.51}P/Ga_{0.99}In_{0.01}As/Ge$ TJSCs, the bandgap energies are 1.9, 1.4, and 0.67 eV, respectively, which means each subcell has its very specific absorption spectrum [11–13]. Hence, the maximum number of photons absorbed by each subcell seems to be expected only if their optical thickness was thick enough. However, the short minority carrier lifetime of a few nanoseconds present in GaInP film limits the typical minority carrier diffusion lengths [14]. Consequently, the thickness of GaInP top-cell is generally allowed in the order of hundreds of nanometers for carrier collection other than its theoretical optical thickness of micrometers, resulting in unsatisfactory absorption of incident solar radiation [15]. Meanwhile, taking the material consumption into account, it also motivates a reduction in thickness of the active region of solar cells [16]. Therefore, a feasible and effective approach should be explored urgently for solving the puzzle between sufficient light absorption and efficient carrier collection in GaInP top-cell.

Most recently, an appealing approach involves the plasmonic nanostructures. The advantages of surface plasmon excitations is attributed to two aspects. The first involves an obviously increased material extinction for incident light arising from an enhanced local electromagnetic field near the nanostructures [17]. The other relates to the extended incident light path owing to a strong scattering of incident light into the active region of solar cells, with a result of increase in absorption and the promotion of photovoltaic conversion efficiency [18]. To date, many studies provide insights into the metal plasmonic nanostructures and, therefore, contribute greatly to practical photovoltaic applications [19–23]. However, few investigations are available on clarifying the feasibility of plasmonic nanostructures for GaInP/GaInAs/Ge TJSCs because the thermal treatment process for fabrication nanoparticles (NPs) is easy to cause the dopant diffusion in the highly-doped tunnel junction [24]. It is well known that annealing metal film is the simplest method to obtain metal nanoparticles. Unfortunately, too low an annealing temperature is not suitable for the formation of metal NPs, and too high an annealing temperature will greatly degrade the performance of TJSCs derived from the diffusion of dopant in TJSCs [25]. However, double hetero-structure tunnel junctions with wider bandgaps and lower diffusion coefficients are known for effectively suppressing impurity diffusion [13], so the problem of impurity diffusion in tunnel junctions is not an insurmountable barrier for improving the photocurrent of GaInP top-cells. Therefore, exploiting appropriate thermal treatment conditions for silver (Ag) NPs without inducing the dopant diffusion is an important strategy needing further study. Herein, the present work was aimed at promoting the photocurrent of GaInP top-cell using plasmonic nanostructures and, therefore, the efficiency of TJSCs.

In this work, Ag NPs were adopted because their extinction spectral range [26] is almost perfectly matched with the absorption range of the GaInP top-cell, and a series of experiments on annealing temperature, exposure time, and heating rate were proceeded to facilitate the light absorption of GaInP top-cell. We found that both the experimental and theoretical study confirmed the ellipsoidal Ag NPs annealed at 350 °C show a better extinction performance than the spherical Ag NPs annealed at 400 °C. The photovoltaic conversion efficiency of the device with ellipsoidal Ag NPs reaches 31.02%, with a nearly 5% relative improvement in comparison with the device without Ag NPs. The findings provide a feasible, cost-effective solution to solve the puzzle between the carrier collection and optical absorption in GaInP top-cells.

2. Experiments

The material growth of TJSCs on a p-type Ge substrate was performed by metal-organic vapor phase epitaxy (MOVPE) and the growth conditions were similar to those described elsewhere [27]. Figure 1a shows a schematic illustration of the material structure of the GaInP/GaInAs/Ge TJSCs evaluated in this work. The overall layer structure for the cell contains three pn-junctions and two tunnel junctions between the subcells. The $Ga_{0.51}In_{0.49}P$ top-cell, $Ga_{0.99}In_{0.01}As$ middle-cell, and Ge bottom-cell are all lattice-matched, which can effectively avoid the formation of dislocations and ensure excellent material quality. The three subcells were series connected with two highly-doped and ultra-thin tunnel junctions, which are favorable for a low resistance and high current density. The device processing on ohmic contacts, wet etching, and anti-reflection coatings were performed after the theoretical design of electrode patterns and anti-reflection coatings. The cell size was 10×11 mm^2 and 1 mm grid pitch of electrode was designed for an optimal electrode distribution under one-sun operation. The performance characterization of current density-voltage (J-V) characteristic curves and external quantum efficiency (EQE) for devices without Ag NPs has been measured. The device parameters of short-circuit current density, open circuit voltage, and efficiency are averaged from 12 individual devices with spherical or ellipsoidal NPs, respectively.

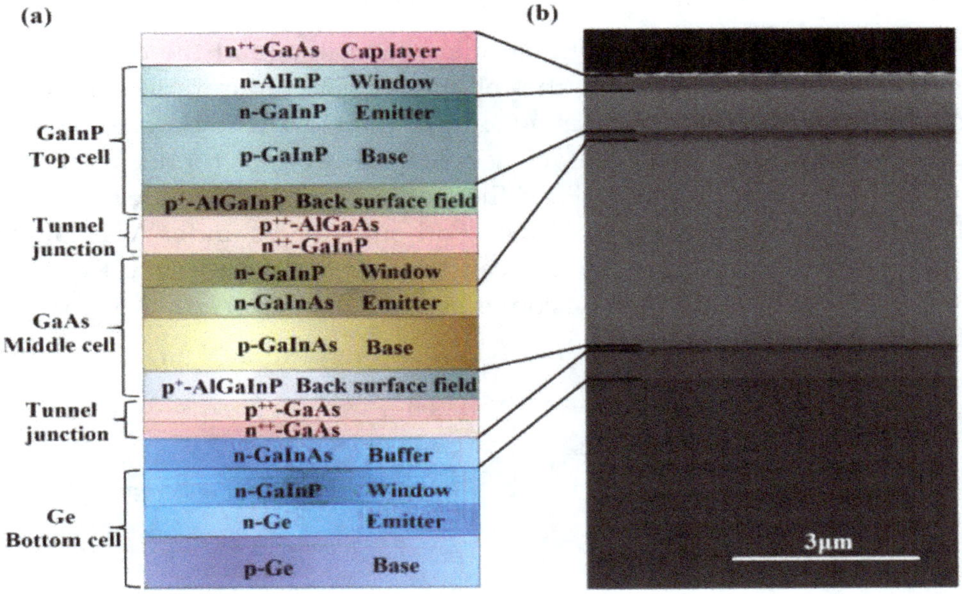

Figure 1. (**a**) Schematic layer structure and (**b**) cross-sectional view scanning electron microscope (SEM) image of the epitaxial structure of GaInP/GaInAs/Ge triple-junction solar cells (TJSCs).

Then, Ag films with 7 nm thickness were deposited on TJSCs at room temperature by magnetron sputtering, and the detailed fabrication processes can be referred to our previous work [18]. In consideration of the growth temperature for epitaxial layers is about 600–700 °C, the annealing temperature and temperature fluctuation for Ag film were strictly controlled. To the former, it is no higher than 450 °C, and to the latter, it is no more than 10 °C according to the computer supervising system. A series of experiments were performed for obtaining an optimal annealing temperature, exposure time, and heating rate. Lastly, the parameters were given as the following: the Ag films were annealed at 350 and 400 °C for 10 min under a nitrogen atmosphere with a heating rate of 150 °C/s. The resulting samples of ellipsoidal and spherical Ag NPs were shown in Figure 2.

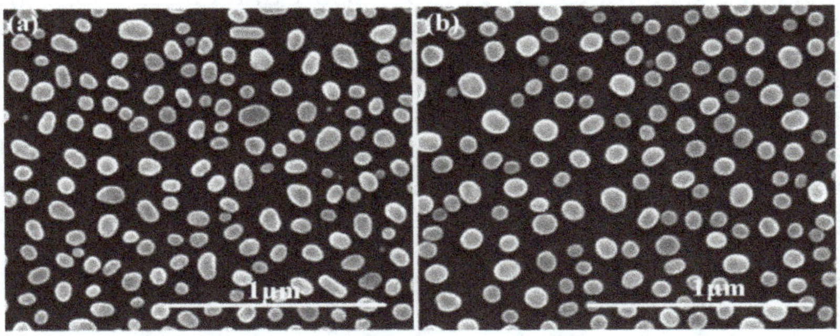

Figure 2. Plane-view SEM images of (**a**) ellipsoidal and (**b**) spherical Ag NPs.

Surface morphologies of Ag NPs were observed by a NOVA NANOSEM 650 scanning electron microscopy (SEM, FEI, Brno, Czech Republic). The optical properties were measured by an ultraviolet-visible spectrophotometer (UV spectrophotometer, Cary 5000, Varian, Palo Alto, CA, USA). The current density versus voltage (*J-V*) characteristic was tested by New-port 92250A Solar Simulator (Newport, Irvine, CA, USA) under one-sun AM1.5 (1000 W/m^2, 25 °C) standard test conditions. The light response property of the devices was further investigated using Qtest-2000 external quantum efficiency (EQE) systems (Crowntech, PA, USA).

3. Results and Discussion

3.1. Epitaxial Structure and Performance of GaInP/GaInAs/Ge TJSCs

Figure 1b demonstrates the cross-sectional view SEM image of the epitaxial structure of the GaInP/GaInAs/Ge TJSCs. As can be seen, the cell has a clear configuration and distinct interfaces, and the epitaxial structure mainly includes a 0.40-μm-thick GaInAs buffer layer, a 3.75-μm-thick middle-cell and a 0.75-μm-thick top-cell, as well as two 0.03-μm-thick tunnel junctions. The thickness of each layer, the doping concentration level and the material composition are based on the theoretical optimization. The *J-V* characteristic curve of TJSCs without Ag NPs under AM 1.5G solar irradiation (100 mW/cm^2) is presented in Figure 3a. As can be seen, the short-circuit current density (J_{sc}), open circuit voltage (V_{oc}), and photovoltaic conversion efficiency (η) are 13.56 mA/cm^2, 2.57 V and 29.54%, respectively.

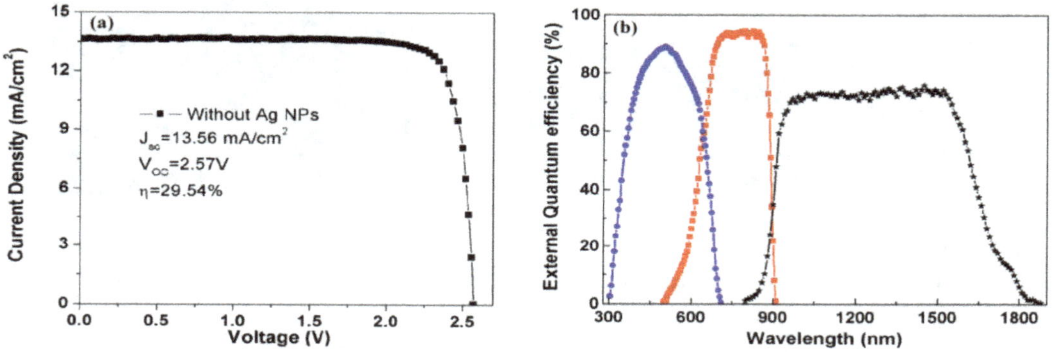

Figure 3. (**a**) Current-voltage (*J-V*) characteristic curve and (**b**) external quantum efficiency of the GaInP/GaInAs/Ge device.

Previous researches demonstrate that the J_{sc} of TJSCs is limited by one of the subcells [28,29], but the J_{sc} of each subcell cannot be directly obtained by *J-V* measurement. Therefore, the light response property of the devices was measured to calculate the J_{sc} of each subcell using EQE. Figure 3b exhibits the EQE of GaInP, GaInAs, and Ge subcells for the TJSCs without Ag NPs. Here, the absorption edges of GaInP, GaInAs, and Ge subcells are 670, 900, and 1800 nm, respectively. The EQE curves of the top- and middle-cell both drop rapidly near their absorption edges *versus* the gradual-decreasing modes of Ge bottom cell when wavelength ranges from 1600 to 1800 nm, which stems from the severe back surface recombination of Ge bottom-cell. The integrated current density for top-, middle- and bottom-cell is 13.70, 14.21, and 18.35 mA/cm^2, respectively, which is in good agreement with the J_{sc} (13.56 mA/cm^2) of the *J-V* measurement. It can be concluded that the J_{sc} of the TJSCs is limited by the GaInP top-cell, and an appropriate approach of increasing photocurrent of the top-cell is indispensable for high-performance TJSCs.

3.2. Morphology and Optical Properties of Ag NPs

Figure 2 presents the plane-view SEM images of the samples of Ag films annealed at 350 and 400 °C, respectively. As seen in Figure 2a, the Ag film was transformed into ellipsoid-like Ag NPs with smooth surface after annealing at 350 °C. The average principal axes length 2*a*–*c* of these ellipsoidal particles are 76, 94, and 76 nm. While the Ag NPs annealed at 400 °C appear very nearly spherical in shape with an average diameter *D* of 88 nm, as shown in Figure 2b. The average sizes of those ellipsoidal and spherical particles in Figure 2a,b are statistically determined by ImageJ software (National Institutes of Health, Bethesda, MD, USA). Hence, the Ag NPs exhibit obviously different morphology features at the two annealing temperatures of 350 and 400 °C, and the shape the Ag NPs tends to be ellipsoid-like at lower annealing temperatures.

The insights into the metal plasmonic nanostructures require a deep understanding of their optical properties. Figure 4 displays both the experimental and theoretical extinction spectra for ellipsoidal and spherical Ag NPs. For comparison, the experimental results are marked with solid lines and the theoretical extinction spectra are marked with dashed lines. Samples of those Ag NPs fabricated on glass with similar process conditions were used to carry out the extinction spectra measurement. The theoretical extinction efficiency was calculated by the discrete dipole approximation (DDA) method [30]. In our calculation, both the perpendicular and parallel polarizations were assumed to have the same proportion of 50%, and the dielectric constant of Ag nanoparticles was extracted from [31].

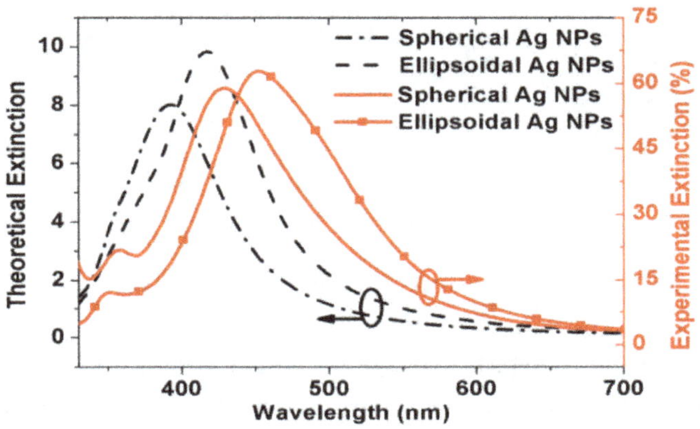

Figure 4. Theoretical (dotted lines) and experimental (solid lines) extinction spectra of spherical Ag NPs with $D = 88$ nm and ellipsoidal Ag NPs with $2a/2b/2c = 76/94/76$ nm.

As shown here, the theoretical calculation is comparatively consistent with the experimental result. The most significant aspect is that both the theoretical and measured extinction spectrum ranges are almost perfectly matched with the absorption wavelength range of GaInP top-cell [26], especially the measured one. The second aspect is that ellipsoidal Ag NPs demonstrate better extinction performance than that of spherical Ag NPs. Finally, the increased extinction ability almost disappears for wavelength beyond 600 nm.

Meanwhile, there are two different features exhibited in Figure 4. Firstly, the theoretical dipole extinction peaks are located at 393 and 418 nm for ellipsoidal and spherical Ag NPs, respectively, and the corresponding experimental dipole extinction peaks are located at 429 and 453 nm. Apparently, a red-shift of about 35-nm, both for the experimental extinction peak of ellipsoidal and spherical Ag NPs, appears, which can be attributed to the retardation effects occurred on the particles [32] and the practical wider size distribution. Secondly, the broadening of full width at half maximum (FWHM) of experimental results can be explicated by the wide size distribution of NPs. Overall, the features of the red-shift, the broader FWHM, and higher peak intensity are beneficial for harvesting more incident sunlight.

3.3. Enhanced Performance of GaInP/GaInAs/Ge TJSCs by Ag NPs

Since the extinction spectra of Ag NPs cover from 300 to 600 nm, we speculate that the plasmonic nanostructures have little effect on the middle- and bottom-cell. To verify the impact of plasmonic nanostructures, the EQE of GaInP/GaInAs/Ge TJSCs with Ag NPs was measured. As expected, only the spectral response characteristic of GaInP top-cell is affected by Ag NPs. Hence, Figure 5 gives the EQE of GaInP top-cell with and without Ag NPs. As shown in Figure 5, the EQEs of devices with Ag NPs are higher than that of device without Ag NPs. The improved EQE for a device with spherical Ag NPs mainly locates from 390 to 490 nm, which is in accordance with the extinction spectra. Similarly, the excellent extinction performance of ellipsoidal Ag NPs leads to a higher EQE of the top-cell in a broader spectral range. The integrated current for devices without and with spherical and ellipsoidal Ag NPs is 13.70, 14.09, and 14.31 mA/cm^2, respectively. Definitely, distinct photocurrent enhancements are obtained both for spherical and ellipsoidal Ag NPs compared with that of the device without Ag NPs, especially for the latter. The results allow the conclusion that ellipsoidal Ag NPs with better extinction property can obviously enhance the performance of TJSCs.

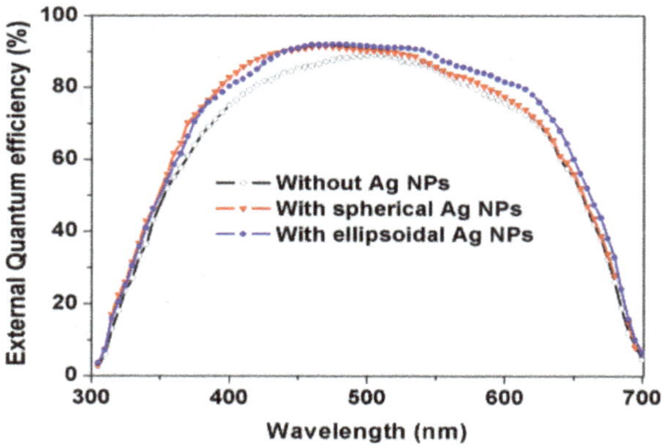

Figure 5. External quantum efficiency of GaInP top-cell in TJSCs.

Figure 6 presents the *J-V* characteristic curves of TJSCs with and without Ag NPs under AM 1.5G solar irradiation. As demonstrated in Figure 6, there is no obvious difference in the V_{oc} for all devices, but their J_{sc} and η increase when Ag NPs are incorporated into the TJSCs. This could be attributed to the increased material extinction of Ag NPs within the spectral range from 300 to 600 nm. For the device without Ag NPs, the J_{sc} and η are 13.56 mA/cm^2 and 29.54%, respectively. The η for devices with spherical and ellipsoidal Ag NPs is improved from 30.27% to 31.02%, respectively. This can be attributed to the key factor of J_{sc}, which increases from 13.94 to 14.17 mA/cm^2 and is in accordance with the EQE results discussed in the previous part. From the results, we conclude that Ag NPs, especially ellipsoidal Ag NPs, are obviously beneficial for the performance of TJSCs, but it is still a challenge to fabricate of metal NPs without inducing the dopant diffusion of tunnel junctions.

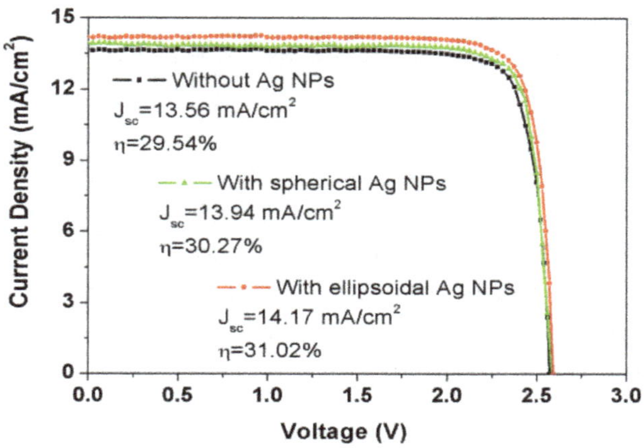

Figure 6. *J-V* characteristics of GaInP/GaInAs/Ge triple-junction solar cells at one-sun AM 1.5G, 25 °C. The measurement was performed on the 1.1 cm^2 solar cell devices.

4. Conclusions

In conclusion, we demonstrated a simple approach for increasing the photocurrent of the top-cell using plasmonic nanostructures for current-matched GaInP/GaInAs/Ge TJSCs. The devices with ellipsoidal Ag NPs display an enhanced photocurrent due to the remarkably increased material extinction. Both the experimental and theoretical results verify the ellipsoidal Ag NPs with the lower annealing temperature of 350 °C show excellent optical properties. Under the illumination of AM 1.5G 100 mW·cm^{-2}, the photovoltaic conversion efficiency of the device with ellipsoidal Ag NPs reaches 31.02%, with a nearly 5% relative improvement in comparison with the device without Ag NPs (29.54%). The findings of this study elucidate that plasmonic nanoparticles located on the illuminated surface of a solar cell is a promising structure for resolving the puzzle between carrier collection and optical absorption in GaInP top-cells and facilitating the light trapping of TJSCs.

Acknowledgments: The authors acknowledge financial support from the National Natural Science Foundation of China (Grant Nos. 61006050, 51573042), the Natural Science Foundation of Beijing (No. 2151004), the Fundamental Research Funds for the Central Universities (Grant No. 2016MS50).

Author Contributions: All the authors together designed the scope of the work. Yiming Bai contributed to the idea, discussion and writing the manuscript. Lingling Yan and Lin Su performed the experiments and analyzed the data. Jun Wang, Zhigang Yin and Yuanyuan Liu contributed to the performance characterization. Nuofu Chen reviewed and finalized the paper.

References

1. Green, M.A.; Emery, K.; Hishikawa, Y.; Warta, W.; Dunlop, E.D. Solar cell efficiency tables (version 46). *Prog. Photovolt. Res. Appl.* **2015**, *23*, 805–812. [CrossRef]
2. Green, M.A.; Keevers, M.J.; Thomas, I.; Lasich, J.B.; Emery, K.; King, R.R. 40% efficient sunlight to electricity conversion. *Prog. Photovolt. Res. Appl.* **2015**, *23*, 685–691. [CrossRef]
3. Li, S.I.; Bi, J.F.; Li, M.Y.; Yang, M.J.; Song, M.H.; Liu, G.Z.; Xiong, W.P.; Li, Y.; Fang, Y.Y.; Chen, C.Q.; *et al.* Investigation of GaInAs strain reducing layer combined with InAs quantum dots embedded in Ga(In)As subcell of triple junction GaInP/Ga(In)As/Ge solar cell. *Nanoscale Res. Lett.* **2015**, *10*. [CrossRef] [PubMed]
4. Paraskeva, V.; Hadjipanayi, M.; Norton, M.; Pravettoni, M.; Georghiou, G.E. Voltage and light bias dependent quantum efficiency measurements of GaInP/GaInAs/Ge triple junction devices. *Sol. Energy Mater. Sol. Cells* **2013**, *116*, 55–60. [CrossRef]
5. Barrigon, E.; Espinet-Gonzalez, P.; Contreras, Y.; Rey-Stolle, I. Implications of low breakdown voltage of component subcells on external quantum efficiency measurements of multijunction solar cells. *Prog. Photovolt. Res. Appl.* **2015**, *23*, 1597–1607. [CrossRef]
6. Geisz, J.F.; Kurtz, S.; Wanlass, M.W.; Ward, J.S.; Duda, A.; Friedman, D.J.; Olson, J.M.; McMahon, W.E.; Moriarty, T.E.; Kiehl, J.T. High-efficiency GaInP/GaAs/InGaAs triple-junction solar cells grown inverted with a metamorphic bottom junction. *Appl. Phys. Lett.* **2007**, *91*. [CrossRef]
7. Dimroth, F.; Grave, M.; Beutel, P.; Fiedeler, U.; Karcher, C.; Tibbits, T.N.D.; Oliva, E.; Siefer, G.; Schachtner, M.; Wekkeli, A.; *et al.* Wafer bonded four-junction GaInP/GaAs//GaInAsP/GaInAs concentrator solar cells with 44.7% efficiency. *Prog. Photovolt. Res. Appl.* **2014**, *22*, 277–282. [CrossRef]
8. Guter, W.; Schone, J.; Philipps, S.P.; Steiner, M.; Siefer, G.; Wekkeli, A.; Welser, E.; Oliva, E.; Bett, A.W.; Dimroth, F. Current-matched triple-junction solar cell reaching 41.1% conversion efficiency under concentrated sunlight. *Appl. Phys. Lett.* **2009**, *94*. [CrossRef]
9. Kurtz, S.; Geisz, J. Multijunction solar cells for conversion of concentrated sunlight to electricity. *Opt. Express* **2010**, *18*, A73–A78. [CrossRef] [PubMed]
10. Pham, N.D.; Kim, J.T.; Jung, T.I.; Han, J.H.; Oh, I. Direct solar water splitting enabled by monolithic III–V triple junction integrated with low-cost catalyst. *Sci. Adv. Mater.* **2016**, *8*, 241–246. [CrossRef]
11. Braun, A.; Katz, E.A.; Gordon, J.M. Basic aspects of the temperature coefficients of concentrator solar cell performance parameters. *Prog. Photovolt. Res. Appl.* **2013**, *21*, 1087–1094. [CrossRef]
12. Meusel, M.; Baur, C.; Létay, G.; Bett, A.W.; Warta, W.; Fernandez, E. Spectral response measurements of monolithic GaInP/Ga(In)As/Ge triple-junction solar cells: Measurement artifacts and their explanation. *Prog. Photovolt. Res. Appl.* **2003**, *11*, 499–514. [CrossRef]

13. Sogabe, T.; Ogura, A.; Okada, Y. Analysis of bias voltage dependent spectral response in $Ga_{0.51}In_{0.49}P/Ga_{0.99}In_{0.01}As/Ge$ triple junction solar cell. *J. Appl. Phys.* **2014**, *115*. [CrossRef]
14. Cotal, H.; Fetzer, C.; Boisvert, J.; Kinsey, G.; King, R.; Hebert, P.; Yoon, H.; Karam, N. III-V multijunction solar cells for concentrating photovoltaics. *Energy Environ. Sci.* **2009**, *2*, 174–192. [CrossRef]
15. Conibeer, G. Third-generation photovoltaics. *Mater. Today* **2007**, *10*, 42–50. [CrossRef]
16. Lee, S.M.; Kwong, A.; Jung, D. High performance ultrathin GaAs solar cells enabled with heterogeneously integrated dielectric periodic nanostructures. *ACS Nano* **2015**, *9*, 10356–10365. [CrossRef] [PubMed]
17. Okamoto, H.; Imura, K. Near-field optical imaging of enhanced electric fields and plasmon waves in metal nanostructures. *Prog. Surf. Sci.* **2009**, *84*, 199–229. [CrossRef]
18. Bai, Y.M.; Gao, Z.; Chen, N.F.; Liu, H.; Yao, J.X.; Ma, S.; Shi, X.Q. Elimination of small-sized Ag nanoparticles via rapid thermal annealing for high efficiency light trapping structure. *Appl. Surf. Sci.* **2014**, *315*. [CrossRef]
19. Yang, L.; Pillai, S.; Green, M.A. Can plasmonic Al nanoparticles improve absorption in triple junction solar cells? *Sci. Rep.* **2015**, *5*. [CrossRef] [PubMed]
20. Zhang, Z.C.; Han, S.; Wang, C.; Li, J.P.; Xu, G.B. Single-walled carbon nanohorns for energy applications. *Nanomaterials* **2015**, *5*, 1732–1755. [CrossRef]
21. Bai, Y.M.; Wang, J.; Yin, Z.G.; Chen, N.F.; Zhang, X.W.; Zhen, F.; Yao, J.X.; Li, N.; Guli, M.N. Ag nanoparticles preparation and their light trapping performance. *Sci. China Technol. Sci.* **2013**, *56*, 109–114. [CrossRef]
22. Stratakis, E.; Kymakis, E. Nanoparticle-based plasmonic organic photovoltaic devices. *Mater. Today* **2013**, *16*, 133–146. [CrossRef]
23. Lee, D.S.; Kim, W.; Cha, B.G.; Kwon, J.; Kim, S.J.; Kim, M.; Kim, J.; Wang, D.H.; Park, J.H. Self-position of Au NPs in perovskite solar cells: Optical and electrical contribution. *ACS Appl. Mater. Interfaces* **2015**, *8*, 449–454. [CrossRef] [PubMed]
24. Keding, R.; Stuwe, D.; Kamp, M.; Kamp, M.; Reichel, C.; Wolf, A.; Woehl, R.; Borchert, D.; Reinecke, H.; Biro, D. Co-diffused back-contact back-junction silicon solar cells without gap regions. *IEEE J. Photovolt.* **2013**, *3*, 1236–1242. [CrossRef]
25. Simrick, N.J.; Kilner, J.A.; Atkinson, A. Thermal stability of silver thin films on zirconia substrates. *Thin Solid Films* **2012**, *520*, 2855–2867. [CrossRef]
26. Hu, P.F.; Cao, Y.L.; Jia, D.Z.; Li, Q.; Liu, R.L. Engineering the metathesis and oxidation-reduction reaction in solid state at room temperature for nanosynthesis. *Sci. Rep.* **2014**, *4*. [CrossRef] [PubMed]
27. Fetzer, C.M.; King, R.R.; Colter, P.C.; Edmondson, K.M.; Law, D.C.; Stavrides, A.P.; Yoon, H.; Ermer, J.H.; Romero, M.J.; Karam, N.H. High-efficiency metamorphic GaInP/GaInAs/Ge solar cells grown by MOVPE. *J. Cryst. Growth* **2004**, *261*, 341–348. [CrossRef]
28. King, R.R.; Law, D.C.; Edmondson, K.M.; Fetzer, C.M.; Kinsey, G.S.; Yoon, H.; Sherif, R.A.; Karam, N.H. 40% efficient metamorphic GaInP/GaInAs/Ge multijunction solar cells. *Appl. Phys. Lett.* **2007**, *90*. [CrossRef]
29. Hoheisel, R.; Fernandez, J.; Dimroth, F.; Bett, A.W. Investigation of radiation hardness of germanium photovoltaic cells. *IEEE Trans. Electron Devices* **2010**, *57*, 2190–2194. [CrossRef]
30. Draine, B.T.; Flatau, P.J. User guide for the discrete dipole approximation code DDSCAT (Version 7.1). 2010. Available online: http://arxiv.org/abs/1002.1505 (accessed on 28 January 2016).
31. Palik, E.D. *Handbook of Optical Constants of Solids*; Academic Press: Washington, DC, USA, 1985.
32. Gao, H.L.; Zhang, X.W.; Yin, Z.G.; Tan, H.R.; Zhang, S.G.; Meng, J.H.; Liu, X. Plasmon enhanced polymer solar cells by spin-coating Au nanoparticles on indium-tin-oxide substrate. *Appl. Phys. Lett.* **2012**, *101*. [CrossRef]

9

Arrays of Plasmonic Nanostructures for Absorption Enhancement in Perovskite Thin Films

Tianyi Shen, Qiwen Tan, Zhenghong Dai, Nitin P. Padture and Domenico Pacifici *

School of Engineering, Brown University, 184 Hope Street, Providence, RI 02912, USA; tianyi_shen@brown.edu (T.S.); qiwen_tan@brown.edu (Q.T.); zhenghong_dai@brown.edu (Z.D.); nitin_padture@brown.edu (N.P.P.)

* Correspondence: domenico_pacifici@brown.edu

Abstract: We report optical characterization and theoretical simulation of plasmon enhanced methylammonium lead iodide ($MAPbI_3$) thin-film perovskite solar cells. Specifically, various nanohole (NH) and nanodisk (ND) arrays are fabricated on gold/$MAPbI_3$ interfaces. Significant absorption enhancement is observed experimentally in 75 nm and 110 nm-thick perovskite films. As a result of increased light scattering by plasmonic concentrators, the original Fabry–Pérot thin-film cavity effects are suppressed in specific structures. However, thanks to field enhancement caused by plasmonic resonances and in-plane interference of propagating surface plasmon polaritons, the calculated overall power conversion efficiency (PCE) of the solar cell is expected to increase by up to 45.5%, compared to its flat counterpart. The role of different geometry parameters of the nanostructure arrays is further investigated using three dimensional (3D) finite-difference time-domain (FDTD) simulations, which makes it possible to identify the physical origin of the absorption enhancement as a function of wavelength and design parameters. These findings demonstrate the potential of plasmonic nanostructures in further enhancing the performance of photovoltaic devices based on thin-film perovskites.

Keywords: perovskite solar cells; surface plasmon polaritons; plasmonic nanostructures; absorption enhancement; FDTD simulations

---

1. Introduction

Hybrid organic–inorganic perovskites have become one of the most popular photovoltaic materials due to their high absorption coefficient [1], long carrier diffusion length [2] as well as low-cost fabrication process [3–7], which make them excellent candidates for high-efficiency solar cells. As one of the most popular members of perovskite materials, methylammonium lead iodide or $CH_3NH_3PbI_3$ ($MAPbI_3$) possesses such a high absorption coefficient that a 400 nm-thick film is sufficient to absorb most of the incident solar spectrum below its bandgap. Although the record power conversion efficiency (PCE) of a single-junction perovskite solar cell has reached values up to 25.2% [8], there is still much room to boost the PCE. Therefore, several methods have been proposed to improve the performance of perovskite solar cells, one of which involves the use of surface plasmons. Plasmonic effects could be employed to improve the performance of perovskite solar cells [9–15] by embedding nanoparticles [16–30] or plasmonic concentrators [31–35] that can increase absorption, especially near the optical band edge of the material. Previous theoretical research has shown that nanohole (NH) or nanodisk (ND) arrays embedded on gold electrodes with a thin layer of $MAPbI_3$ deposited on them can significantly improve the solar cell PCE by up to $\sim$10%. The reduced film thickness has the implied benefits of reducing the amount of toxic materials, improving electronic performance and enabling fabrication on flexible substrates [36].

In this work, we report the experimental fabrication, structural and optical characterization, as well as theoretical simulations of plasmon enhanced $MAPbI_3$ perovskite solar cell with NH and ND arrays on gold substrates, as a function of geometry parameters and varying thickness of $MAPbI_3$ deposited on top. Four geometry features, height h, diameter D, pitch P and $MAPbI_3$ thickness t, are systematically studied by three-dimensional finite-difference time-domain (3D FDTD) simulations to identify the physical effects responsible for the absorption and PCE enhancement.

2. Materials and Methods

An adhesion layer of 4 nm-thick titanium and a layer of 200 nm-thick gold was sequentially deposited on a 1 mm-thick quartz substrate using electron beam evaporation. Subsequently, NH or ND arrays of varying geometry parameters were fabricated on the thick gold film using focused ion beam (FIB) milling.

Following the FIB milling, the substrates were treated with UV-ozone for 45 min to enhance the wettability. The $MAPbI_3$ precursor solution was prepared by dissolving 159 mg of methylammonium iodide (Greatcell, Queanbeyan, Australia) and 461 mg of PbI_2 (Sigma-Aldrich, St. Louis, MO, USA) in 78 mg dimethyl sulfoxide (Sigma-Aldrich, St. Louis, MO, USA) and 1368 mg of N,N-dimethylformamide (Acros organics, NJ, USA) to obtain a 30 wt% solution. To deposit the $MAPbI_3$ layer, the solution was spin-coated at 4000 rpm for 30 s with an acceleration of 1300 rpm/s in a nitrogen-filled glove box. At the 10th second of spinning, 250 μL diethyl ether (Sigma-Aldrich, St. Louis, MO, USA) was dripped onto the substrate. The as-coated film was then annealed at 100 °C for 20 min to obtain the $MAPbI_3$ thin film. The film thickness can be controlled by adjusting the amount of N,N-dimethylformamide. Schematic illustrations of the fabricated NH and ND array coated with $MAPbI_3$ thin film are displayed in Figure 1a,b, respectively.

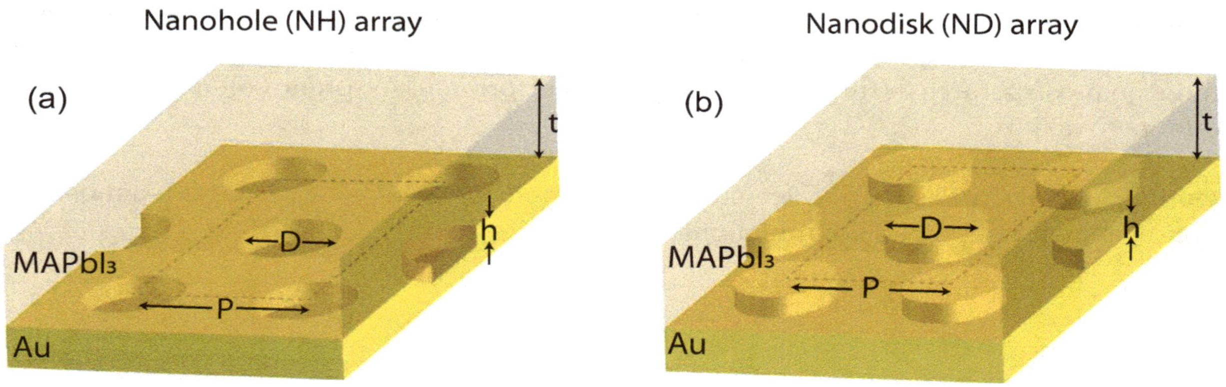

Figure 1. Schematic illustrations of methylammonium lead iodide ($MAPbI_3$) thin-film over nanostructured gold (Au) surface with (**a**) NH or (**b**) ND array, respectively. t represents the perovskite film thickness on the top of gold surface. P represents the triangular array pitch, while h and D respectively represent the height and diameter of a plasmonic concentrator. A negative h indicated NH, while a positive h corresponded to ND.

The thickness of the synthesized $MAPbI_3$ films were characterized using a variable angle spectroscopic ellipsometer (J.M. Woollam, Lincoln, NE, USA, M-2000). The reflectance spectra were characterized using an inverted microscope (Nikon Instruments, Melville, NY, USA, Eclipse Ti) with a small numerical aperture objective (10×, NA = 0.3) to mimic normal incidence. A randomly polarized broadband light source was used to illuminate the sample. The reflected light from the samples was coupled into a spectrograph (Princeton Instruments, Acton, MA, USA, Acton SpectraPro SP-2300) and was detected using a charge-coupled device camera (Princeton Instruments, Acton, MA, USA, Pixis 100). A piece of flat silicon wafer was used as a calibration reference during the characterization.

Apart from experimentally characterizing the absorptance spectra, simulations are also performed using FDTD method. A commerical-grade 3D electromagnetic simulator from Lumerical Inc.

(Vancouver, BC, Canada) was used [37]. Anti-symmetric and symmetric boundary conditions were used in the lateral directions while perfectly matched layer boundary conditions were used in the vertical directions. The real and imaginary parts of refractive indices for gold and $MAPbI_3$ were obtained from literature [38,39]. Considering the balance between computational cost and accuracy for different structures, the mesh size varied among 4 nm $\times$ 4 nm $\times$ 2 nm, 4 nm $\times$ 4 nm $\times$ 1 nm and 3 nm $\times$ 3 nm $\times$ 2 nm in the Cartesian coordinates.

3. Results and Discussion

Figure 2a,b,e,f show the tilted and top-view scanning electron microscope (SEM) micrographs of example NH and ND arrays prior to the $MAPbI_3$ synthesis. To investigate the absorption enhancement brought by the plasmonic nanostructure, the corresponding reflectance spectra $R(\lambda)$ were measured, and the measured absorptance spectra $A(\lambda) = 1 - R(\lambda)$ are reported in Figure 2c,g. Absorptance of nanostructured gold surface (black solid lines) shows a similar spectral shape as that of flat gold surface (black dashed lines) for both NH and ND arrays, while the NH array exhibits relatively higher absorption. This is because a portion of the scattered light by plasmonic concentrators is not collected by the finite NA of the objective lens, therefore a lower measured reflectance (i.e., higher absorption). By contrast, ND array in Figure 2e,f has smaller feature size and larger array pitch. Therefore, a smaller portion of light is scattered and a higher portion of reflected light is collected by the objective (i.e., lower absorption). The scattered light can be efficiently absorbed by adding a layer of $MAPbI_3$ on the top. After being coated with a 110 nm-thick $MAPbI_3$ film, the absorptance spectrum of the flat-interface structure (red dashed lines) shows a peak at wavelength λ = 526 nm. This results from the Fabry–Pérot cavity effect supported by the perovskite thin-film. The absorption decreases with increasing wavelength because the perovskite layer no longer supports any Fabry–Pérot cavity mode and the material has weaker absorbing property at longer wavelength. With NH or ND nanostructures (red solid lines), the absorption gets significantly enhanced at longer wavelengths, as is shown as red fillings in Figure 2d,h. For NH array, the spectral peak at λ = 526 nm gets weaker. This is because the structure disrupts the interference in the film [36]. In contrast, ND array still preserves that spectral peak due to its smaller feature size and lower nanostructure concentration. The absorptance spectra of perovskite/nanostructured gold are normalized to that of perovskite/flat gold and are displayed in Figure 2d,h. The blue filling at lower wavelengths represents the suppression of Fabry–Pérot cavity mode while the red filling represents the absorption enhancement brought by the plasmonic nanostructure. It is obvious that the example NH could not preserve the cavity effect as well as ND array (larger blue filling area), but the greater absorption enhancement by the plasmonic effects at longer wavelengths compensates for the loss at lower wavelengths (larger red filling area). As a comparison, for the example ND array, both blue and red filling areas are much smaller in Figure 2h, indicating negligible suppression at short wavelengths but also less significant enhancement at longer wavelengths.

A variety of nanostructured gold surfaces (NH or ND arrays) were fabricated based on optimal design considerations [36]. The top row of Figure 3 shows the SEM micrographs of the NH (left) and ND arrays (right). The measured and normalized absorptance spectra of these NH and ND arrays are shown in Figure 3a–d. For NH arrays (left column), as pitch P increases from 200 nm (circle symbols) to 250 nm (triangle symbols), the spectral peaks redshift from λ = 632 nm to λ = 666 nm. This results from the pitch-dependent collective surface plasmon polariton (SPP) resonances in the hexagonal lattice [36,40]. In addition, with hole depth increasing from 40 nm (solid symbols) to 70 nm (open symbols), the Fabry–Pérot cavity mode gets disrupted, resulting in a relatively lower absorption at λ = 526 nm, but there is more significant absorption enhancement at longer wavelengths. A similar behavior is also observed in ND arrays: structures suppressing the Fabry–Pérot thin-film interference effects show stronger absorption enhancement near the $MAPbI_3$ band edge.

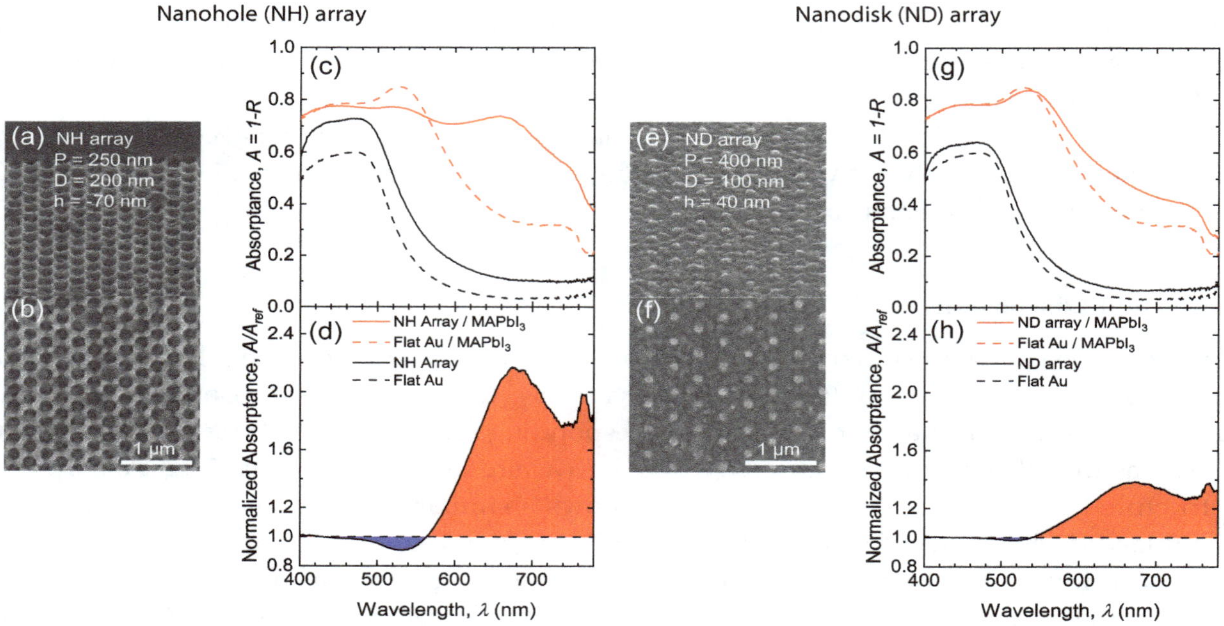

Figure 2. (**a**,**b**,**e**,**f**) display the tilted-view (**a**,**e**) and top-view (**b**,**f**) SEM micrographs of the fabricated nanohole (NH) array with P = 250 nm, D = 200 nm, h = −70 nm (**a**,**b**) and nanodisk (ND) array with P = 400 nm, D = 100 nm, h = 40 nm (**e**,**f**). (**c**,**g**) The measured absorptance spectra $A = 1 - R$ for nanostructured (solid lines) or flat (dashed lines) gold with (red lines) or without (black lines) 110nm-thick $MAPbI_3$ thin-film. (**d**,**h**) show the absorptance spectra for 110nm-thick $MAPbI_3$ thin-films on nanostructured gold normalized to that of equal-thickness $MAPbI_3$ on flat gold. The red filling ($A/A_{ref} > 1.0$) represents absorption enhancement while the blue filling ($A/A_{ref} < 1.0$) represents absorption suppression.

The absorption enhancement with different perovskite layer thicknesses is also investigated experimentally. $MAPbI_3$ layers with different thicknesses (t = 75, 110, 300 nm) were synthesized on the same nanostructure and the corresponding absorptance spectra were characterized and displayed in Figure 4. The spectral peaks and dips in the flat-interface cases (dashed lines) again result from the Fabry–Pérot cavity modes supported by different perovskite thicknesses at different wavelengths. The SEM micrographs of nanostructured surfaces are displayed in the top panels. Low-density and small-feature-size nanostructures and the corresponding absorptance spectra are shown in the left column. For different thicknesses t, these two structures can preserve the Fabry–Pérot cavity modes, as well as bring additional absorption enhancement. However, the absorption enhancement becomes much less significant as t increases to 300 nm, because most of the incident light is absorbed by the perovskite material before interacting with the plasmonic concentrators. For NH or ND arrays with larger feature sizes and higher concentration (right column), Fabry–Pérot cavity effects are significantly suppressed at the resonant wavelengths. In addition, both high-density and large-feature-size arrays exhibit stronger absorption enhancement at other wavelength ranges comparing to the low-density and small-feature-size arrays. Moreover, obvious absorption enhancement is observed even when t increases to 300 nm.

To further evaluate how the plasmonic effects and absorption enhancement will affect the overall performance of real photovoltaic devices, the PCE can be obtained using the measured absorptance spectra for different active layer thicknesses. With the detailed balance assumption, PCE can be expressed as

$$\mathrm{PCE} = \frac{\int_{\lambda<\lambda_g} \mathrm{AM1.5}(\lambda)\frac{2\pi\lambda}{\hbar c}A(\lambda)E_g d\lambda}{1\ sun} \tag{1}$$

where λ_g is the bandgap wavelength of $MAPbI_3$, AM1.5(λ) is the wavelength-dependent air mass

1.5 solar radiation spectrum, $\hbar$ is the reduced Planck constant and 1 *sun* is the 1000 W/m^2 incident solar power density [41]. The experimentally measured absorptance spectra includes the absorption in both gold and $MAPbI_3$, but the portion absorbed by gold is relatively small according to previously reported study [36]. For NH and ND arrays in Figure 4 coated with multiple active layer thicknesses, the calculated PCE and PCE enhancement is shown in Figure 5.

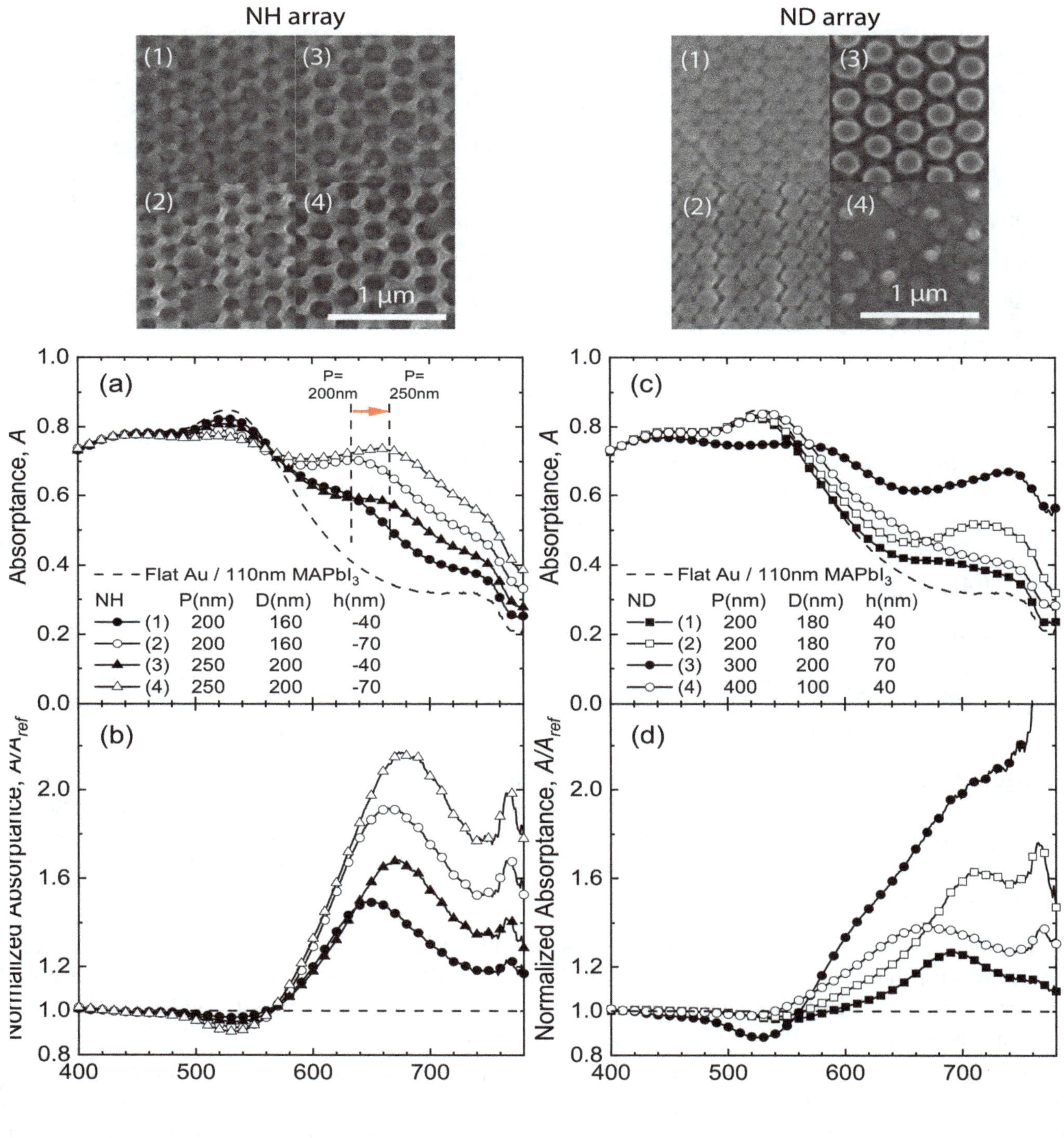

Figure 3. Top-view SEM micrographs (top), experimental absorptance spectra (**a**,**c**) and normalized absorptance spectra (**b**,**d**) for 110nm-thick $MAPbI_3$ on flat (dashed lines) and nanostructured (solid lines with symbols) gold NH (left column) and ND (right column) arrays. The vertical dashed lines in (**a**) show the absorption spectral redshift as a function of pitch P (increasing from 200 nm to 250 nm).

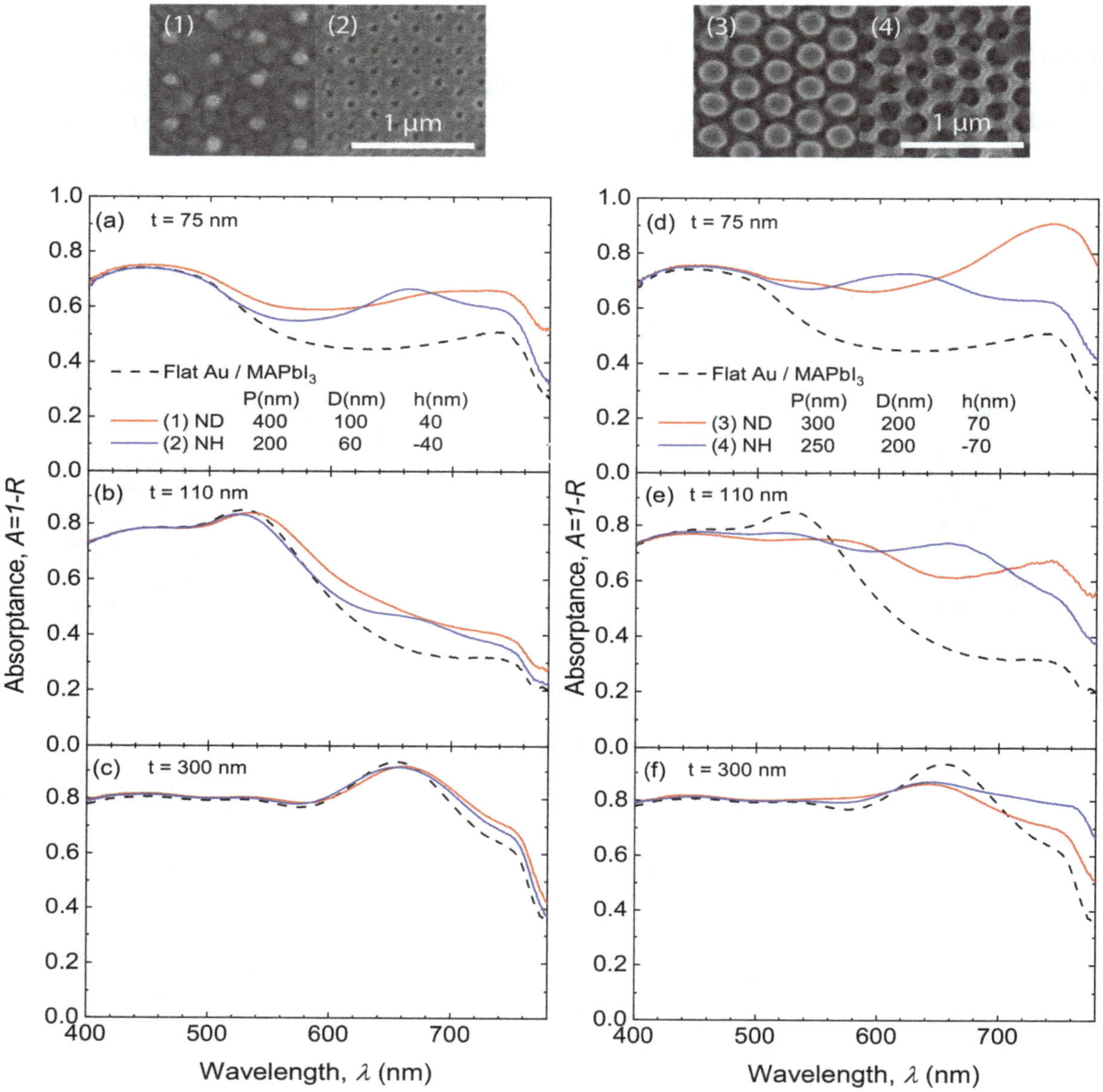

Figure 4. Measured absorptance spectra of flat (dashed lines) and nanostructured (solid lines) gold surfaces coated with varying-thickness $MAPbI_3$ with $t = 75$ nm (**a**,**d**); 110 nm (**b**,**e**); 300 nm (**c**,**f**). The left (right) column corresponds to nanostructured gold surfaces with low (high) surface density of optical scatterers. The top panels show top-view SEM micrographs of gold surfaces with low (1–2) and high-concentration (3–4) ND or NH arrays.

The PCE of flat structure $PCE_{flat}(t)$ (dashed line) is obtained using the multilayer interference model. For all three thicknesses ($t = 75$, 110 and 300 nm), all four nanostructure arrays exhibit PCE enhancement. Specifically, for active layer thickness $t = 75$ nm, high-density NH and ND arrays (red symbols) increase the PCE by 26.6% and 45.5%, respectively.

As a comparison, low-density nanostructure arrays (blue symbols) do not enhance the PCE as much. As the thickness increases to 300 nm, the PCE enhancement becomes much less significant due to the weaker plasmonic effect. The performance difference narrows between arrays of different concentration and feature sizes, enhancing the PCE by 3.7–7.0%.

As demonstrated in the experiment, plasmonic nanostructures with various geometry parameters exhibit different absorption enhancement behavior. Therefore, FDTD simulations were performed to further investigate the influence of different geometry parameters on the absorption enhancement.

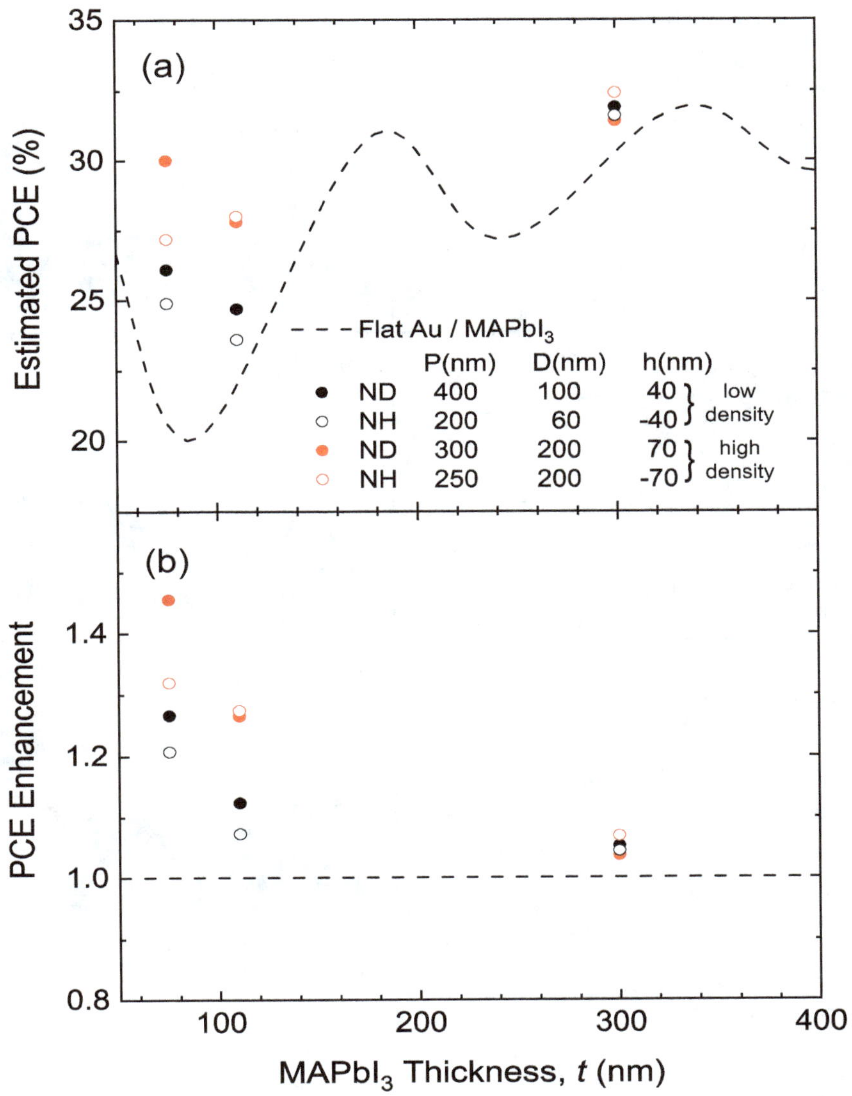

Figure 5. (**a**) Power conversion efficiency (PCE) and (**b**) PCE enhancement estimated from experimental absorptance spectra for varying $MAPbI_3$ thickness values, t = 75, 110 and 300 nm with low (blue symbols) and high (red symbols) surface density of optical scatterers in NH (open symbols) and ND arrays (solid symbols). The dashed lines represent the PCE of flat-interface structures as a function of $MAPbI_3$ thickness. A strong PCE enhancement is clearly demonstrated for high-density nanostructures at t = 75 and 110 nm (red symbols).

Figure 6a–d display the simulated absorptance spectra of an example NH array (P = 250 nm, D = 200 nm, h = −70 nm) coated with 110 nm-thick $MAPbI_3$ film. In the left column, each of the four parameters are individually varied, meanwhile the other three parameters are kept fixed. When each parameter is varied, specific spectra with spectral peaks of interest are selected and respectively exhibited in Figure 6e–h and the insets show the electric field intensity distribution at the peak wavelengths.

First, the role of NH depth $|h|$ is investigated. It is noticeable that the spectral peak at λ = 545 nm redshifts when $|h|$ increases from 10 nm (black solid line) to 70 nm (yellow line). This peak results from the shallow hole, not disrupting the Fabry–Pérot effect (the bright horizontal band in inset (1)). When $|h|$ increases to 50 nm, the peak redshifts due to a localized surface plasmon resonance (LSPR) arising inside the hole (inset (2) in Figure 6e). When $|h|$ further increases to 150 nm, the constructive interference near the $MAPbI_3$/gold interface is restored, but is not as strong (faint horizontal band in inset (3) in Figure 6e).

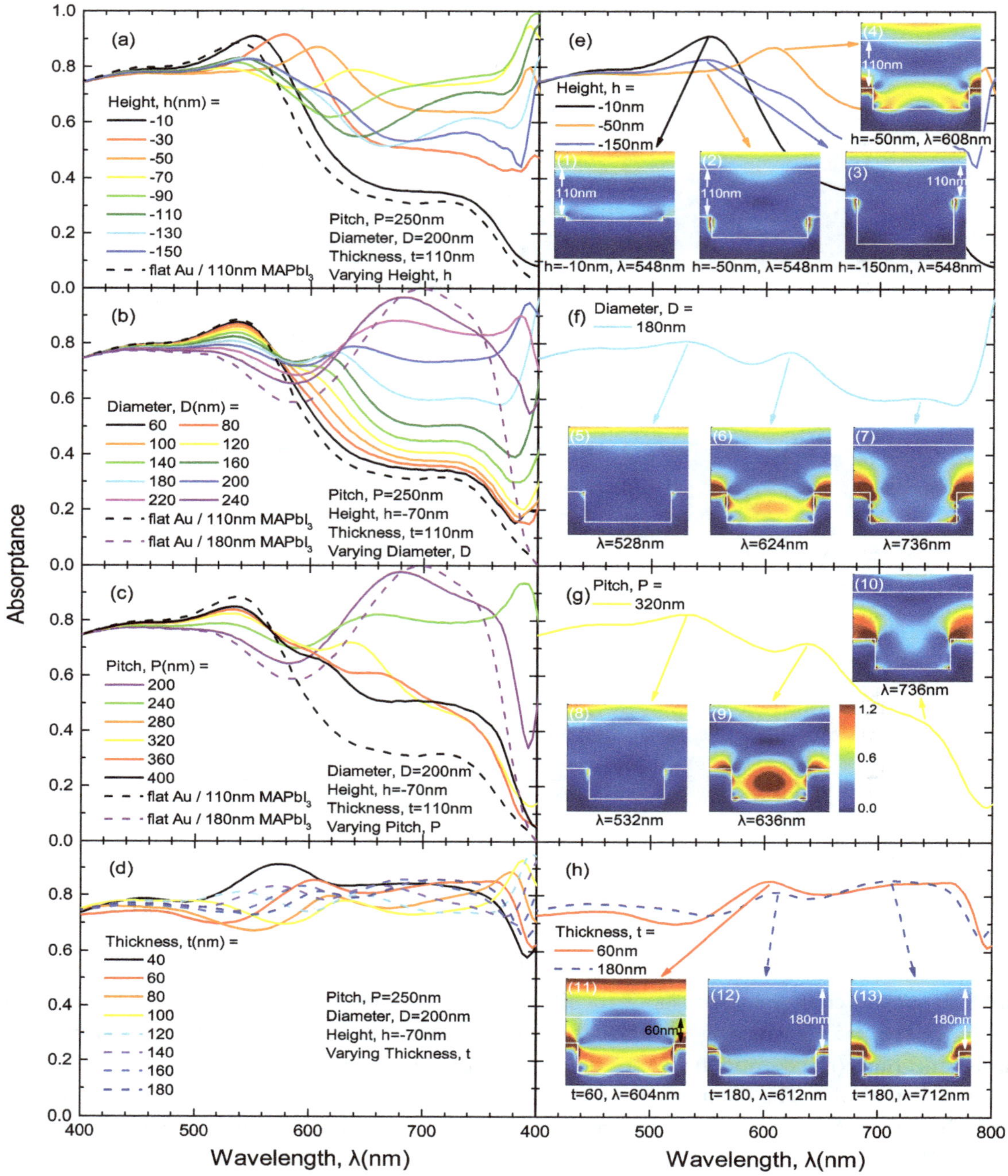

Figure 6. (**a–d**) Simulated absorptance spectra of representative NH arrays coated with a 110 nm-thick $MAPbI_3$ film. Each of the geometry parameters (i.e., height, h, diameter, D, pitch, P and $MAPbI_3$ thickness, t) is individually varied while keeping the others constant. (**e–h**) Representative absorptance spectra from (**a–d**) showing spectral peaks of interest. The color maps reported as insets show cross sections of electric field intensity simulated at each corresponding spectral peak wavelength by using 3D finite-difference time-domain (FDTD).

Figure 6b,c respectively demonstrate the absorptance spectra change with varying pitch P and diameter D. The black and purple dashed lines correspond to absorptance spectra of 110 nm and (110+$|h|$) nm-thick $MAPbI_3$ film on flat gold, respectively. With small feature sizes (small diameter D) or low NH concentration (large pitch P), the absorptance spectra are similar to that of 110 nm-thick $MAPbI_3$ on flat gold (black dashed lines). As the NH size or concentration increases, the absorptance spectral shape starts to shift towards that of 180 nm-thick $MAPbI_3$ on flat gold (purple dashed lines), which results from the Fabry–Pérot mode between the bottom of the holes and top of $MAPbI_3$ film. In addition, another peak near 630 nm shows up in some structures. Insets (6, 9) in Figure 6f,g show that these spectral peaks attribute to the LSPR inside the NH structure. Moreover, an additional spectral peak is observed at $\lambda = 736$ nm. These peaks results from the LSPR between neighboring NH structures, as is shown in insets (7, 10) in Figure 6f,g. Figure 6d demonstrates the thickness dependence of the absorptance spectra. With increasing thickness, the multiple peaks corresponds to the LSPR inside the holes or between neighboring holes. In addition, the resonances gets weaker as t increases due to less light–matter interaction.

Likewise, the role of the ND geometry parameters in absorption enhancement is also investigated. Figure 7 shows the simulated absorptance spectra of an example ND array ($P = 400$ nm, $D = 100$ nm, $h = 40$ nm) coated with $MAPbI_3$ thin film of $t = 110$ nm.

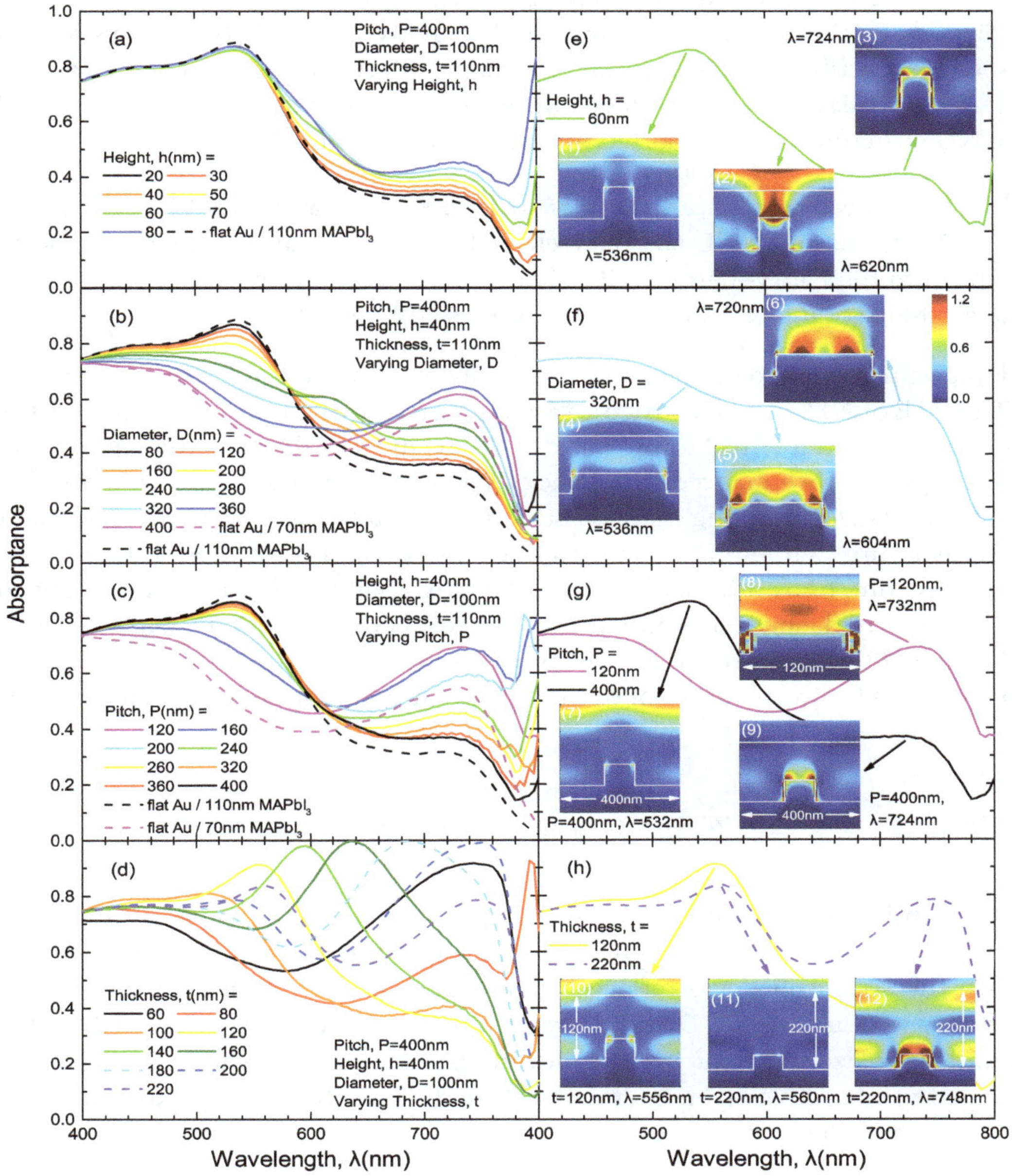

Figure 7. (**a**–**d**) Simulated absorptance spectra of representative ND arrays coated with a 110 nm-thick $MAPbI_3$ film. The geometry parameters (i.e., height, h, diameter, D, pitch, P and $MAPbI_3$ thickness, t) are individually varied while the other three parameters are kept constant; (**e**–**h**) Representative absorptance spectra from (**a**–**d**) showing spectral peaks of interest. The color maps reported as insets show cross sections of electric field intensity simulated at each corresponding spectral peak wavelength by using 3D FDTD.

While varying h (Figure 7a), the absorptance spectra shape mainly follows that of flat-interface structure (dashed black line), showing peaks at $\lambda = 530$ nm, which corresponds to the Fabry–Pérot effect of 110 nm-thick $MAPbI_3$ on gold. Insets (1, 3) in Figure 7e show a discontinuous horizontal band, indicating that the existence of ND disrupts the cavity effect, but the spectral peaks still persists. Additionally, the ND also brings additional plasmonic resonances, as is shown in inset (3), enhancing the absorption with increasing height. Moreover, for a specific structure ($h = 60$ nm), another resonance is observed on the top of the disk structure, as is shown in inset (2) in Figure 7e.

The diameter D and array pitch P play a similar role in ND arrays: smaller feature sizes and low concentration preserve cavity effects, but also contribute to less plasmonic effect, as is shown in Figure 7b,c. For $D = 320$ nm (Figure 7f), specifically, the suppression of Fabry–Pérot effect is observed (inset (4) in Figure 7f). Similar to NH arrays, high-density (small P) ND arrays also suppress the cavity effects, while contributing to significant plasmonic effects (insets (7–9) in Figure 7g). In addition, multiple orders of localized standing-wave surface plasmonic resonances are observed on the top surface of the ND structure, as are reported in insets (5, 6) in Figure 7f.

Figure 7d shows the absorptance spectra of structures with varying $MAPbI_3$ thickness t coated on the ND arrays. Due to the small ND size, as well as the relatively large array pitch, the spectra peaks generally result from the Fabry–Pérot cavity effects (horizontal bands in insets (10–12) in Figure 7h). In spite of the existence of the ND structure, the modes still persist.

The PCE is calculated based on the previous simulation results, and is reported as black circle lines in Figure 8. The PCE of NH and ND structure are respectively displayed in the left and right column. As is expected, when coated with 110nm-thick $MAPbI_3$, most of the structures with nanostructured surfaces exhibit significantly enhanced PCE (black circles) compared to their flat-interface counterparts (red dashed lines). It is worth noting that NH structures are adding extra perovskite material volume to the 110 nm-thick film, while ND structures have less perovskite material than the flat structure. To take this into account, two additional references are exhibited in Figure 8: (1) the blue dashed lines in Figure 8a–c and gray dashed lines in Figure 8e–g respectively show $\text{PCE}_{flat}(\text{t} = 110 \pm |h| \text{ nm})$ for NH and ND arrays; (2) the red triangle lines represent the PCE of flat structures with a combined thickness that accounts for the presence of two effective Fabry–Pérot cavities, which can be respectively expressed as

$$\text{PCE}_{combined} = \frac{\pi D^2}{2\sqrt{3}P^2}\text{PCE}_{flat}(\text{t=}110 \pm |h|\text{nm}) + (1 - \frac{\pi D^2}{2\sqrt{3}P^2})\text{PCE}_{flat}(\text{t=110nm}) \quad (2)$$

where the plus (minus) sign refers to NH (ND) arrays. For most of the structures, the nanostructured-surface PCE is greater than $\text{PCE}_{combined}$, which further proves that the plasmonic effects improve the solar cell performance. In addition, the difference between these two PCEs generally increases with increasing D and decreasing P, i.e., the cell performance benefits more from high-density and large-size nanostructures. In addition, with small thickness t, a larger portion of light penetrates the perovskite active layer, interacting strongly with the nanostructure array, and therefore also results in a larger difference.

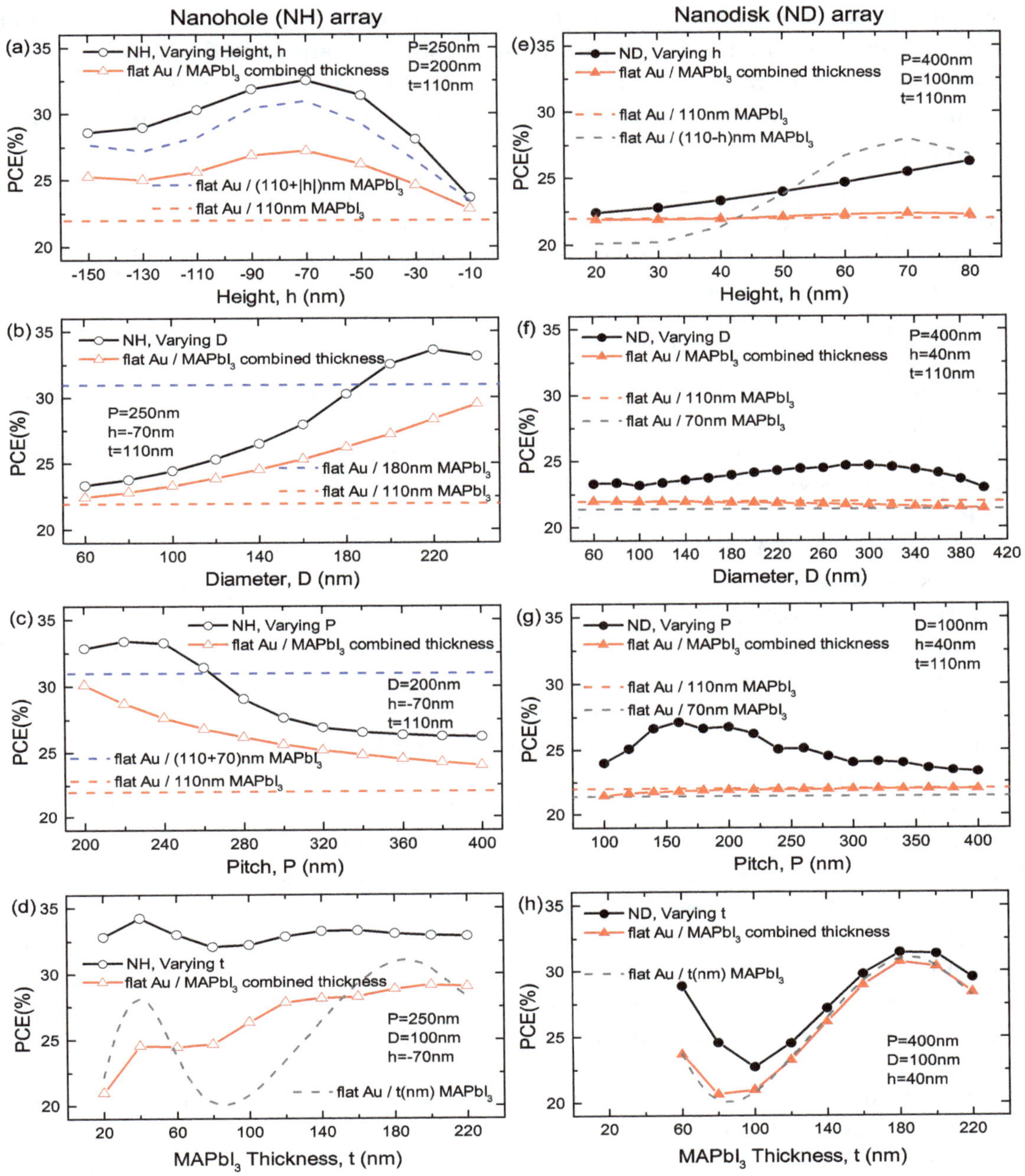

Figure 8. Calculated PCE for NH (**a–d**) and ND (**e–h**) arrays as a function of varying geometry parameters (i.e., height, h, diameter, D, array pitch, P and $MAPbI_3$ thickness, t). Dashed lines show the PCE of flat structures with t = 110 nm or $110 \pm |h|$ nm, while red triangle lines show the PCE of flat structures with a combined thickness that accounts for the presence of two effective Fabry–Pérot cavities in the nanostructured structure.

4. Conclusions

In conclusion, various plasmonic nanostructure arrays are experimentally fabricated on gold surfaces coated with perovskite films with varying thickness and their optical absorption properties are theoretically calculated by using 3D FDTD simulations. It is shown that, although thin-film interference effects can be negatively affected by the presence of nanoscatterers, overall, the plasmonic concentrators allow for significant absorption enhancement as the result of a combination of physical effects, i.e., (1) localized optical resonances (including surface plasmon resonances); (2) plasmonic modes in vertical cavities (such as the space within nanoholes); (3) in-plane constructive interference of surface plasmon polaritons that propagate along the metal/dielectric interface. Structures with higher surface density of nanoscattereres and with reduced film thickness generally show larger absorption and calculated PCE enhancements. The experimental results are supported by 3D FDTD simulations that help identify the individual physical effects and elucidate their interplay in redistributing the incident field intensity, which in turns determine the observed absorption and the calculated PCE enhancements.

Author Contributions: Data curation, T.S. and Q.T.; formal analysis, T.S. and Q.T.; resources, Z.D.; supervision, N.P.P. and D.P.; writing—original draft, T.S. and Q.T.; writing—review and editing, Z.D., N.P.P. and D.P. All authors have read and agreed to the published version of the manuscript.

Acknowledgments: The authors thank Min Chen for fruitful discussions.

References

1. Huang, J.; Yuan, Y.; Shao, Y.; Yan, Y. Understanding the physical properties of hybrid perovskites for photovoltaic applications. *Nat. Rev. Mater.* **2017**, *2*, 1–19. [CrossRef]
2. Dong, Q.; Fang, Y.; Shao, Y.; Mulligan, P.; Qiu, J.; Cao, L.; Huang, J. Electron-hole diffusion lengths > 175 μm in solution-grown $CH_3NH_3PbI_3$ single crystals. *Science* **2015**, *347*, 967–970. [CrossRef] [PubMed]
3. Yang, S.; Fu, W.; Zhang, Z.; Chen, H.; Li, C.Z. Recent advances in perovskite solar cells: efficiency, stability and lead-free perovskite. *J. Mater. Chem. A* **2017**, *5*, 11462–11482. [CrossRef]
4. Elumalai, N.K.; Mahmud, M.A.; Wang, D.; Uddin, A. Perovskite solar cells: Progress and advancements. *Energies* **2016**, *9*, 861. [CrossRef]
5. Chen, M.; Ju, M.G.; Carl, A.D.; Zong, Y.; Grimm, R.L.; Gu, J.; Zeng, X.C.; Zhou, Y.; Padture, N.P. Cesium titanium (IV) bromide thin films based stable lead-free perovskite solar cells. *Joule* **2018**, *2*, 558–570. [CrossRef]
6. Song, Z.; Watthage, S.C.; Phillips, A.B.; Heben, M.J. Pathways toward high-performance perovskite solar cells: review of recent advances in organo-metal halide perovskites for photovoltaic applications. *J. Photon. Energy* **2016**, *6*, 022001. [CrossRef]
7. Zhou, Y.; Zhou, Z.; Chen, M.; Zong, Y.; Huang, J.; Pang, S.; Padture, N.P. Doping and alloying for improved perovskite solar cells. *J. Mater. Chem. A* **2016**, *4*, 17623–17635. [CrossRef]
8. Best Research–Cell Efficiencies. Available online: https://www.nrel.gov/pv/assets/pdfs/best-research-cell-efficiencies.20200406.pdf (accessed on 8 July 2020).
9. Ferry, V.E.; Sweatlock, L.A.; Pacifici, D.; Atwater, H.A. Plasmonic nanostructure design for efficient light coupling into solar cells. *Nano Lett.* **2008**, *8*, 4391–4397. [CrossRef] [PubMed]
10. Atwater, H.A.; Polman, A. Plasmonics for improved photovoltaic devices. In *Materials For Sustainable Energy: A Collection of Peer-Reviewed Research and Review Articles from Nature Publishing Group*; World Scientific: Singapore, 2011; pp. 1–11.
11. Polman, A.; Atwater, H.A. Photonic design principles for ultrahigh-efficiency photovoltaics. *Nat. Mater.* **2012**, *11*, 174–177. [CrossRef] [PubMed]
12. Gan, Q.; Bartoli, F.J.; Kafafi, Z.H. Plasmonic-enhanced organic photovoltaics: Breaking the 10% efficiency barrier. *Adv. Mater.* **2013**, *25*, 2385–2396. [CrossRef]
13. Yang, J.; You, J.; Chen, C.C.; Hsu, W.C.; Tan, H.r.; Zhang, X.W.; Hong, Z.; Yang, Y. Plasmonic polymer tandem solar cell. *ACS Nano* **2011**, *5*, 6210–6217. [CrossRef] [PubMed]
14. Ding, I.K.; Zhu, J.; Cai, W.; Moon, S.J.; Cai, N.; Wang, P.; Zakeeruddin, S.M.; Grätzel, M.; Brongersma, M.L.; Cui, Y.; et al. Plasmonic dye-sensitized solar cells. *Adv. Energy Mater.* **2011**, *1*, 52–57. [CrossRef]
15. Spinelli, P.; Ferry, V.; Van de Groep, J.; Van Lare, M.; Verschuuren, M.; Schropp, R.; Atwater, H.; Polman, A. Plasmonic light trapping in thin-film Si solar cells. *J. Opt.* **2012**, *14*, 024002. [CrossRef]
16. Jiménez-Solano, A.; Carretero-Palacios, S.; Míguez, H. Absorption enhancement in methylammonium lead iodide perovskite solar cells with embedded arrays of dielectric particles. *Opt. Express* **2018**, *26*, A865–A878. [CrossRef] [PubMed]
17. Miranda-Muñoz, J.M.; Carretero-Palacios, S.; Jiménez-Solano, A.; Li, Y.; Lozano, G.; Míguez, H. Efficient bifacial dye-sensitized solar cells through disorder by design. *J. Mater. Chem. A* **2016**, *4*, 1953–1961. [CrossRef]
18. Carretero-Palacios, S.; Jiménez-Solano, A.; Míguez, H. Plasmonic nanoparticles as light-harvesting enhancers in perovskite solar cells: a user's guide. *ACS Energy Lett.* **2016**, *1*, 323–331. [CrossRef]
19. Carretero-Palacios, S.; Calvo, M.E.; Míguez, H. Absorption enhancement in organic–inorganic halide perovskite films with embedded plasmonic gold nanoparticles. *J. Phys. Chem. C* **2015**, *119*, 18635–18640. [CrossRef]

20. Kang, S.M.; Jang, S.; Lee, J.K.; Yoon, J.; Yoo, D.E.; Lee, J.W.; Choi, M.; Park, N.G. Moth-Eye TiO_2 Layer for Improving Light Harvesting Efficiency in Perovskite Solar Cells. *Small* **2016**, *12*, 2443–2449. [CrossRef]
21. Mali, S.S.; Shim, C.S.; Kim, H.; Patil, P.S.; Hong, C.K. In situ processed gold nanoparticle-embedded TiO_2 nanofibers enabling plasmonic perovskite solar cells to exceed 14% conversion efficiency. *Nanoscale* **2016**, *8*, 2664–2677. [CrossRef]
22. Ghahremanirad, E.; Olyaee, S.; Hedayati, M. The Influence of Embedded Plasmonic Nanostructures on the Optical Absorption of Perovskite Solar Cells. *Photonics* **2019**, *6*, 37. [CrossRef]
23. Mohsen, A.A.; Zahran, M.; Habib, S.; Allam, N.K. Refractory plasmonics enabling 20% efficient lead-free perovskite solar cells. *Sci. Rep.* **2020**, *10*, 1–12. [CrossRef] [PubMed]
24. Deng, W.; Yuan, Z.; Liu, S.; Yang, Z.; Li, J.; Wang, E.; Wang, X.; Li, J. Plasmonic enhancement for high-efficiency planar heterojunction perovskite solar cells. *J. Power Sour.* **2019**, *432*, 112–118. [CrossRef]
25. Wang, B.; Zhu, X.; Li, S.; Chen, M.; Liu, N.; Yang, H.; Ran, M.; Lu, H.; Yang, Y. Enhancing the Photovoltaic Performance of Perovskite Solar Cells Using Plasmonic Au@ Pt@ Au Core-Shell Nanoparticles. *Nanomaterials* **2019**, *9*, 1263. [CrossRef]
26. Yao, K.; Zhong, H.; Liu, Z.; Xiong, M.; Leng, S.; Zhang, J.; Xu, Y.X.; Wang, W.; Zhou, L.; Huang, H.; et al. Plasmonic Metal Nanoparticles with Core–Bishell Structure for High-Performance Organic and Perovskite Solar Cells. *ACS Nano* **2019**, *13*, 5397–5409. [CrossRef] [PubMed]
27. Chen, L.C.; Tien, C.H.; Lee, K.L.; Kao, Y.T. Efficiency Improvement of $MAPbI_3$ Perovskite Solar Cells Based on a $CsPbBr_3$ Quantum Dot/Au Nanoparticle Composite Plasmonic Light-Harvesting Layer. *Energies* **2020**, *13*, 1471. [CrossRef]
28. Lu, Z.; Pan, X.; Ma, Y.; Li, Y.; Zheng, L.; Zhang, D.; Xu, Q.; Chen, Z.; Wang, S.; Qu, B.; et al. Plasmonic-enhanced perovskite solar cells using alloy popcorn nanoparticles. *RSC Adv.* **2015**, *5*, 11175–11179. [CrossRef]
29. Zhang, W.; Saliba, M.; Stranks, S.D.; Sun, Y.; Shi, X.; Wiesner, U.; Snaith, H.J. Enhancement of perovskite-based solar cells employing core–shell metal nanoparticles. *Nano Lett.* **2013**, *13*, 4505–4510. [CrossRef]
30. Bayles, A.; Carretero-Palacios, S.; Calió, L.; Lozano, G.; Calvo, M.E.; Míguez, H. Localized surface plasmon effects on the photophysics of perovskite thin films embedding metal nanoparticles. *J. Mater. Chem. C* **2020**, *8*, 916–921. [CrossRef]
31. Ostfeld, A.; Pacifici, D. Plasmonic concentrators for enhanced light absorption in ultrathin film organic photovoltaics. *Appl. Phys. Lett.* **2011**, *98*, 113112. [CrossRef]
32. Mao, J.; Sha, W.E.; Zhang, H.; Ren, X.; Zhuang, J.; Roy, V.A.; Wong, K.S.; Choy, W.C. Novel direct nanopatterning approach to fabricate periodically nanostructured perovskite for optoelectronic applications. *Adv. Funct. Mater.* **2017**, *27*, 1606525. [CrossRef]
33. Wang, H.; Haroldson, R.; Balachandran, B.; Zakhidov, A.; Sohal, S.; Chan, J.Y.; Zakhidov, A.; Hu, W. Nanoimprinted perovskite nanograting photodetector with improved efficiency. *ACS Nano* **2016**, *10*, 10921–10928. [CrossRef] [PubMed]
34. Long, M.; Chen, Z.; Zhang, T.; Xiao, Y.; Zeng, X.; Chen, J.; Yan, K.; Xu, J. Ultrathin efficient perovskite solar cells employing a periodic structure of a composite hole conductor for elevated plasmonic light harvesting and hole collection. *Nanoscale* **2016**, *8*, 6290–6299. [CrossRef] [PubMed]
35. Paetzold, U.W.; Qiu, W.; Finger, F.; Poortmans, J.; Cheyns, D. Nanophotonic front electrodes for perovskite solar cells. *Appl. Phys. Lett.* **2015**, *106*, 173101. [CrossRef]
36. Shen, T.; Siontas, S.; Pacifici, D. Plasmon-enhanced thin-film perovskite solar cells. *J. Phys. Chem. C* **2018**, *122*, 23691–23697. [CrossRef]
37. Lumerical Inc. Available online: https://www.lumerical.com/products/ (accessed on 8 July 2020).
38. Löper, P.; Stuckelberger, M.; Niesen, B.; Werner, J.; Filipic, M.; Moon, S.J.; Yum, J.H.; Topič, M.; De Wolf, S.; Ballif, C. Complex refractive index spectra of $CH_3NH_3PbI_3$ perovskite thin films determined by spectroscopic ellipsometry and spectrophotometry. *J. Phys. Chem. Lett.* **2014**, *6*, 66–71. [CrossRef]
39. Haynes, W.M. *CRC Handbook of Chemistry and Physics*; CRC Press: Boca Raton, FL, USA, 2014.
40. Ng, R.J.; Goh, X.M.; Yang, J.K. All-metal nanostructured substrates as subtractive color reflectors with near-perfect absorptance. *Opt. Express* **2015**, *23*, 32597–32605. [CrossRef]
41. Würfel, P.; Würfel, U. *Physics of Solar Cells: From Basic Principles to Advanced Concepts*; John Wiley & Sons: Hoboken, NJ, USA, 2016.

10

Numerical Study of Complementary Nanostructures for Light Trapping in Colloidal Quantum Dot Solar Cells

Jue Wei [1,2], Qiuyang Xiong [1], Seyed Milad Mahpeykar [1] and Xihua Wang [1,*]

1 Department of Electrical and Computer Engineering, University of Alberta, Edmonton, AB T6G 2V4, Canada; weijue@hotmail.com (J.W.); qiuyang@ualberta.ca (Q.X.); mahpeyka@ualberta.ca (S.M.M.)
2 Key Laboratory of Coherent Light and Atomic and Molecular Spectroscopy of Ministry of Education, College of Physics, Jilin University, Changchun 130012, China
* Correspondence: xihua@ualberta.ca

Academic Editor: Guanying Chen

Abstract: We have investigated two complementary nanostructures, nanocavity and nanopillar arrays, for light absorption enhancement in depleted heterojunction colloidal quantum dot (CQD) solar cells. A facile complementary fabrication process is demonstrated for patterning these nanostructures over the large area required for light trapping in photovoltaic devices. The simulation results show that both proposed periodic nanostructures can effectively increase the light absorption in CQD layer of the solar cell throughout the near-infrared region where CQD solar cells typically exhibit weak light absorption. The complementary fabrication process for implementation of these nanostructures can pave the way for large-area, inexpensive light trapping implementation in nanostructured solar cells.

Keywords: solar cells; nanostructures; light trapping

1. Introduction

In the last few years, colloidal quantum dot (CQD) solar cells have received a great deal of attention due to their potential for large-area, high-throughput, and low-cost manufacturing [1]. Despite all the achievements in CQD synthesis, surface treatment and film deposition technologies [2], the power conversion efficiency of this type of solar cell continues to lag behind traditional silicon solar cells. Because of the lack of long diffusion lengths for photo-generated carriers in CQD films, a CQD film capable of taking advantage of all the incident solar power would be too thick to extract all the generated carriers, leading to an absorption-extraction trade-off [3–5]. Light trapping, or effectively increasing optical path lengths in the absorbing material through structuring without any change in light absorbing material's thickness, is one option to overcome the aforementioned trade-off [6].

Periodic nanostructured gratings have been extensively explored for light trapping in various types of thin-film solar cells and various silicon or metamaterial-based structures have been proposed, such as nanopillar, nanowire, nanohole, and pyramid arrays [7–14]. The downside, however, is that these structures are difficult to fabricate due to their complicated structures or material compositions. Metallic gratings have also been considered for light harvesting enhancement by taking advantage of surface plasmons [15–17]. However, parasitic light absorption intrinsic to metallic structures which can compete against useful absorption in light absorbing layer has severely limited the application of metallic nanostructures in photovoltaic devices [18].

Recently, we reported on periodic nano-branch indium-doped tin oxide (ITO) electrodes as diffraction gratings for light absorption enhancement in CQD solar cells [19]. Using numerical simulations, a significant polarization-independent broadband light absorption enhancement was observed for two-dimensional ITO nano-branch gratings and the absorption enhancement was demonstrated to be almost independent of common fabrication flaws in nano-branch structure. On the other hand, current fabrication technologies are unable to implement the fabrication of such nanostructures due to the difficulty of keeping the periodicity of the structure over a large area and also incorporation of CQDs into such porous structure.

In order to be able to apply a periodic nanostructure for practical and tangible light trapping in solar cells, maintaining the periodicity of the structure over a large area is the key requirement. Therefore, the fabrication process used to impose the periodic pattern must be able to produce periodic patterns over large areas and also inexpensive at the same time. Recent advances in large-scale nanofabrication techniques have allowed sophisticated nanostructures to be employed in solar cells and photodetectors with impressive results [20,21]. In this work, in order to fully satisfy the large-area requirement of light trapping structures, we consider two experimentally available nanostructures (nanocavity and nanopillar) for light absorption enhancement in CQD solar cells due to their potential for easy large-area fabrication and CQD incorporation. A facile fabrication process is performed to achieve large-area periodic pattern of the nanocavity and nanopillar structures. The simulation results show that compared to the reference flat structure, our proposed structures can achieve relatively high absorption enhancement in CQD solar cells.

2. Results and Discussion

2.1. Structure Design

The main reason that nanocavity and nanopillar arrays are proposed as nanostructured electrodes to improve absorption enhancement in PbS CQD solar cells is that such nanostructures can be easily fabricated over large areas by a low-cost pattern transfer method known as nanosphere lithography (NSL). As a proof of concept, PDMS nanopillar and silver nanocavity arrays were fabricated utilizing nanosphere lithography. The proposed process steps for fabrication of nanocavity and nanopillar arrays is presented schematically in Figure 1a. Firstly, colloidal nanosphere mask is deposited on a Si substrate. After nanosphere mask formation, oxygen plasma is employed to shrink the nanospheres to ideal diameter needed for the intended structure through reactive ion etching (RIE).

The next step is to deposit the desired material, in this case silver (Ag), on the sample covered by colloidal mask. The final step is to lift off the nanosphere mask, which is usually done in ultrasonic bath with organic solvents such as acetone, after which the nanocavity array is formed on the substrate. In addition to being a standalone light trapping structure, the fabricated nanocavity array can be utilized for fabrication of nanopillar array through PDMS casting and peel off. This is possible because the nanocavity array can act as a mold for formation of nanopillar array. The Scanning Electron Microscope (SEM) images of the cavity and pillar arrays fabricated using the described fabrication process are shown in Figure 1b. We believe a similar procedure with minimal modification can be used to fabricate well-defined ITO nanocavity and nanopillar electrodes. Substituting the silver with ITO in material deposition step will easily lead to ITO nanocavity structure and depositing ITO on top of fabricated PDMS nanopillars can form the desired ITO nanopillars suitable for light absorption enhancement in CQD solar cells.

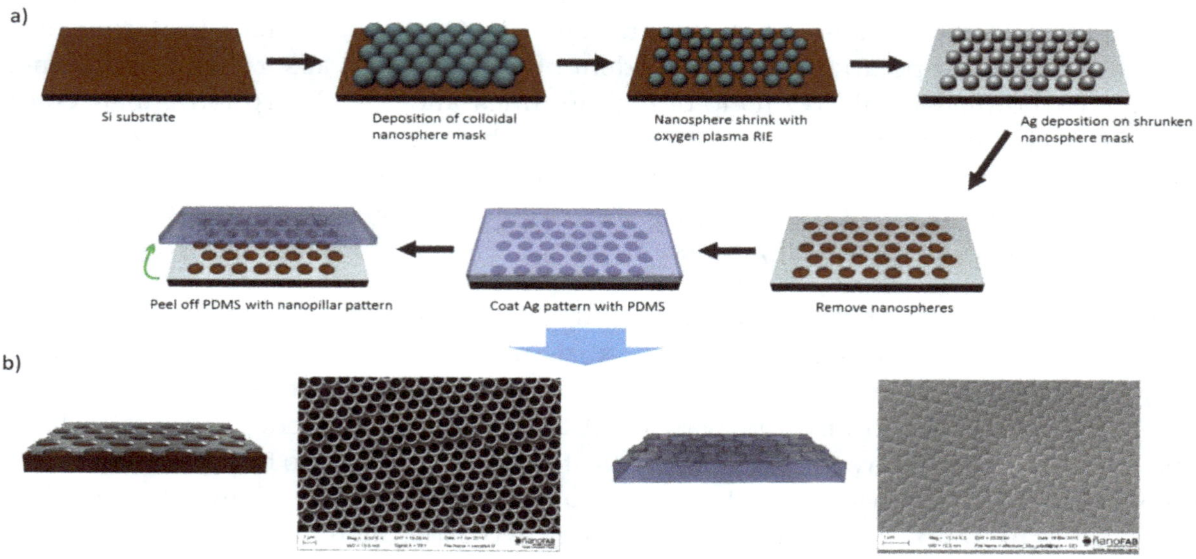

Figure 1. (**a**) The proposed process flow for fabrication of nanocavity and nanopillar arrays. (**b**) Top view scanning electron microscope (SEM) images and cross-sectional schematic of the nanocavity (left) and nanopillar (right) arrays fabricated using the proposed process.

The periodic ITO electrodes proposed in this work are designed for a typical depleted heterojunction CQD solar cell structure. The depleted heterojunction architecture utilizes a TiO_2 layer as the n-side of the junction and p-type PbS quantum dots as the p-side. The bottom contact to the junction is formed on a glass substrate and consists of a thin transparent conductive ITO layer. The top contact employs a deep work function metal such as gold to collect photo-excited holes and also reflect back any unabsorbed photons into the light absorbing layer. A conformal layer of TiO_2 with thickness of 50 nm was considered as a layer between ITO electrode and active layer. The designed periodic nanostructures are implemented at the interface between the ITO bottom contact and PbS QDs. Figure 2a shows the schematic of the structure of PbS QD solar cell with patterned ITO electrode used for simulation. As depicted in the figure, when the light is normally incident on ITO diffraction gratings through the transparent substrate, forward diffraction of light can induce light trapping by effectively increasing optical path lengths inside the absorbing material especially for higher diffracted orders supported by the grating structure. Optical constants of the materials used in the simulation model are shown in Figure 2b.

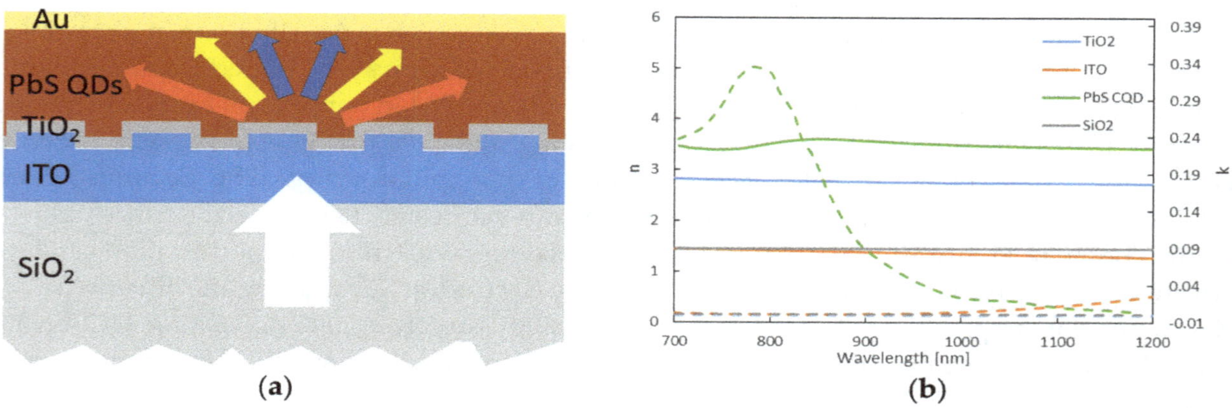

Figure 2. (**a**) Schematic of light diffraction in PbS quantum dot (QD) solar cell with patterned indium-doped tin oxide (ITO) electrode. (**b**) Optical constants of the materials used in the simulation model.

2.2. Light Trapping Analysis

Grating far-field projection analysis [22] was firstly used to analyze the diffraction behavior of the proposed periodic grating structures in PbS CQD solar cells. The resulting transmission efficiencies of the simulated patterned structures are illustrated in Figure 3. As is clear from the figure, both cavity and pillar structures demonstrate high transmission efficiency in wavelength range of 700 nm to 900 nm. On the other hand, the amount of transmitted power is not significant beyond 900 nm. This trend, however, is broken for both nanocavity and nanopillar arrays at around the wavelength of 950 nm with a strong increase in the amount of transmitted power. The same behavior is also observed in the case of the cavity array at wavelength of 1080 nm, the intensity of which, however, is not as strong as the peak at 950 nm. This sudden increase in transmitted power can be attributed to the resonant coupling of the incident light into wave guiding modes supported by the PbS CQD layer located adjacent to the ITO grating structures due to the periodic nature of their structure [19]. Although throughout the spectrum, a portion of incident light is not diffracted (shown as the order (0,0)), by paying close attention to the total transmission values and their difference with order (0,0), it is obvious that a significant amount of energy is diffracted into higher orders, especially at resonance wavelengths. This can greatly contribute to light absorption enhancement in CQD layer by increasing the optical path length of the light inside the layer or light trapping through resonant coupling with the incident light [19]. The transmission efficiencies of two of the strongest diffracted orders (1,1) and (2,0) are plotted in Figure 3.

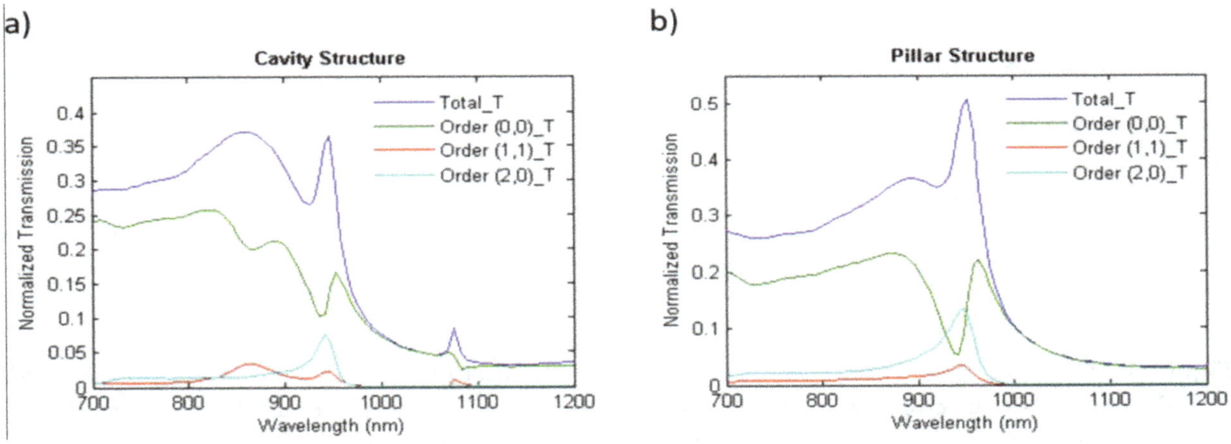

Figure 3. The normalized transmission spectra of simulated patterned ITO structures: (**a**) nanocavity, (**b**) nanopillar. The plot shows the relative power transmitted into different diffracted orders and the net total transmitted power normalized to the simulation source power. Two of the strongest diffracted orders (1,1) and (2,0) are plotted. (0,0) represents the part of incident power not being diffracted by the structures.

Figure 4 depicts the simulated light absorption spectra for PbS CQD layer of the modelled depleted heterojunction solar cell normalized to the AM1.5G solar spectrum (Figure 4a) and simulation light source (Figure 4b). In order to be able to compare the effect of proposed structures on absorption enhancement in PbS CQD layer, a flat ITO layer was considered as the reference. The available power from AM1.5G spectrum is also included in the figure for comparison. As is obvious from the figure, both nanocavity and nanopillar arrays can induce more light absorption in CQD layer than the flat ITO layer within most parts of the near-infrared region. It is also noticeable that both proposed structures have achieved almost perfect absorption in the range of 720 nm to 850 nm by absorbing all the power available from the sun in this range. As for beyond this range, especially in the case of cavity structure, the resonant coupling of the incident light into guided modes supported by CQD layer is the major responsible for strong but narrowband absorption enhancements at resonant peaks, previously predicted by grating projection analysis. This is possible because of the major difference in refractive indices of the CQD layer and the ITO layer which can form an efficient waveguide in the middle of the cell's structure.

In order to have an overall evaluation for the light absorption enhancement performance of the proposed structures, the average absorption enhancement of the structures over the entire simulated spectrum was measured against the flat reference structure using the following equation:

$$\text{Absorption Enhancement}\,(\%) = (P_g - P_r)/P_r \times 100 \tag{1}$$

where P_g depicts the total power absorbed by the cell with grating structure and P_r denotes the power absorbed by the reference flat structure. According to the equation, the calculated absorption enhancement factors for nanocavity and nanopillar grating structures compared to the flat structure are 15.0% and 13.6%, respectively. This amount of absorption enhancement can significantly boost charge carrier generation and thus short-circuit current density of a CQD solar cell which can ultimately lead to remarkable improvement in power conversion efficiency of the cell.

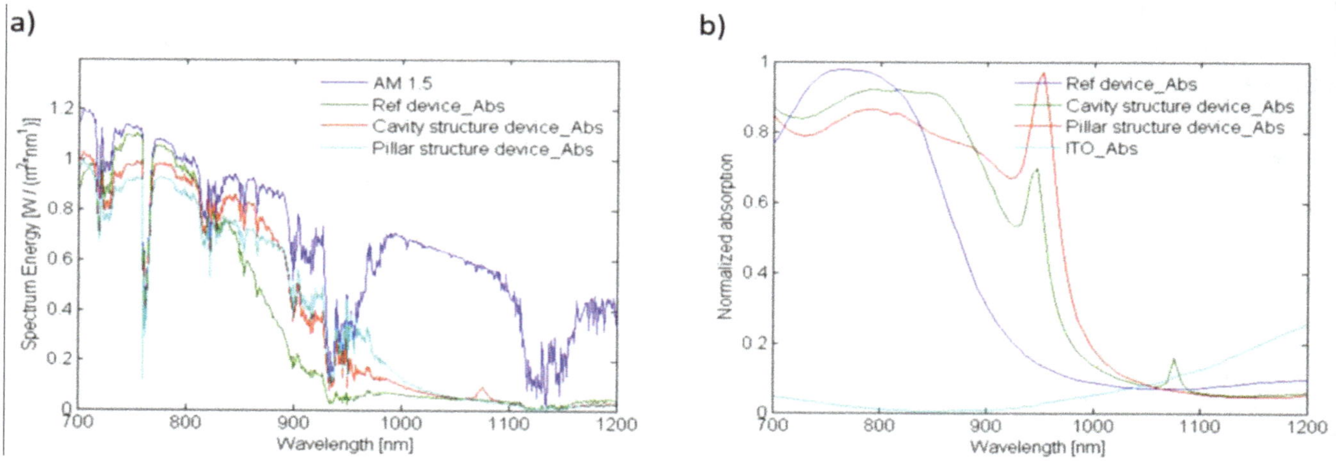

Figure 4. The light absorption spectra for PbS colloidal quantum dot (CQD) layer incorporated into different ITO structures normalized to (**a**) AM1.5G spectra and (**b**) simulation light source. The absorption enhancement for both cavity and pillar structures over the reference flat structure is obvious especially at resonance wavelengths of 950 nm for both structures and 1080 nm for cavity arrays. A slight absorption loss by ITO layer was also observed, as shown in Figure 4b.

To clearly demonstrate the influence of nanocavity and nanopillar resonance effect on absorption enhancement in CQD layer, the electric field distributions inside the PbS CQD layer with patterned structures were investigated and are shown in Figure 5. The on resonance profiles for nanocavity and nanopillar structures are plotted at wavelength of 950 nm and the off resonant profile wavelength is chosen at 1000 nm for both structures. The on resonance profiles for both structures reveal various absorption hot spots for on-resonance wavelengths where as in the case of off resonance profiles, no hot spot is visible at off-resonance wavelengths. It is obvious that the presence of high intensity E-field spots (hot spots) indicates the occurrence of strong absorption inside PbS CQD layer. In addition, the periodic pattern of the hot spots observed in the obtained profiles discloses the type of resonance to be the guided mode kind usually excited by periodic dielectric nanostructures [6]. The difference in field distribution observed between on and off resonance profiles implies the impressive light trapping performance of the proposed structures at resonance wavelengths. This confirms the superiority of the proposed structures for absorption enhancement in CQD solar cells through resonant coupling of the incident light with supported waveguide modes inside the CQD layer.

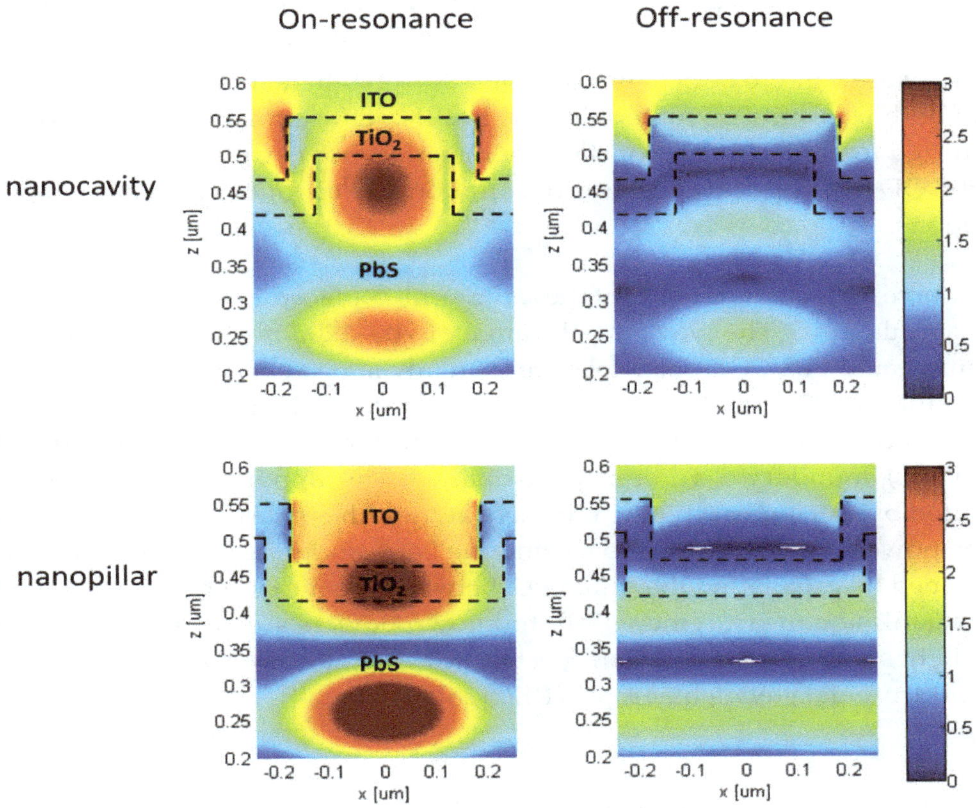

Figure 5. Simulated electric field distributions inside the PbS QDs layer with patterned structures. The hot spots present at resonance wavelengths (950 nm for both structures) with high field intensity indicate strong absorption inside PbS CQD. No hot spots are observed at off resonance wavelengths (1000 nm for both structures) suggesting the importance of resonant coupling of the incident into CQD layer for significant absorption enhancement.

3. Materials and Methods

3.1. Complementary Structure Fabrication

Self-assembly of nanospheres in Si substrate was accomplished by the air/water interface self-assembly process [23]. The etching of deposited nanospheres was done by using a 20 W O_2 plasma at 5 sccsm oxygen flow for 1 min. E-beam evaporation was used for Silver deposition. The nanosphere mask was removed through sonication in acetone for 10 min. PDMS nanopillars were fabricated by mixing silicone elastomer with curing agent from a Sylgard 184 kit (Dow Corning, Midland, MI, USA) in 10 wt % ratio. The mixture was then degassed in a desiccator for 30 min and was spin-coated on the nanocavity structures at 200 rpm for 30 s. The resulting film was then cured on a hot plate at 80 °C for 2 h after which it was peeled off using a doctor blade. The SEM images were obtained by a Ziess EVO Scanning Electron Microscope (Carl Zeiss, Oberkochen, Germany).

3.2. Simulation Methods

The Lumerical FDTD Solutions software (Lumerical Solutions Inc, Vancouver, Canada) was used for simulations in this work. A period of 500 nm was chosen for both cavity and pillar structures which is same as the diameter of nanospheres used for pattern generation. A cavity depth and pillar height of 80 nm and a diameter of 360 nm for both structures was found to be an optimum value. The PbS QD layer was considered to be a quasi-bulk homogeneous film (QDs were not considered as individual particles) without any voids and its thickness (excluding the nanostructure) is chosen to be 300 nm, which is usually considered the maximum thickness for efficient photo-generated carrier collection. The TiO_2 layer was assumed to be 50 nm thick. Both gold and SiO_2 glass layers are considered with infinite thickness for ease of modeling. The optimum thickness of ITO layer (excluding the nanostructure) was found to be 500 nm. It should be noted here that the periodic grating structure layer consists of both ITO and PbS materials. The multi-coefficient fitting tool inside the simulation software was utilized to model optical constants of materials from available experimental data [24–27]. In the case of PbS QDs, the optical constants of commonly used QDs with a bandgap of 1.3 eV were used for simulations.

The light source was considered a planewave source placed inside the substrate (SiO_2 layer) to simplify the simulations. The wavelength range of 700–1200 nm was chosen as the simulation wavelength span because PbS CQD solar cells are currently in need of absorption enhancement mostly in this region of sunlight [27]. For directions perpendicular to the incident light propagation direction, Bloch boundary conditions and for directions parallel to the light propagation direction, perfectly matched layer (PML) boundary conditions were defined. The amount of absorption inside the CQD layer was measured by placing two power monitors at the either sides of the layer. This configuration can calculate the power flow entering and exiting the layer and thus the power absorbed inside the layer can be obtained by calculating the difference between the measurements from the two monitors. In the case of transmission measurements, all the shown powers are normalized with respect to the power from the light source. For absorption spectra, however, the absorbed powers are normalized to AM1.5G solar spectrum data available from NREL [28].

4. Conclusions

In this work, ITO nanocavity and nanopillar diffraction gratings are proposed as light trapping structures in CQD solar cells to realize absorption enhancement and power conversion efficiency improvement. A facile fabrication process is demonstratedfor patterning these periodic nanostructures over a large area which is a critical requirement of practical light trapping structures designed for photovoltaic devices. The simulation results show that both proposed periodic structures can effectively increase the light absorption in CQD layer of the solar cell throughout the near-infrared region where CQD solar cells typically exhibit weak light absorption. The overall absorption enhancement of 15.0% and 13.6% was achieved for nanocavity and nanopillar structures, respectively. It is verified that the two nanostructures are useful for enhancing the efficiency of photovoltaics as large-area, inexpensive light trapping structures.

Abbreviations

The following abbreviations are used in this manuscript:

CQD	Colloidal Quantum Dot
ITO	Indium-doped Tin Oxide
NSL	Nano Sphere Lithography
RIE	Reactive Ion Etching
SEM	Scanning Electron Microscope
PDMS	Polydimethylsiloxane
FDTD	Finite Difference Time Domain

Acknowledgments: This work was supported by the Natural Sciences and Engineering Research Council of Canada (NSERC) Discovery Grant, the University of Alberta Start-Up Fund, and the IC-IMPACTS Centres of Excellence. J.W. also gratelfully acknowledges financial support from China Scholarship Council.

Author Contributions: J.W. and X.W. conceived and designed the study; J.W. and Q.X. performed the experiments; J.W. and S.M.M. performed and analyzed the simulations; All the authors contributed to the preparation of the manuscript.

References

1. Chang, L.Y.; Lunt, R.R.; Brown, P.R.; Bulovic, V.; Bawendi, M.G. Low-temperature solution-processed solar cells based on PbS colloidal quantum dot/CdS heterojunctions. *Nano Lett.* **2013**, *13*, 994–999. [CrossRef] [PubMed]
2. Tang, J.; Sargent, E.H. Infrared colloidal quantum dots for photovoltaics: Fundamentals and recent progress. *Adv. Mater.* **2011**, *23*, 12–29. [CrossRef] [PubMed]
3. Clifford, J.P.; Konstantatos, G.; Johnston, K.W.; Hoogland, S.; Levina, L.; Sargent, E.H. Fast, sensitive and spectrally tuneablecolloidal-quantum-dot photodetectors. *Nat. Nanotechnol.* **2009**, *4*, 40–44. [CrossRef] [PubMed]
4. Zhitomirsky, D.; Voznyy, O.; Hoogland, S.; Sargent, E.H. Measuring charge carrier diffusion in coupled colloidal quantum dot solids. *ACS Nano* **2013**, *7*, 5282–5290. [CrossRef] [PubMed]
5. Labelle, A.J.; Thon, S.M.; Kim, J.Y.; Lan, X.; Zhitomirsky, D.; Kemp, K.W.; Sargent, E.H. Conformal fabrication of colloidal quantum dot solids for optically enhanced photovoltaics. *ACS Nano* **2015**, *9*, 5447–5453. [CrossRef] [PubMed]
6. Brongersma, M.L.; Cui, Y.; Fan, S. Light management for photovoltaics using high-index nanostructures. *Nat. Mater.* **2014**, *13*, 451–460. [CrossRef] [PubMed]
7. Cao, S.; Yu, W.; Wang, T.; Xu, Z.; Wang, C.; Fu, Y.; Liu, Y. Two-dimensional subwavelength meta-nanopillar array for efficient visible light absorption. *Appl. Phys. Lett.* **2013**, *102*, 161109. [CrossRef]
8. Han, Q.; Jin, L.; Fu, Y.Q.; Yu, W.X. Si substrate-based metamaterials for ultrabroadbandperfect absorption in visible regime. *J. Nanomater.* **2014**, *2014*, 893202.
9. Han, S.E.; Chen, G. Optical absorption enhancement in silicon nanoholearrays for solar photovoltaics. *Nano Lett.* **2010**, *10*, 1012–1015. [CrossRef] [PubMed]
10. Lin, C.; Povinelli, M.L. Optical absorption enhancement in silicon nanowire arrays with a large lattice constant for photovoltaic applications. *Opt. Express* **2009**, *17*, 19371–19381. [CrossRef] [PubMed]
11. Adachi, M.M.; Labelle, A.J.; Thon, S.M.; Lan, X.; Hoogland, S.; Sargent, E.H. Broadband solar absorption enhancement via periodic nanostructuring of electrodes. *Sci. Rep.* **2013**, *3*. [CrossRef] [PubMed]
12. Fu, Y.; Dinku, A.G.; Hara, Y.; Miller, C.W.; Vrouwenvelder, K.T.; Lopez, R. Modeling photovoltaic performance in periodic patterned colloidal quantum dot solar cells. *Opt. Express* **2015**, *23*, A779–A790. [CrossRef] [PubMed]
13. Kim, J.; Koh, J.K.; Kim, B.; Kim, J.H.; Kim, E. Nanopatterning of mesoporous inorganic oxide films for efficient light harvesting of dye-sensitized solar cells. *Angew. Chem. Int. Ed.* **2012**, *51*, 6864–6869. [CrossRef] [PubMed]
14. Na, J.; Kim, Y.; Park, C.; Kim, E. Multi-layering of a nanopatterned TiO_2 layer for highly efficient solid-state solar cells. *NPG Asia Mater.* **2015**, *7*, E217. [CrossRef]
15. Atwater, H.A.; Polman, A. Plasmonics for improved photovoltaic devices. *Nat. Mater.* **2010**, *9*, 205–213. [CrossRef] [PubMed]
16. Ding, I.-K.; Zhu, J.; Cai, W.; Moon, S.-J.; Cai, N.; Wang, P.; Zakeeruddin, S.M.; Grätzel, M.; Brongersma, M.L.; Cui, Y.; *et al.* Plasmonic dye-sensitized solar cells. *Adv. Energy Mater.* **2011**, *1*, 52–57. [CrossRef]
17. Min, C.; Li, J.; Veronis, G.; Lee, J.-Y.; Fan, S.; Peumans, P. Enhancement of optical absorption in thin-film organic solar cells through the excitation of plasmonic modes in metallic gratings. *Appl. Phys. Lett.* **2010**, *96*, 133302. [CrossRef]
18. Raman, A.; Yu, Z.; Fan, S. Dielectric nanostructures for broadband light trapping in organic solar cells. *Opt. Express* **2011**, *19*, 19015–19026. [CrossRef] [PubMed]
19. Mahpeykar, S.M.; Xiong, Q.; Wang, X. Resonance-induced absorption enhancement in colloidal quantum dot solar cells using nanostructured electrodes. *Opt. Express* **2014**, *22*, 1576–1588. [CrossRef] [PubMed]
20. Hall, A.S.; Friesen, S.A.; Mallouk, T.E. Wafer-scale fabrication of plasmoniccrystals from patterned silicon templates prepared by nanospherelithography. *Nano Lett.* **2013**, *13*, 2623–2627. [CrossRef] [PubMed]
21. Yang, S.; Lapsley, M.I.; Cao, B.; Zhao, C.; Zhao, Y.; Hao, Q.; Kiraly, B.; Scott, J.; Li, W.; Wang, L.; *et al.* Large-scale fabrication of three-dimensional surface patterns using template-defined electrochemical deposition. *Adv. Funct. Mater.* **2013**, *23*, 720–730. [CrossRef]
22. Lumerical Solutions Inc. Grating Projections. Available online: http://docs.lumerical.com/en/solvers_grating_projections.html (accessed on 15 February 2016).

23. Yu, J.; Geng, C.; Zheng, L.; Ma, Z.; Tan, T.; Wang, X.; Yan, Q.; Shen, D. Preparation of high-quality colloidal mask for nanospherelithography by a combination of air/water interface self-assembly and solvent vapor annealing. *Langmuir* **2012**, *28*, 12681–12689. [CrossRef] [PubMed]
24. Synowicki, R.A. Spectroscopic ellipsometry characterization of indium tin oxide film microstructure and optical constants. *Thin Solid Films* **1998**, *313–314*, 394–397. [CrossRef]
25. Palik, E.D. *Handbook of Optical Constants of Solids*; Academic Press: Cambridge, MA, USA, 1998; pp. 12–24.
26. SOPRA Refractive Index Database. Available online: http://www.soprasteria.com (accessed on 15 February 2016).
27. Wang, X.; Koleilat, G.I.; Tang, J.; Liu, H.; Kramer, I.J.; Debnath, R.; Brzozowski, L.; Barkhouse, D.A.R.; Levina, L.; Hoogland, S.; *et al.* Tandem colloidal quantum dot solar cells employing a graded recombination layer. *Nat. Photon.* **2011**, *5*, 480–484. [CrossRef]
28. Reference Solar Spectral Irradiance: ASTM G-173. Available online: http://rredc.nrel.gov/solar/spectra/am1.5/ASTMG173/ASTMG173.html (accessed on 15 February 2016).

11

Enhancing Solar Cell Efficiency using Photon Upconversion Materials

Yunfei Shang [1], Shuwei Hao [1,2], Chunhui Yang [1,2,*] and Guanying Chen [1,3,*]

1 School of Chemical Engineering and Technology, Harbin Institute of Technology, Harbin 150001, China; shangyunfei@hit.edu.cn (Y.S.); haosw@hit.edu.cn (S.H.)
2 Harbin Huigong Technology Co., Ltd., Harbin 150001, China
3 Institute for Lasers, Photonics, and Biophotonics, University at Buffalo, State University of New York, Buffalo, NY 14260, USA
* Authors to whom correspondence should be addressed; yangchh@hit.edu.cn (G.Y.); chenguanying@hit.edu.cn or guanying@buffalo.edu (G.C.).

Academic Editor: Jiye Fang

Abstract: Photovoltaic cells are able to convert sunlight into electricity, providing enough of the most abundant and cleanest energy to cover our energy needs. However, the efficiency of current photovoltaics is significantly impeded by the transmission loss of sub-band-gap photons. Photon upconversion is a promising route to circumvent this problem by converting these transmitted sub-band-gap photons into above-band-gap light, where solar cells typically have high quantum efficiency. Here, we summarize recent progress on varying types of efficient upconversion materials as well as their outstanding uses in a series of solar cells, including silicon solar cells (crystalline and amorphous), gallium arsenide (GaAs) solar cells, dye-sensitized solar cells, and other types of solar cells. The challenge and prospect of upconversion materials for photovoltaic applications are also discussed.

Keywords: photovoltaic; upconversion; efficiency

---

1. Introduction

Fossil fuels (coal, oil, and natural gas) form the major energy source for meeting current human needs [1–4], yet cause a range of serious environmental issues. Moreover, the ever-growing consumption rate outpaces their regeneration rate, endangering the exhaustion of fossil fuels on earth. Dealing with the energy crisis is an urgent need [5,6]. Among all new competing energy sources (biomass, wind, hydroelectricity, geothermal energy, and nuclear energy), solar energy is considered to be the most abundant, renewable, and environment-friendly energy form [7,8]. The total solar power that strikes the Earth's surface is about 100,000 terawatts, which is 10,000 times more than that consumed globally [9,10]. If just 0.1% of the sunlight that reaches the Earth's surface could be converted by photovoltaic (PV) devices, with an average conversion efficiency of 10%, the amount would be sufficient to meet our current energy demands. The achievement of PV technology in recent decades has been tremendous [11–13]. The high-cost per kilowatt delivered by PV stations, however, limits its competitiveness with other sources. This is caused mainly by the inferior power conversion efficiencies of PV devices. The inability to absorb infrared (IR) light (700–2500 nm), which constitutes 52% of the energy of the entire solar spectrum, forms the major energy loss mechanism of conventional solar cells. This fundamental issue is set by the sizable bandgap of light-absorbing materials in PV devices. Crystalline silicon (c-Si) photovoltaic (PV) cells are the most used among all types of solar cells on the market, representing about 90% of the world's total PV cell production in 2008 [14]. However, even for single crystalline silicon (Si) PV cells with a rather small semiconductor band-gap

(1.12 eV, corresponding to a wavelength of ~1100 nm), the transmission loss of sub-band-gap photons can still amount to about 20% of the sun's energy irradiated onto the Earth's surface [15]. For PV cells with a larger band-gap, such as amorphous Si (1.75 eV) solar cells, which are limited to absorb sunlight with wavelengths below 708 nm, they manifest even higher near infrared transmission losses.

Photon upconversion (UC) provides a means to circumvent transmission loss by converting two sub-band-gap photons into one above-band-gap photon, where the PV cell has high light responsivity. This technology enables us to break the Shockley–Queisser limit of a single-junction PV cell (about 31% for non-concentrated sunlight irradiation for a semiconductor material with an optimized band-gap of around 1.35 eV) by transforming the solar spectrum [16]. Indeed, Trupke *et al.* demonstrated through a detailed balance model that modification of the solar spectrum with an up-converter could elevate the upper theoretical efficiency limit of a single-junction crystalline silicon PV cell to be as high as 40.2% under non-concentrated sunlight irradiation [17]. This value is far beyond the Shockley–Queisser limit for crystalline silicon solar cells (~1.1 eV bandgap) of approximately 30%. Figure 1 schematically illustrates the use of upconversion processes to convert the solar spectrum in the IR-Near IR (NIR)-short visible range into the peak (~500 nm) of sun radiation. There are three typical photon upconversion materials under investigation now outlined. (1) Rare-earth-doped micro- and nano-crystals (RED-UC), which usually work with wavelengths above 800 nm, but also possibly below 800 nm with appropriate composition design [18–24]. The abundant electronic states of trivalent lanthanide ions enable upconverting a range of IR wavelengths by selecting varied types of rare earth ions [25,26]. (2) Triplet–triplet annihilation upconversion (TTA-UC, response range $\lambda < 800$ nm) [27–29], whereby the triplet states of two organic molecules interact with each other, exciting one molecule to its emitting state to produce fluorescence. The excitation power density required for TTA UC is quite low, a few mW/cm^2, which is comparable to that of sun radiation. (3) Upconversion in Quantum Nanostructures (QN-UC, response range $\lambda < 800$ nm) [30–32]. This process leans on the use of a unique design comprising a compound semiconductor nanocrystal, which incorporates two quantum dots with different bandgaps separated by a tunneling barrier. Upconversion occurs by excitation of an electron in the lower energy transition, followed by intra-band absorption of the hole, allowing it to cross the barrier to a higher energy state. Detailed mechanisms of these three types of upconversion materials are discussed in Section 2. These upconversion materials are now emerging in use to improve PV efficiency in the real word.

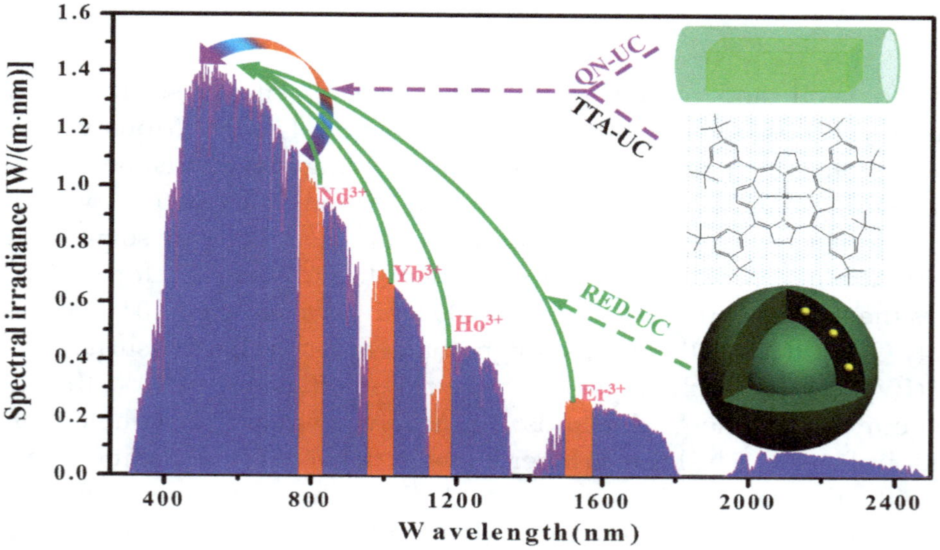

Figure 1. The absorption and emission range of three types of upconversion materials in reference to AM 1.5 spectrum (QN-UC (purple): upconversion in quantum nanostructures; TTA-UC (purple): triplet-triplet annihilation upconversion; RED-UC (green): Rare-earth-doped upconversion materials).

2. Upconversion Materials

2.1. *Rare-Earth-Doped Upconversion Materials*

The rare-earth family is comprised of 17 elements, which includes 15 lanthanide elements (from La to Lu) plus the elements of yttrium (Y) and scandium (Sc). The trivalent lanthanide ions possess a $4f^n5s^25p^6$ electronic structure with 14 available orbitals ($0 < n < 14$), offering 14 possible electronic group configurations. The quantum interaction of involved electrons endows lanthanide elements with abundant energy levels covering a spectral range of NIR, visible and ultraviolet (UV) [33–37]. In addition, the perfect shielding of $4f$ electrons by outer complete $5s$ and $5p$ shells enables electronic transitions to occur with limited influence from the surrounding environment, thus exhibiting high resistance to processes of photobleaching and photochemical degradation. As the symmetries of involved quantum states are identical, the intra-$4f$ electronic transitions of lanthanide ions are electric-dipole forbidden, yet can be relaxed due to local-crystal-field-induced intermixing of the f states with higher electronic configurations. The primary forbidden nature yields metastable energy levels of lanthanide ions (lifetime can be as long as tens of milliseconds), thus favoring the occurrence of sequential excitations in excited states of a single lanthanide ion as well as permitting favorable ion-ion interactions in excited states to allow energy transfers between two or more lanthanide ions.

There are four main basic mechanisms for rare-earth ions based upconversion processes, comprising of excited state absorption (ESA), energy transfer upconversion (ETU), photon avalanche (PA) and energy migration-mediated upconversion (EMU) (Figure 2). Excited state absorption (ESA) takes the form of successive absorption of pump photons by a single ion utilizing the ladder-like structure of a simple multi-level system as portrayed in Figure 2a. Figure 2a illustrates a simplified three level system for two sequential photon absorption processes. The ETU process involves at least two types of ions, namely a sensitizer and an activator. In this process, ion I known as the sensitizer is firstly excited from the ground state to its metastable level by absorbing a pump photon; it then successively transfers its harvested energy to the ground state and the first excited state of ion II, known as the activator, exciting ion II to its upper emitting state, which is followed by radiative decay to its ground state. The PA in Figure 2c is a looping process that involves an efficient cross relaxation mechanism between ion I in the ground state and ion II in the second excited state, resulting in generation of two ion IIs in the metastable state. The population of ion II in the second excited state is created through absorption of laser photons at its metastable state (the first excited state), which is initially populated through non-resonant weak ground state absorption. When the looping process ensues, an avalanche population of ion II will be created at its metastable state, producing avalanche upconverted luminescence from the emitting state. The generation of PA UC typically occurs above a certain threshold of excitation density. Below the threshold, very little up-converted fluorescence is produced, while the luminescence intensity increases by orders of magnitude above the pump threshold. In addition, the looping nature enables the evoked UC luminescence to be strongly dependent on the laser pump power, especially around the threshold laser power. The lanthanide ions designed for realizing energy migration-mediated upconversion (EMU) comprise four types: the sensitizers (type I), the accumulators (type II), the migrators (type III), and the activators (type IV; Figure 2d). A sensitizer ion is used to harvest excitation photons and subsequently promotes a neighboring accumulator ion to its excited states. A migrator ion extracts the excitation energy from high-lying energy states of the accumulator, followed by random energy hopping through the migrator ion sublattice and trapping of the migrating energy by an activator ion that produces luminescence by decaying to the ground state. The excitation density for RED-UC is typically in the range of 10^{-1}–10^2 W/cm^2.

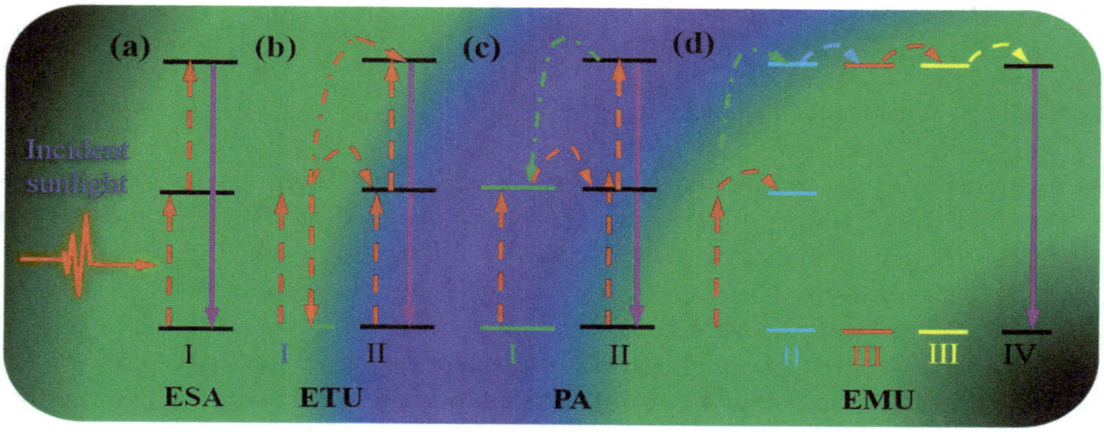

Figure 2. A schematic illustration of four typical upconversion processes: (**a**) Excited state absorption (ESA); (**b**) Energy transfer upconversion (ETU); (**c**) Photon avalanche (PA); (**d**) Energy migration-mediated upconversion (EMU).

Rare-earth-doped upconversion materials typically consist of an appropriate dielectric host matrix and doped Ln^{3+} ions that are dispersed as the guest in the lattice of the host matrix. Host materials with low phonon energy are able to produce upconversion luminescence at high efficiency, as multiphonon-assisted nonradiative relaxations between the closely spaced energy levels can be minimized, thus yielding increased lifetime of intermediate energy levels. Investigated low phonon energy host materials typically include fluorides, chlorides, iodides, and bromides [38], while high phonon energy host materials such as silicates, borates, and phosphates are also under study [39]. In general, host materials with low phonon energy are hygroscopic, while the high phonon energy ones are robust even under acute environment (strong acid, base, high temperature, *etc.*). Yet, the type of fluoride host material is unique and has attracted a lot of attentions in recent years. This is because fluoride host lattice not only has low phonon energy but also shows excellent chemical stability. In particular, hexagonal $NaYF_4$ lattice is considered to be one of the most efficient host materials to date [40–42]. Interestingly, even for cubic phase $NaYF_4$, a well-defined distribution of Na^+ and Y^{3+} ions in the crystal lattice can enable ultrahigh upconversion luminescence [43].

The Ln^{3+} dopants provide light harvesting ability as well as upconverting ability for RED UC. Among Ln^{3+} ions, the research for enhancing the efficiency of solar cells explores the use of single Ln^{3+} doping such as Er^{3+}, Ho^{3+}, Tm^{3+}, and Pr^{3+} to upconvert IR light. Meanwhile, a utilization of Yb^{3+} ions as co-dopants can provide new, strong absorption at ~980 nm ($^2F_{7/2} \rightarrow {}^2F_{5/2}$). The Yb^{3+} ions are able to sensitize most lanthanide activator ions, typically, Er^{3+}, Ho^{3+}, Tm^{3+}, resulting in intense upconversion when excited. Recently, Nd^{3+} ion has been proposed as another extraordinary sensitizer. It provides a new absorption at 808 nm and is able to sensitize the Yb^{3+} ion. This further broadens the absorption range of RED upconversion for PV application. Table 1 summarizes some selected upconverting materials which can be excited, utilizing light with wavelength longer than 800 nm. (1) Single Er^{3+} doped UC materials can convert light at 1523 nm to green (550 nm) and red (650 nm) (Figure 3a) [44–47]; (2) Yb^{3+}/Er^{3+} codoped system can produce green (525 nm, 542 nm), red (655 nm), as well as purple (415 nm, weak) emissions under 980 nm laser excitation [42,48–53]; (3) Yb^{3+}/Tm^{3+} pairs are able to convert light at 980 nm into UV (345 nm), blue (480 nm) and NIR (800 nm) emissions [54–58]; (4) $Nd^{3+}/Yb^{3+}/Ln^{3+}$ (Ln = Er, Tm, Ho, *etc.*) tri-doped UC materials can convert 808 nm NIR light to visible luminescence [57–62]. As an example, the energy transfer processes for Nd^{3+}, Yb^{3+}, and Er^{3+} are depicted in Figure 3b for illustration. The Er^{3+} ion emits in the green and red range after absorbing the excitation energy by either the Nd^{3+} or the Yb^{3+} ions. Simultaneous use of Nd^{3+}, Yb^{3+} and Er^{3+} ions enables light harvesting at ~800 nm, ~980 nm, as well as ~1523 nm, covering broader spectral range for upconversion.

Table 1. Some selected upconverters excitable with wavelength longer than 800 nm.

Dopant Ion	Host Material	Excitation (nm)	Emission (nm)	References
Er^{3+}	$NaYF_4$	1523	550, 660, 800, 980	[63]
Yb^{3+}-Er^{3+}	$NaYbF_4$	980	520, 540, 654	[64]
Yb^{3+}-Er^{3+}	$NaYF_4$	980	410, 522, 540, 650	[65–67]
Yb^{3+}-Er^{3+}	$NaYF_4$@$NaYF_4$	980	510~570, 640~680	[68]
Yb^{3+}-Er^{3+}	$NaYF_4$@$NaYF_4$:Nd^{3+}	808/980	520, 540, 655	[69]
Yb^{3+}-Tm^{3+}	$NaYF_4$	980	375, 450, 475, 679, 800	[70,71]
Yb^{3+}-Ho^{3+}	$NaYF_4$	980	545, 650	[72]
Yb^{3+}-Er^{3+}-Nd^{3+}	$NaYF_4$	808/980	410, 520, 545, 650	[60]

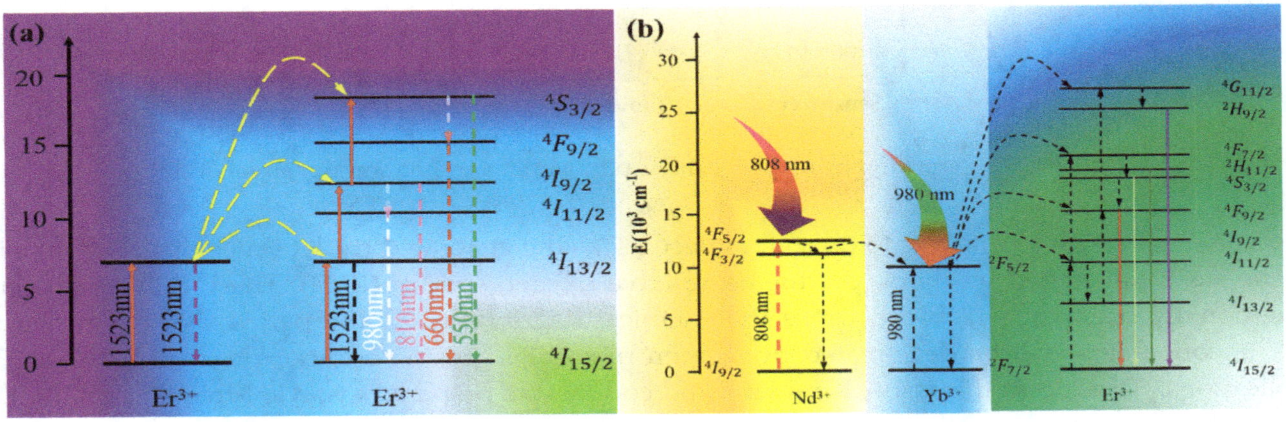

Figure 3. The upconversion mechanisms for (**a**) single Er^{3+} doped upconversion particles (UCNPs) excited under 1523 nm laser exciation; (**b**) Nd^{3+}-Yb^{3+}-Er^{3+} tridoped system under 808 nm or 980 nm laser exciation.

2.2. *Triplet-Triplet Annihilation Upconversion Materials*

Triplet-triplet annihilation (TTA) up-conversion is a process occurring in a pair of sensitizer-annihilator dyes, as portrayed in Figure 4. Light harvesting is enabled by the sensitizer, followed by populating its singlet excited state ($S_0 \rightarrow S_1$). The intersystem crossing process (ISC, $S_1 \rightarrow T_1$) allows the triplet excited state of the sensitizer (organo-metallic type) to be efficiently populated thanks to the heavy atom effect of the transition metal atom. Subsequently, a transfer of the energy in the triplet to a neighboring acceptor (annihilator) at S_0 state *via* a Dexter energy transfer (DET) process, can excite the acceptor (annihilator) to its triplet state. Two nearby acceptor (annihilator) molecules in the triplet states collide with each other, and results in one acceptor molecule being excited to the higher singlet state (S_1), while the other one returns to the ground singlet state (S_0). A radiative decay from the generated singlet excited state of the acceptor produces an upconverted fluorescence, which is called triplet-triplet annihilation (TTA) upconversion. Furthermore, the wavelength of light harvesting can be varied by selecting the sensitizer, while the emission wavelength of TTA up-conversion can be tuned by selection of the triplet acceptor (annihilator). Since sun radiation can be utilized in a direct way to induce TTA UC, this type of upconversion is promising for applications in PV solar cells. Yet, the inability to produce efficient TTA UC in long visible as well as in the IR range, limits its impact on PV technology. To address this challenge, Kimizuka *et al.* [73] and Bardeen *et al.* [74] extended the absorption wavelength beyond 850 nm by using metallonaphthalocyanines and semiconductor nanocrystals as triplet sensitizers. However, the upconversion quantum efficiency of TTA-UC remains rather low. To address the low efficiency problem, F. Meinardi and co-workers [75] combined organic TTA with fluorescent semiconductor nanocrystals that not only broaden the absorption but also lower the excitation intensity. The same group [76] synthesized a series of sensitizers with tuned absorption and achieved a conversion efficiency of 10% under broadband

AM 1.5 irradiation. Kimizuka *et al.* [77] demonstrated a novel metal-organic framework, in which donor-acceptor position could be precisely tailored to achieve a high UC efficiency even under a weak excitation (less than that of sunlight).

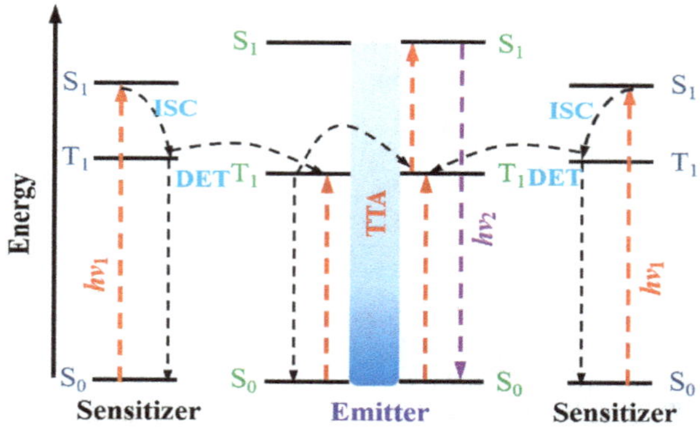

Figure 4. The working principle of the triplet-triplet annihilation upconversion.

TTA UC is highly sensitive to oxygen molecules, as the generated triplet states of TTA dye pairs can be easily quenched by them, producing reactive oxygen species (ROS). These ROS species are exceptionally reactive, which can cause the chemical degradation of the paired dyes. As a result, TTA upconversion will be deactivated by oxygen molecules, posing another problem for their PV applications. To address this issue, Kimizuka and co-workers prepared a solvent-free liquid TTA UC system that can function well even under aerated conditions [78]. Subsequently, they reported a series of air-stable TTA upconversion in supramolecular organogel [79], supramolecular self-assemblies [80], and ionic liquids [81], providing a new approach to solve the problem of the sensitivity to oxygen molecules. Alternatively, Weder *et al.* demonstrated TTA upconversion in molecular glasses and organogels where oxygen-induced upconversion fluorescence quenching could also be prevented [82,83].

2.3. Upconversion in Quantum Nanostructures

Upconversion in quantum nanostructures is realized through a design of a compound semiconductor nanocrystal, which incorporates two quantum dots with different bandgaps separated and connected by a tunneling barrier (a semiconductor rod). The implementation of upconversion is through a two-step absorption of two subsequent photons. The first photon generates an electron-hole pair via interband absorption in the lower-energy core (small band-gap dots), leaving a confined hole and a delocalized electron in the compound semiconductor nanocrystal. The second absorbed photon can lead, either directly or indirectly, to further excitation of the hole, enabling it to cross the barrier layer. The direct way is through an intraband absorption of the photon by the confined hole at the lower-energy core (Figure 5a), while the indirect way is via an Auger mediated energy transfer process (Figure 5b).The Auger process describes a recombination of the second electron-hole pair, generated by absorbing the second photon, while simultaneously allowing nonradiative energy transfer to the confined hole at the lower energy core (small band-gap dots), empowering it to cross the barrier to the higher energy quantum dot (large band-gap dots). This, in turn, is followed by a radiative recombination with the delocalized electron, producing upconverted luminescence. This system combines the stability of an inorganic crystalline structure, with the spectral tunability afforded by quantum confinement. Since the absorption, emission, and lifetime of the semiconductor nanocrystals can be controlled by variation of their size, shape, as well as composition, upconversion in quantum nanostructures holds prime promise for applications in PV devices. However, a relative high excitation density (~10^4 W/cm^2) is needed to activate this type of upconversion. Lowering the excitation density to the range of sun irradiation (~10^{-1} W/cm^2) is an inviting direction, which could produce a pronounced impact for PV applications in a straightforward way.

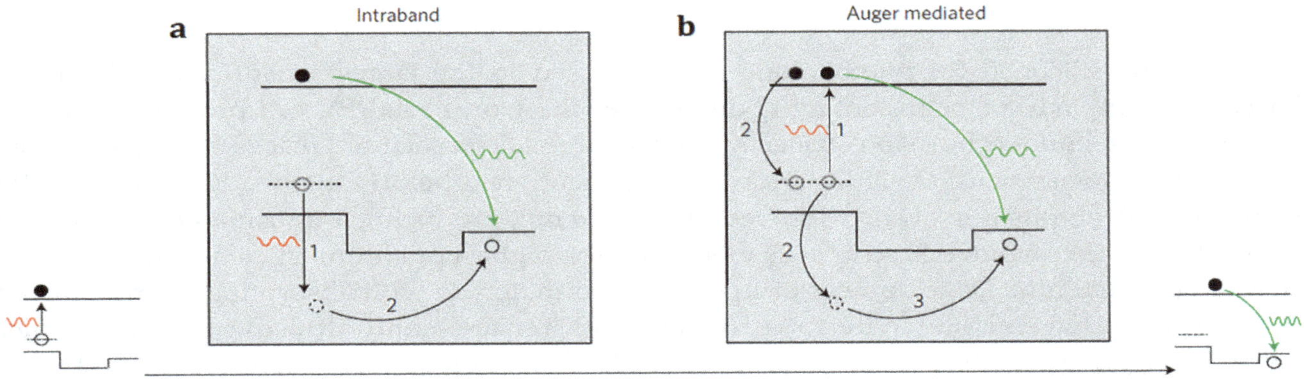

Figure 5. Mechanism of the upconversion in quantum nanostructures: (**a**) Direct intraband hole absorption mechanism of upconversion; (**b**) Auger-mediated upconversion (adapted with permission from [30]; Copyright Nature Publishing Group, 2013).

3. PV Applications

The solar cell, which directly converts solar energy into electricity, is one of the most attractive solutions to the growing energy demands, due to the omnipresence and abundance of solar energy. Light harvesting is the first but also the most important step, which determines how much sun irradiation can be absorbed. The standard solar irradiation spectrum (AM 1.5) covers the wavelength region from UV to IR (300–2500 nm). However, most single-junction PV devices are unable to absorb light quanta in the long wavelength range, typically NIR and IR, creating the performance-limiting issue of transmission loss. Upconversion materials are in play to circumvent this issue by spectral conversion, which has been applied to a wide range of solar cells. These PV devices include c-Si, amorphous Si thin-film, GaAs and dye sensitized solar cells (DSSCs). Two typical configurations have been exploited for applications in PV cells, as portrayed in Figure 6, to upconvert sub-bandgap light into above-bandgap luminescence. Structure One involves the use of a reflection layer to increase the optical path of NIR or IR light within the upconverting layer for an increase of the upconversion luminescence output, while Structure Two does not use any reflection layer, typically employed by DSSCs.

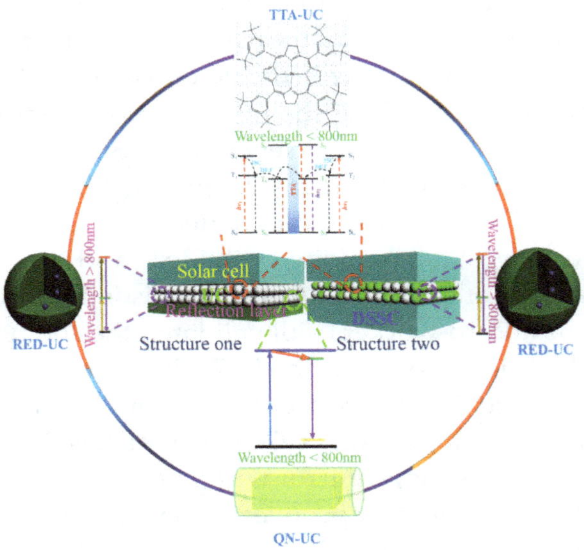

Figure 6. A schematic illustration of the usage of three typical upconversion materials in solar cells using two configurations. (Rare-earth upconversion, RED-UC; Upconversion in quantum nanostructure, QN-UC; Triplet-triplet annihilation upconversion, TTA-UC).

3.1. c-Si Solar Cells

Crystalline silicon (c-Si) photovoltaic cells are used in the largest quantity of all types of solar cells on the market, representing about 90% of the world total PV cell production in 2008. The recorded maximum conversion efficiency of crystalline silicon solar has reached 25% [84], close to the Schockley–Queisser limit (~30%). However, as mentioned before, the bandgap of crystalline silicon (~1.12 eV) limits it to absorb light less than 1100 nm, producing a transmission loss as high as 20%. On the other hand, c-Si solar cells work with the highest quantum efficiency in the spectral region of 800–1100 nm. Development of upconversion materials with absorption below 1100 nm, and the upconverted emission in the range of ~800–1100 nm are highly attractive for improvement of the efficiency of a c-Si PV cell. Yb^{3+}-sensitized or Nd^{3+}-sensitized RED UC materials are beyond consideration for use in c-Si PV cells, as they provide absorption at ~980 nm and ~800 nm where the semiconductor silicon has existing high spectral responses.

Single Er^{3+}-doped RED UC materials are attractive for c-Si solar cells, as they display absorption at 145–1580 nm with sub-bandgap energy of silicon, and emit strong luminescence at ~980 nm, ~540 nm, and 650 nm that can be very useful for silicon to produce excitons. In 2005, Shalav *et al.* [63] exploited the use of $NaYF_4$:20% Er^{3+} microcrystals as an upconverter in bifacial c-Si solar cells. In their geometric arrangement of implementation of upconversion, microsized $NaYF_4$:20% Er^{3+} phosphors were dispersed into an acrylic adhesive medium with matched refractive index, and then this mixture was deposited as a thick film on the rear of a bifacial c-Si solar cell. An impressive increase of external quantum efficiency of 2.5% was accomplished under excitation at 1523 nm. Fischer *et al.* [47] also investigated the potential use of microsized or bulk $NaYF_4$:20% Er^{3+} materials (~3 μm) to improve the conversion performance of c-Si solar cells. The upconversion efficiency of $NaYF_4$:20% Er^{3+} microcrystals was quantified to be 5.1% when irradiated with 1523 nm laser with a power density of 1880 W/m^2. Shalav *et al.* showed that a c-Si solar cell device can have an external quantum efficiency of 0.34% under sub-bandgap irradiation at 1522 nm with a power density of 1090 W/m^2.

Alongside Er^{3+} ions, the Ho^{3+} ions provide a possibility to harvest sub-bandgap energy of silicon at a new wavelength range (1150–1230 nm). The centroids of upconverted luminescence of Ho^{3+} ions are at ~910 nm (corresponding to the $^5I_5 \rightarrow {}^5I_8$ transition) in the NIR range, and at ~650 nm (corresponding to the $^5F_5 \rightarrow {}^5I_8$ transition) in the visible range. Both radiations have energy above the bandgap of silicon. It should be noted that the intensity of sun radiation at 1170 nm is approximately twice than that at 1520 nm, potentially delivering more effective improvement of solar cell efficiency. Lahoz *et al.* [85] first reported on the use of single Ho^{3+} ion doped upconverting glass ceramics in a c-Si solar cell. They successfully demonstrated that the c-Si solar cell responds to light irradiation at ~1170 nm due to the upconverted visible emission (650 nm) and NIR emission (910 nm) in the glass ceramics. In another work of theirs, Ho^{3+}-Yb^{3+} codoped upconversion materials were investigated to improve the conversion efficiency of a c-Si solar cell [86]. They found that the Ho^{3+}-Yb^{3+} codoped upconverter produced much stronger NIR emission intensity than that of the single Ho^{3+}-doped counterpart. This is because the excited Ho^{3+} ions (at the $^5I_5$ state) can sensitize Yb^{3+} ions (at the ground state $^2F_{7/2}$ state), making both Ho^{3+} and Yb^{3+} luminescence. This, in turn, result in an improved NIR response of Si solar cells in comparison to the one using single Ho^{3+}-doped glass ceramics. Moreover, Lahoz and co-workers [86] validated the combined use of single-Er^{3+} doped RED UC material and single Ho^{3+} doped oxyfluoride RED UC material to encompass broader sunlight harvesting in the NIR range (Figure 7). This concept was demonstrated by placing the Er^{3+} doped RED UC layer to the back of the Ho^{3+} doped RED UC layer, which is then attached to the back of a c-Si solar cell. To enhance NIR light harvesting, a mirror was placed at the bottom of the cell to reflect the unabsorbed sub-bandgap sunlight back to the upconverting layers. From a material development point of view, Chen *et al.* [87] prepared a core-shell-shell structure of $NaGdF_4$:Er^{3+}@$NaGdF_4$:Ho^{3+}@$NaGdF_4$ nanocrystals, in which the Er^{3+} and the Ho^{3+} ions were separately doped into the core and the first shell layer (Figure 8). Intense UC emissions from both Er^{3+} and Ho^{3+} were shown in the same core-shell-shell nanoparticle. The intensities of luminescence bands from both ions are all much stronger than that from Er^{3+}/Ho^{3+}-codoped nanocrystals due to a spatial isolation of Er^{3+} and Ho^{3+} ions, which can avoid the detrimental cross relaxation between these ions. However, because of high doping concentration of Er^{3+} and Ho^{3+}, adverse relaxations could still induce quenching beyond the interfaces, even though the dopants were in different layers. To circumvent this, a multi-layer core/shell design of $NaYF_4$:10%Er^{3+}@$NaYF_4$@$NaYF_4$:10% Ho^{3+}@$NaYF_4$@$NaYF_4$:1% Tm^{3+}@$NaYF_4$ nanoparticles were reported by us [88], in which the inert $NaYF_4$ layers in between upconverting domains were utilized to efficiently suppress the detrimental cross-relaxation processes between different types of lanthanide ions, yielding about two times more efficient upconversion photoluminescence than the counterpart $NaYF_4$:10% Er^{3+}@$NaYF_4$:10% Ho^{3+}@$NaYF_4$:1% Tm^{3+}@$NaYF_4$ without the inert $NaYF_4$ layers. Moreover, these core/multishell nanoparticles can be excited at ~1120–1190 nm (due to Ho^{3+}), ~1190–1260 nm (due to Tm^{3+}), and ~1450–1580 nm (due to Er^{3+}), collectively covering a broad spectral range of ~270 nm in the infrared range that can be very useful for infrared photosensitization of c-Si solar cells.

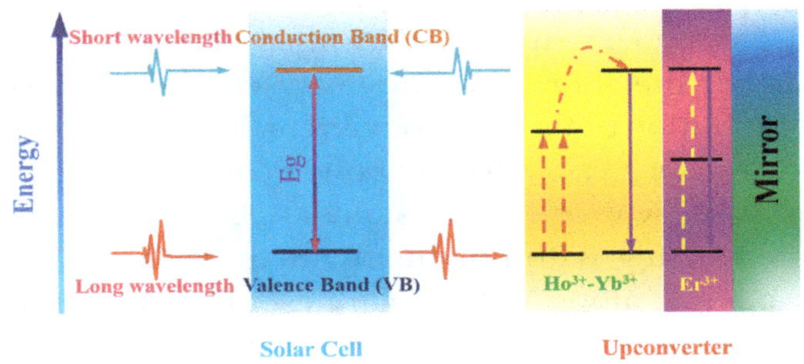

Figure 7. An operating mechanism for a c-Si solar cell with two upconverters, one co-doped with Ho^{3+}–Yb^{3+} and the other one single doped with Er^{3+}. Photons with short wavelength can be absorbed directly by the solar cell. The transmitted sub-bandgap light can be upconverted into high-energy photons, which would be absorbed by c-Si. The mirror behind the upconverter increases the probability of absorption of sub-bandgap light in the upconverter layer.

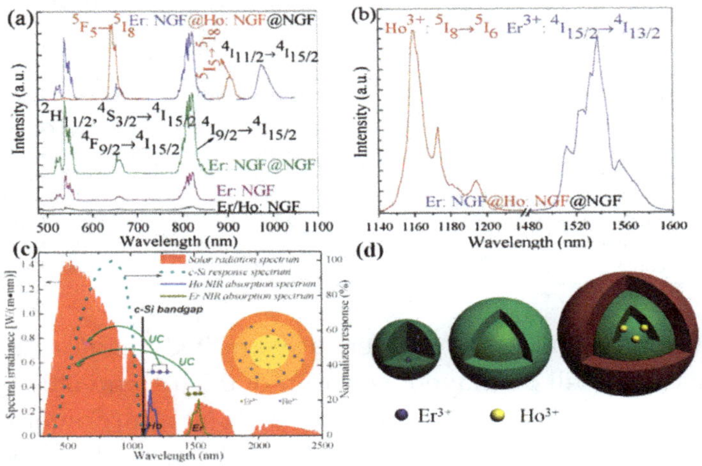

Figure 8. (**a**) Upconversion emission spectra of the core $NaGdF_4$:Er^{3+}/Ho^{3+} nanocrystals, the core $NaGdF_4$:Er^{3+} nanocrystals, the core/shell $NaGdF_4$:Er^{3+}@$NaGdF_4$ nanocrystals, and the $NaGdF_4$:Er^{3+}@$NaGdF_4$:Ho^{3+}@$NaGdF_4$ core/shell/shell nanocrystals; (**b**) Upconversion excitation spectrum of the $NaGdF_4$:Er^{3+}@$NaGdF_4$:Ho^{3+}@$NaGdF_4$ core/shell/shell nanocrystals; (**c**) A schematic illustration of the upconversion luminescence of $NaGdF_4$:Er^{3+}@$NaGdF_4$:Ho^{3+}@$NaGdF_4$ nanocrystals *versus* the bandgap of c-for Si solar cell, and the spectrum of AM 1.5 sun irradiation; (**d**) A schematic of illustration of different nanostructures: the $NaGdF_4$:Er^{3+} core, the $NaGdF_4$:Er^{3+}@$NaGdF_4$ core/shell, and the $NaGdF_4$:Er^{3+}@$NaGdF_4$:Ho^{3+}@$NaGdF_4$ core/shell/shell (adapted with permission from [87]; Copyright Royal Society of Chemistry, 2012).

In addition, Alexander Dobrovolsky and co-workers [89] designed a system based on GaNP nanowires that can harvest IR light through a two-step two-photon absorption process and emits visible light. Though it is claimed to be suitable for application in third-generation Si-based solar cells, no demonstrations have yet been shown.

3.2. Amorphous Silicon Solar Cells

Amorphous Si solar cells have been considered as a promising substitute of c-Si cells due to their low cost, easy preparation, and excellent chemical stability. The band-gap of amorphous Si is larger than that of c-Si at about 1.75 eV, which confines it to absorb NIR light shorter than 700 nm. This means that upconverting materials with absorption above 700 nm should be appealing for uses in amorphous Si solar cells to circumvent the transmission loss.

Table 2 lists a range of investigated and selected single/co-doped RED UC materials in the literature that can be utilized towards this purpose. Zhang *et al.* [90] investigated the use of $NaYF_4$:18% Yb^{3+}, 2% Er^{3+} nanocrystals as upconverter in an amorphous Si solar cell. This kind of upconverter shows visible emissions at around 655 nm (red), 525 nm and 540 nm (green) after absorbing light at 980 nm. They showed that the short circuit density can be increased by 6.25% (from 16 to 17 $mA \cdot cm^{-2}$) due to the contribution of these upconverters. In analogy, De Wild *et al.* [91,92] utilized β-$NaYF_4$:Yb^{3+}/Er^{3+} as upconverter to enhance the power conversion efficiency of an amorphous Si solar cell. The upconverter layer with a thickness of 200–300 μm was deposited at the rear of the amorphous Si solar cell after mixing with polymethylmethacrylate. The maximum photocurrent was increased from 2.1 to 6.2 μA, along with a manifestation of an external quantum efficiency of 0.02% at 980 nm. Yb^{3+}/Er^{3+}-codoped TeO_2-PbF_2 oxyfluoride tellurite glass [93] as well as $Gd_2(MoO_4)_3$:Er^{3+}/Yb^{3+} UC materials [94] have also been investigated in a similar way by applying them at the back of amorphous silicon cells. The broader NIR light harvesting ability, due to the effect of host matrix, was supposed to yield a significant improvement in the solar efficiency. Unfortunately, there was only a tiny improvement obtained when co-excited by AM 1.5 and 980 nm laser radiation. The underlying mechanism remains unclear, and deserves further investigations.

Table 2. Typical dopant ions, the main emissions, and the corresponding energy transitions for upconversion nanocrystals reported in the literature.

Dopant Ion	Host Material	Excitation (nm)	Emission (nm)	Reference
Er^{3+}	$NaYF_4$	1523	550, 660, 800, 980	[63]
Er^{3+}	YF_3	1490	410, 530, 550, 660, 810, 980	[95]
Er^{3+}	CaF_2	1510	410, 550, 660, 980	[96]
Er^{3+}	Y_2O_3	1538	562, 659, 801, 987	[97]
Er^{3+}	Fluoride glasses	1538	550, 660, 820, 980	[98]
Er^{3+}	$BaCl_2$	1535	410, 550, 660, 980	[99]
Ho^{3+}	Glass ceramics containing PbF_2 nanocrystals	1170	650, 910	[85]

In addition, Er^{3+} doped β-$NaYF_4$ powders were also applied to amorphous Si solar cells by Chen and co-workers [100]. This solar cell displayed a current of 0.3 μA and 0.01 μA when irradiated with a laser at 980 nm (power density of 60 mW/cm^2) and at 1560 nm (power density of 80 mW/cm^2), respectively. An enhanced current of 0.54 μA in solar cells was observed when irradiated with 980 nm (60 mW/cm^2) and 1560 nm (100 mW/cm^2) simultaneously. This was associated with a broadened NIR response due to simultaneous absorption at both ~980 nm and ~1560 nm. UC materials with broad band NIR absorption are more effective to circumvent transmission losses. In this way, TTA upconversion is in play which exploits the strong and broad absorption of dye molecules. Recently, a TTA-upconverter was investigated in hydrogenated amorphous silicon (a-Si:H) solar cells, producing a high efficiency of 10.1% with an absorption threshold of 700 nm [84]. Cheng and co-workers carried out a great deal of research work on optimizing the design of amorphous silicon solar cells combined with TTA-UC materials [101–107]. For example, an amorphous silicon solar cell was shown to achieve an obvious increase of current of $2.40 \times 10^{-3} \cdot mA/cm^2$, on equipping with a hybrid-emitter TTA system (namely, a filled cuvette with Ag-coated glass beads) as a back-reflecting medium. The use of TTA-UC to increase the light harvest efficiency of amorphous Si solar cells is still underway.

3.3. *GaAs Solar Cells*

Gibart *et al.* in 1995 firstly reported an application of RED UC material on a GaAs solar cell, demonstrating that it is useful to enhance the solar cell efficiency by improving the harvest of unavailable IR light [108]. In their subsequent experiment [108], a substrate-free GaAs solar cell (a band-gap of 1.43 eV) was coupled to a 100 μm thick vitroceramic upconverting materials which contained Yb^{3+} and Er^{3+} ions. The efficiency of the GaAs solar cell was increased quadratically with the power of excitation due to the nonlinear nature of the upconversion process. A power conversion efficiency as high as 2.5% was achieved under 891 nm (1.391 eV) illumination with an irradiance of 25.6 W/cm^2. In a similar way, Lin *et al.* [109] adhered a 300 μm thick UC phosphor layer of $Y_{5.86}W_2O_{15}$:0.05Yb^{3+}, 0.09Er^{3+} to the rear of a GaAs solar cell. The maximum output power of 0.339×10^{-6} W was obtained with a 973 nm laser irradiation at 145.65 W/cm^2. These results clearly indicate that GaAs solar cells can work effectively under sub-bandgap light irradiance. However, the involved light irradiances are much higher than that of sun radiance (10^{-2}–10^{-1} W/cm^2). Development of efficient UC materials under low irradiance is required.

3.4. *Dye-Sensitized Solar Cells*

Dye-sensitized solar cells (DSSCs) are third-generation PV cells, which have brought revolutionary innovation to PV technology since the first report by Grätzel in 1991 [110]. Owing to their low cost, simple fabrication methodology, environmental friendliness, as well as flexible structure, DSSCs have been considered as the most promising alternative for Si-based solar cells [111]. However, due to the limited absorption range of investigated dyes, the improvement of the conversion efficiency of a DSSC is always challenging [112–115]. The band-gap of typically used dyes in DSSC, such as N3, N719, and N749, *etc.*, is usually higher than 1.8 eV. This means that they are only able to absorb photons with wavelengths shorter than 700 nm, leading to unharvesting of approximately 52% of the solar energy in the IR range (from 700 to 2500 nm) [116–118]. Efforts to develop panchromatic sensitizers for DSSCs have been hampered by poor electron injection efficiency and competing charge recombination when the absorption spectrum of the sensitizer is extended to the IR region, adversely affecting the efficiency. Moreover, the photostability of an IR absorbing dye is known to be very poor [119,120]. Photon upconversion provides an alternative approach by converting unavailable IR photons into high-energy photons that can be absorbed by the sensitizing dyes with a typical structure as presented in Figure 9.

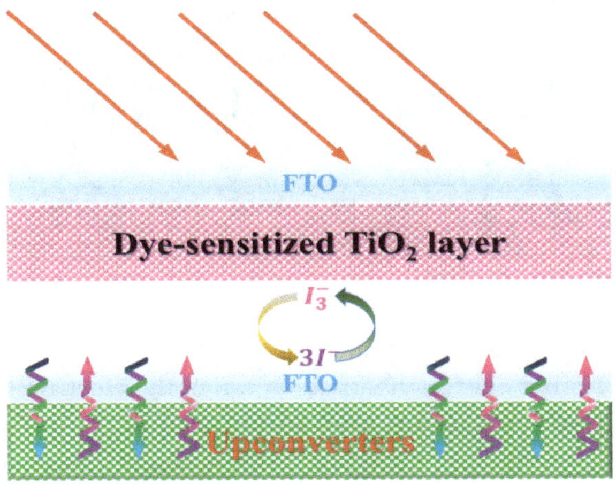

Figure 9. A typical schematic configuration of a DSSC equipped with upconverters.

The first DSSC device, based on rare-earth-doped UC materials, was presented by Demopoulos *et al.*, in which LaF_3:Yb^{3+}/Er^{3+}-TiO_2 nanocomposites were used to form a multilayer electrode structure [121]. As a proof of concept demonstration, the response of DSSC at NIR light (~980 nm) was shown. Later, they deposited a layer of microsized β-$NaYF_4$:Yb^{3+}/Er^{3+} particles on the rear side of a counter electrode [122], which was able to provide both light reflection and NIR light harvesting without apparent charge recombination at interfaces. An improvement of NIR response was accomplished. In parallel, Yuan *et al.* [123] in 2012 proposed and validated the use of colloidal β-$NaYF_4$:2% Er^{3+}/20% Yb^{3+} nanoparticles in a DSSC. As the size is less than 20 nm, they are allowed to diffuse in the mesoporous TiO_2 layer of the DSSC, enabling them to efficiently interact with sensitization dyes. The advantage of this approach is that these nanoparticles can be utilized in the way of using sensitizing dyes without modifications of device preparation procedure. Following that, Yang and co-workers attached YF_3:Yb^{3+}/Er^{3+} particles to the surface of the porous TiO_2 thin film, the power conversion efficiency was increased from 5.18% (blank DSSC without RED UC materials) to 6.76% [124].

Researchers also placed a lot of effort in the application of RED-UC structures in photoanode layers. Wang *et al.* [125] achieved an efficiency increase of 23% when doping YOF:Yb^{3+}/Er^{3+} particles into the TiO_2 photoanode layer. Zhang *et al.* exploited the use of $NaYF_4$:Er^{3+}/Yb^{3+}@TiO_2 core-shell composite as a photoanode. The conversion efficiency was enhanced by a factor of 1.23 [126]. To avoid electron recombination losses, Zhao *et al.* separated the upconversion and TiO_2 layers by growing the middle SiO_2 layer between them, achieving an improvement of 29.4% in efficiency [127]. In another way, Demopoulos *et al.* [128] investigated the utilization of β-$NaYF_4$:Yb^{3+}/Er^{3+}@TiO_2 submicro particles as both a light harvesting and an IR energy relay layer. An optimization of the layer of the DSSC structure resulted in a 16% relative increase of power conversion efficiency. On the other hand, there are a number of studies using a range of RED UC materials doped with Yb^{3+}-Er^{3+} [68,129,130], Yb^{3+}-Tm^{3+} [29], Er^{3+} [131], for improving the NIR response of DSSCs, and the following illustration (Figure 10) presents the energy transfer mechanisms in a DSSC equipped with a upconverter codoped with Yb^{3+} and Er^{3+}.

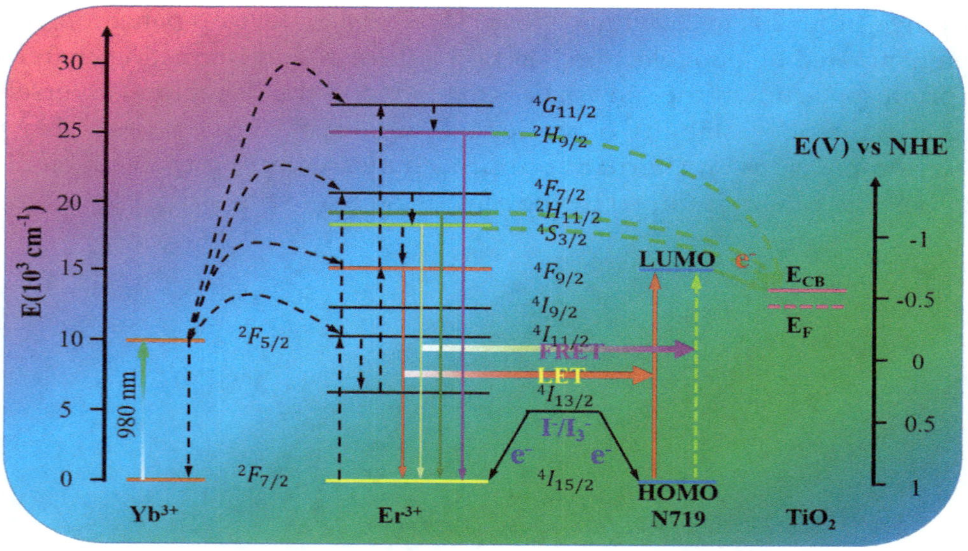

Figure 10. A cartoon illustration of energy transfer mechanisms in upconversion nanoparticles (UCNPs)-Doped DSSCs. This figure shows the absorption and conversion of near infrared (NIR) photons into visible light of higher energy via upconversion; the electron transfer from UCNPs to the conduction band of TiO_2, as well as the electron transfer from I^-/I_3^- to the ground state of UCNPs. FRET and LET are fluorescence resonance energy transfer and luminescence-mediated energy transfer, respectively. HOMO and LUMO are highest occupied molecular orbital and lowest unoccupied molecular orbital, respectively.

Narrow absorption and high excitation threshold of RED-UC are two main limits for efficiency improvement in DSSCs. To mitigate this, Bai and co-workers [132] enhanced light harvesting by 14% due to the usage of CeO_2:Er^{3+}/Yb^{3+} nanofibers that exploited host effect to broaden the light absorption ability. Chen *et al.* [133] provided a method to broaden the NIR absorption of upconversion nanoparticles by utilization of a hexagonal core-shell structure of β-$NaYbF_4$:2% Er^{3+}@$NaYF_4$:30% Nd^{3+}. This allows simultaneous use of the absorption of Nd^{3+} (~800 nm), Yb^{3+} (~980 nm), and possibly Er^{3+} (~1550 nm) ions for light harvesting, while the core/shell structure enables a spatial isolation of Nd^{3+} from Yb^{3+} and Er^{3+} ions to avoid detrimental cross relaxations to entail a high upconversion efficiency. Zhao *et al.* [134] achieved an efficiency of 8.32% (with a noticeable enhancement of 14.78%) by putting core-shell-structured β-$NaYF_4$:Yb^{3+}/Er^{3+}@SiO_2@Au nanocomposites on top of a mesoporous TiO_2 layer. Yang's group [124,135,136] also did a lot research on enhancing the performance of DSSCs using different lanthanide ion doped nanoparticles. For example, the design of Ho^{3+}-Yb^{3+}-F^- tri-doped TiO_2 nanoparticles enabled a 37% improvement in the power conversion efficiency. Despite recent achievements, improvement of DSSCs efficiency is still limited by the upconversion efficiency as well as the narrow and low absorption ability of RED-UC. Overcoming these two problems would lead to an epidemic use of RED-UC in DSSCs.

3.5. Other Types of Solar Cells

Apart from the solar cells discussed above, organic solar cells are considered to be one of the most promising alternatives for Si-based solar cells, due to their advantages of being flexible, low-cost, light-weight, of simple fabrication and large-scale production for the PV industry [137–141]. However, limited by the spectral mismatch of the absorption of organic molecules with sun irradiation, the improvement in efficiency remains a daunting task. Bulk heterojunction-based PV cells with a large bandgap of organic molecules are only able to absorb visible sunlight. Currently, some organic solar cells have been designed to harvest 800–900 nm sunlight by using low-band-gap polymer materials such as PCPDTBT (poly-[*N*-9′-heptadecanyl-2,7-carbazole-alt-5,5(4′,7′-di-2-thienyl-2′,1′,3′-benzothiadia-zole]) and its derivatives [18,142–144]. The poly (3-hexylthiophene) (P3HT) and the fullerenederivative [6,6]-phenyl-C61-butyric acid methyl ester (PCBM) have been exploited to increase the optical response of organic solar cells to 650–700 nm [145].

Upconversion materials have been dedicated to further extend the NIR spectral response. Wang and co-workers utilized LaF_3:Yb^{3+}/Er^{3+} phosphors to improve the NIR response of P3HT:PCBM organic solar cells. An upconverted photocurrent density of ~16.5 $\mu A/cm^2$ was obtained under an excitation density of 25 mW/cm^2 at 975 nm [146]. In another work of theirs, they demonstrated that the UC material (MoO_3:Yb^{3+}/Er^{3+}), incorporated into P3HT:PCBM organic solar cells, contributed about 1% to the improvement of short-circuit current under one-sun (AM 1.5) illumination [147]. Wu *et al.* [148] in 2012 added $NaYF_4$:Yb^{3+}, Er^{3+} nanoparticles on the rear of a P3HT:PCBM organic solar cell, the short-circuit current was enhanced by 0.5 μA when illuminated by a 980 nm light.

To utilize the NIR solar spectrum more efficiently, Adikaari *et al.* [149] reported an application of Y_2BaZnO_5:Yb^{3+}, Ho^{3+} UC particles in PCDTBT:PCBM organic solar cells. Two different layout designs are presented in Figure 11. The PCDTBT:PCBM active layer absorbs photons of wavelength shorter than 700 nm, while Y_2BaZnO_5:Yb^{3+}/Ho^{3+} shows an intense absorption in the NIR region of 870–1030 nm due to the $^2F_{7/2} \rightarrow {}^2F_{5/2}$ (Yb^{3+}) transition. Moreover, the UC emission peak at 545 nm from Ho^{3+}, corresponding to the $^5S_2/^5F_4 \rightarrow {}^5I_8$ transition, matches well with the absorption band of PCDTBT:PCBM active layer. Illuminated by a 986 nm laser with an excitation density of ~390 mW/cm^2, a maximum photocurrent density of 16 $\mu A/cm^2$ and a conversion of 0.45% were obtained, showing the ability to expand the response into the NIR range. Guo *et al.* [150] used a similar method to improve the performance of both NIR harvesting and light scattering of inverted polymer BHI solar cells through mixing $NaYF_4$:Yb^{3+}/Er^{3+} with PCDTBT:PCBM.

Instead of using UC materials as upconverters, the porphyrin-based TTA-sensitizers have also been utilized to efficiently increase the NIR response of solar cells [151]. Schulze's group carried out a comprehensive study to explore the application of TTA-UC in organic solar cells [104,106], among which the photocurrent was increased up to 0.2% under a moderate concentration of 19 suns [105]. In this system, a TTA-UC unit was conjugated to the inverted organic cell, to avoid parasitic optical losses [102].

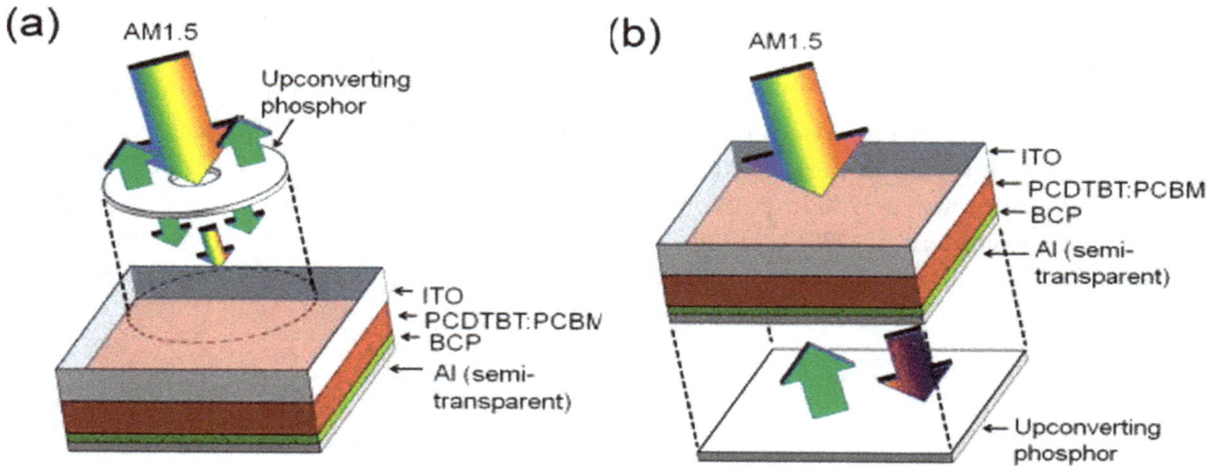

Figure 11. Schematic design of an organic PV device with upconversion phosphors placed (**a**) in front of or (**b**) behind the device (adapted with permission from [149]; Copyright AIP Publishing, 2012).

4. Conclusions

In this review, we have summarized photon upconversion materials including rare-earth-doped upconversion materials, triplet-triplet annihilation upconversion materials, and quantum nanostructure upconversion materials as spectral modifiers for various types of photovoltaic solar cells to circumvent their major energy loss mechanism of transmission loss. The ability to convert the transmitted sub-band-gap photons into above-band-gap light where solar cells typically have high quantum efficiency, enables the Schockley-Queisser limit of single junction PV devices to be broken. The inorganic rare-earth-doped upconversion materials are able to work above 800 nm, while the organic triplet-triplet annihilation upconversion dye pairs typically work below 700 nm. Moreover, quantum nanostructures emerge as a new type of upconversion materials, where the size- and shape-induced quantum effects can be exploited to upconvert at therequired wavelength ranges. Indeed, the use of these upconversion materials has improved the performance of c-Si, amorphous Si, GaAs, DSSCs as well as other PV devices with varying bandgaps. Despite recent progress, the increase of efficiency remains less than 2%, far below the theoretical predication of ~10% as in the case of c-Si PV cells when irradiated with unconcentrated sun light [17]. This discrepancy is ascribed to several problems of current photon upconversion materials. (i) The established energy conversion efficiency of long wavelength rare-earth-doped upconversion materials is less than 3% [152], thus limiting the contribution of spectral conversion of absorbed photons for improvement of PV efficiency. To improve the upconversion efficiency, various approaches can be utilized such as non-luminescent impurity doping, using photonic crystals to tailor the excitation field [153], architecture of a core-shell structure to suppress surface-related quenching mechanisms, utilization of metallic structures for surface plasmon enhanced upconversion [154], *etc.*; (ii) The low and narrow absorption of rare-earth ions results in harvesting of only a small fraction of sunlight for rare earth upconversion. Design of hierarchical nanostructures to incorporate a range of rare earth ions, without introducing deleterious cross relaxations, collectively can produce intense broad band upconversion [61,155]. Alternatively, external sensitizers that have strong and broad absorption, such as organic dyes [156], quantum dots [157], and transition metal ions [158] can be utilized to sensitize rare earth ions to entail upconversion; (iii) Photon upconversion is a nonlinear optical process, defining the strong dependence of upconversion luminescence on light irradiance. Most photon upconversion materials thus have limited luminescent upconverting efficiency in the range of sun irradiance (~1000 W/m^2). Triplet-triplet annihilation upconversion (TTA UC) has impressive efficacy under sun irradiation, favorable for uses in solar cells but with wide band-gap [101,159]. This is because the involved absorption is typically less than 700 nm [27,29,101,159–162]. Development of TTA UC with wavelengths in the IR range is appealing [102]; (iv) The upconversion in quantum nanostructures has been demonstrated at a visible wavelength, but not shown in the broad spectral range, in particular, the IR range under sunlight irradiance. This limits its uses in PV devices. Future works to overcome this problem are attractive. In all, development of upconversion materials with broad-band, strong, and tunable absorption as well as high upconverting efficiency is required, which will significantly boost solar cell efficiency by upconverison of transmitted sub-band-gap photons.

Acknowledgments: This work was supported in part by the National Science Foundation of China (No. 51102066), the National Science Fund for Distinguished Young Scholars (No. 51325201), the International Cooperation Project in the Ministry of Science and Technology (No. 2014DFA50740), the Program for Basic Research Excellent Talents in Harbin Institute of Technology, China (BRETIII 2012018), and the Fundamental Research Funds for the Central Universities, China (AUGA 5710052614).

References

1. Kalyanasundaram, K.; Gratzel, M. Themed issue: Nanomaterials for energy conversion and storage. *J. Mater. Chem.* **2012**, *22*, 24190–24194. [CrossRef]
2. Nozik, J.; Miller, J. Introduction to solar photon conversion. *Chem. Rev.* **2010**, *110*, 6443–6445. [CrossRef] [PubMed]
3. Luna-Rubio, R.; Trejo-Peria, M.; Vargas-Vázquez, D.; Ríos-Moreno, G.J. Optimal sizing of renewable hybrids energy systems: A review of methodologies. *Sol. Energy* **2012**, *86*, 1077–1088. [CrossRef]
4. Muller-Furstenberger, G.; Wagner, M. Exploring the environmental Kuznets hypothesis: Theoretical and environmental problem. *Ecol. Econ.* **2007**, *62*, 648–660. [CrossRef]
5. Simmons, M.R. Twilight in the Desert. In *The Coming Saudi Oil Shock and the World Economy*; John Wiley & Sons: Hoboken, NJ, USA, 2005.
6. Olah, G.O.; Goepert, A.; Prakash, G.K.S. *Beyond Oil and Gas: The Methanol Economy*; Wiley-VCH: Weinheim, Germany, 2006.
7. Scholes, G.D.; Fleming, G.R.; Olaya-Castro, A.; van Grondelle, R. Lessons from nature about solar light harvesting. *Nat. Chem.* **2011**, *3*, 763–774. [CrossRef] [PubMed]
8. Lewis, N.S. Toward cost-effective solar energy use. *Science* **2007**, *315*, 798–801. [CrossRef] [PubMed]
9. Águas, H.; Ram, S.K.; Araújo, A.; Gaspar, D.; Vicente, A.; Filonovich, S.A.; Fortunato, E.; Martins, R.; Ferreira, I. Silicon thin film solar cells on commercial tiles. *Energy Environ. Sci.* **2011**, *4*, 4620–4632. [CrossRef]
10. Grätzl, M. Photovoltaic and photoelectrochemical conversion of solar energy. *Philos. Trans. R. Soc. A* **2007**, *365*, 993–1005. [CrossRef] [PubMed]
11. Lewis, N.S.; Nocera, D.G. Powering the planet: Chemical challenges in solar energy utilization. *Proc. Natl. Acad. Sci. USA* **2006**, *103*, 15729–15735. [CrossRef] [PubMed]
12. Morton, O. Solar energy: A new day dawning? Silicon Valley Sunrise. *Nature* **2006**, *443*, 19–22. [CrossRef] [PubMed]
13. Van der Ende, B.M.; Aarts, L.; Andries, M. Lanthanide ions as spectral converters for solar cells. *Phys. Chem. Chem. Phys.* **2009**, *11*, 11081–11095. [CrossRef] [PubMed]
14. Saga, T. Advances in crystalline silicon solar cell technology for industrial mass production. *NPG Asia Mater.* **2010**, *2*, 96–102. [CrossRef]
15. Chen, G.Y.; Seo, J.; Yang, C.H.; Prasad, P.N. Nanochemistry and nanomaterials for photovoltaics. *Chem. Soc. Rev.* **2013**, *42*, 8304–8338. [CrossRef] [PubMed]
16. Schockley, W.; Queisser, H.J. Detailed Balance Limit of Efficiency of p-n Junction Solar Cells. *J. Appl. Phys.* **1961**, *32*, 510–519. [CrossRef]

17. Trupke, T.; Shalav, A.; Richards, B.S.; Würfel, P.; Green, M.A. Efficiency enhancement of solar cells by luminescent up-conversion of sunlight. *Sol. Energy Mater. Sol. Cells* **2006**, *90*, 3327–3338. [CrossRef]
18. Zhou, J.; Liu, Q.; Feng, W.; Sun, Y.; Li, F. Upconversion Luminescent Materials: Advances and Applications. *Chem. Rev.* **2015**, *115*, 395–465. [CrossRef] [PubMed]
19. Haase, M.; Schäfer, H. Upconverting nanoparticles. *Angew. Chem. Int. Ed.* **2011**, *50*, 5808–5829.
20. Sun, L.D.; Wang, Y.F.; Yan, C.H. Paradigms and challenges for bioapplication of rare earth upconversion luminescent nanoparticles: Small size and tunable emission/excitation spectra. *Acc. Chem. Res.* **2014**, *47*, 1001–1009. [CrossRef] [PubMed]
21. Zhou, J.; Liu, Z.; Li, F.Y. Upconversion nanophosphors for small-animal imaging. *Chem. Soc. Rev.* **2012**, *41*, 1323–1349. [CrossRef] [PubMed]
22. Gorris, H.H.; Wolfbeis, O.S. Photon-upconverting nanoparticles for optical encoding and multiplexing of cells, biomolecules, and microspheres. *Angew. Chem. Int. Ed.* **2013**, *52*, 3584–3600.
23. Liu, Q.; Feng, W.; Li, F. Water-soluble lanthanide upconversion nanophosphors: Synthesis and bioimaging applications *in vivo*. *Chem. Rev.* **2014**, *273*, 100–110. [CrossRef]
24. Feng, W.; Han, C.M.; Li, F.Y. Upconversion-Nanophosphor-Based Functional Nanocomposites. *Adv. Mater.* **2013**, *25*, 5287–5303. [CrossRef] [PubMed]
25. Wang, F.; Banerjee, D.; Liu, Y.S.; Chen, X.Y.; Liu, X.G. Upconversion Nanoparticles in Biological Labeling, Imaging and Therapy. *Analyst* **2010**, *135*, 1839–1854. [CrossRef] [PubMed]
26. Wang, F.; Liu, X.G. Recent advances in the chemistry of lanthanide-doped upconversion nanocrystals. *Chem. Soc. Rev.* **2009**, *38*, 976–989. [CrossRef] [PubMed]
27. Zhao, J.Z.; Ji, S.M.; Guo, H.M. Triplet-triplet annihilation based upconversion: From triplet sensitizers and triplet acceptors to upconversion quantum yields. *RSC Adv.* **2011**, *1*, 937–950. [CrossRef]
28. Soo, H.L.; Mathieu, A.A.; Roberto, V.; Christoph, W.; Yoan, C.S. Light upconversion by triplet–triplet annihilation in diphenylanthracene-based copolymers. *Polym. Chem.* **2014**, *5*, 6898–6904.
29. Singh-Rachford, T.N.; Castellano, F.N. Photon upconversion based on sensitized triplet-triplet annihilation. *Coord. Chem. Rev.* **2010**, *254*, 2560–2573. [CrossRef]
30. Deutsch, Z.; Neeman, L.; Oron, D. Luminescence upconversion in colloidal double quantum dots. *Nat. Nanotechnol.* **2013**, *8*, 649–653. [PubMed]
31. Deutsch, Z.; Schwartz, O.; Tenne, R.; Popovitz-Biro, R.; Oron, D. Two-color antibunching from band-gap engineered colloidal semiconductor nanocrystals. *Nano Lett.* **2012**, *12*, 2948–2952. [CrossRef] [PubMed]
32. Xing, G.; Liao, Y.; Wu, X.; Chakrabortty, S.; Liu, X.; Yeow, E.K.L.; Chan, Y.; Sum, T.C. Ultralow-threshold two-photon pumped amplified spontaneous emission and lasing from seeded CdSe/CdS nanorod heterostructures. *ACS Nano* **2012**, *6*, 10835–10844. [CrossRef] [PubMed]
33. Bunzli, J.C.G. Benefiting from the Unique Properties of Lanthanide Ions. *Acc. Chem. Res.* **2006**, *39*, 53–61. [CrossRef] [PubMed]
34. Bunzli, J.C.G. Lanthanide luminescence for biomedical analyses and imaging. *Chem. Rev.* **2010**, *110*, 2729–2755. [CrossRef] [PubMed]
35. Carlos, L.D.; Ferreira, R.A.S.; Bermudez, V.D.; Julian-Lopez, B.; Escribano, P. Progress on lanthanide-based organic-inorganic hybrid phosphors. *Chem. Soc. Rev.* **2011**, *40*, 536–549. [PubMed]
36. Kar, A.; Patra, A. Impacts of core-shell structures on properties of lanthanide-based nanocrystals: Crystal phase, lattice strain, downconversion, upconversion and energy transfe. *Nanoscale* **2012**, *4*, 3608–3619. [CrossRef] [PubMed]
37. Goesmann, H.; Feldmann, C. Nanoparticulate Functional Materials. *Angew. Chem. Int. Ed.* **2010**, *49*, 1362–1395. [CrossRef] [PubMed]
38. Ohwaki, J.; Wang, Y. Efficient 1.5mm to visible upconversion in Er^{3+}-doped halide phosphors. *Jpn. J. Appl. Phys.* **1994**, *33*, 334–337. [CrossRef]
39. Auzel, F.; Pecile, D.; Morin, D. Rare earth doped vitroceramics: New, efficient, blue and green emitting materials for infrared upconversion. *J. Electrochem. Soc.* **1975**, *122*. [CrossRef]
40. Menyuk, N.; Dwight, K.; Pierce, J.W. $NaYF_4$:Yb,Er—An efficient upconversion phosphor. *Appl. Phys. Lett.* **1972**, *21*, 159–161. [CrossRef]
41. Suyver, J.F.; Grimm, J.; Krämer, K.W.; Güdel, H.U. Highly Efficient Near-Infrared to Visible Up-Conversion Process in $NaYF_4$:Er^{3+},Yb^{3+}. *J. Lumin.* **2005**, *114*, 53–59. [CrossRef]

42. Krämer, K.W.; Biner, D.; Frei, G.; Güdel, H.U.; Hehlen, M.P.; Lüthi, S.T. Hexagonalsodium yttrium fluoride based green and blue emitting upconversion phosphors. *Chem. Mater.* **2004**, *16*, 1244–1251. [CrossRef]
43. Wang, L.; Li, X.; Li, Z.; Chu, W.; Li, R.; Lin, K.; Qian, H.; Wang, Y.; Wu, C.; Li, J.; *et al.* A New Cubic Phase for a $NaYF_4$ Host Matrix Offering High Upconversion Luminescence Efficiency. *Adv. Mater.* **2015**, *27*. [CrossRef]
44. Chen, G.Y.; Ohulchanskyy, T.Y.; Kachynski, A.; Agren, H.; Prasad, P.N. Intense Visible and Near-Infrared Upconversion Photoluminescence in Colloidal $LiYF_4:Er^{3+}$ Nanocrystals under Excitation at 1490 nm. *ACS Nano* **2011**, *5*, 4981–4986. [CrossRef] [PubMed]
45. Kumar, G.A.; Pokhrel, M.; Sardar, D.K. Intense visible and near infrared upconversion in M_2O_2S:Er (M = Y, Gd, La) phosphor under 1550 nm excitation. *Mater. Lett.* **2012**, *68*, 395–398. [CrossRef]
46. Zheng, K.Z.; Zhao, D.; Zhang, D.S.; Liu, N.; Qin, W.P. Ultraviolet upconversion fluorescence of Er^{3+} induced by 1560 nm laser excitation. *Opt. Lett.* **2010**, *35*, 2442–2444. [CrossRef] [PubMed]
47. Fischer, S.; Goldschmidt, J.C.; Loper, P.; Bauer, G.H.; Bruggemann, R.; Kramer, K.; Biner, D.; Hermle, M.; Glunz, S.W. Enhancement of silicon solar cell efficiency by upconversion: Optical and electrical characterization. *J. Appl. Phys.* **2010**, *108*, 044912. [CrossRef]
48. Li, C.; Lin, J. Rare earth fluoride nano-/microcrystals: Synthesis, surface modification and application. *J. Mater. Chem.* **2010**, *20*, 6831–6847. [CrossRef]
49. Chen, G.Y.; Zhang, Y.G.; Somesfalean, G.; Zhang, Z.G.; Sun, Q.; Wang, F.P. Two-color upconversion in rare-earth-ion-doped ZrO_2 nanocrystals. *Appl. Phys. Lett.* **2006**, *89*, 163105. [CrossRef]
50. Chen, G.Y.; Somesfalean, G.; Liu, Y.; Zhang, Z.G.; Sun, Q.; Wang, F.P. Upconversion mechanism for two-color emission in rare-earth-ion-doped ZrO_2 nanocrystals. *Phys. Rev. B* **2007**, *75*, 195204. [CrossRef]
51. Boyer, J.C.; Johnson, N.J.J.; van Veggel, F.C.J.M. Upconverting Lanthanide-Doped $NaYF_4$-PMMA Polymer Composites Prepared by in Situ Polymerization. *Chem. Mater.* **2009**, *21*, 2010–2012.
52. Boyer, J.C.; Cuccia, L.A.; Capobianco, J.A. Synthesis of Colloidal Upconverting $NaYF_4:Er^{3+}/Yb^{3+}$ and Tm^{3+}/Yb^{3+} Monodisperse Nanocrystals. *Nano Lett.* **2007**, *7*, 847–852. [CrossRef] [PubMed]
53. Heer, S.; Kompe, K.; Gudel, H.U.; Haase, M. Highly Efficient Multicolour Upconversion Emission in Transparent Colloids of Lanthanide-Doped $NaYF_4$ Nanocrystals. *Adv. Mater.* **2004**, *16*, 2102–2105. [CrossRef]
54. Chen, D.; Zhou, Y.; Wan, Z.; Huang, P.; Yu, H.; Lu, H.; Ji, Z. Enhanced upconversion luminescence in phase-separation-controlled crystallization glass ceramics containing Yb/Er(Tm):$NaLuF_4$ nanocrystals. *J. Eur. Ceram. Soc.* **2015**, *35*, 2129–2137. [CrossRef]
55. Cheng, E.; Yin, W.; Bai, S.; Qiao, R.; Zhong, Y.; Li, Z. Synthesis of vis/NIR-driven hybrid photocatalysts by electrostatic assembly of $NaYF_4$:Yb, Tm nanocrystals on g-C_3N_4 nanosheets. *Mater. Lett.* **2015**, *146*, 87–90. [CrossRef]
56. Chen, G.Y.; Shen, J.; Ohulchanskyy, T.Y.; Patel, N.J.; Kutikov, A.; Li, Z.P.; Song, J.; Pandey, R.K.; Agren, H.; Prasad, P.N.; *et al.* (α-$NaYbF_4:Tm^{3+}$)/CaF_2 Core/Shell Nanoparticles with Efficient Near-Infrared to Near-Infrared Upconversion for High-Contrast Deep Tissue Bioimaging. *ACS Nano* **2012**, *6*, 8280–8287. [CrossRef] [PubMed]
57. Li, Q.; Lin, J.; Wu, J.; Lan, Z.; Wang, Y.; Peng, F.; Huang, M. Enhancing photovoltaic performance of dye-sensitized solar cell by rare-earth doped oxide of Lu_2O_3:(Tm^{3+},Yb^{3+}). *Electrochim. Acta* **2011**, *56*, 4980–4984. [CrossRef]
58. Chen, G.Y.; Ohulchanskyy, T.Y.; Kumar, R.; Agren, H.; Prasad, P.N. Ultrasmall Monodisperse $NaYF_4:Yb^{3+}/Tm^{3+}$ Nanocrystals with Enhanced Near-Infrared to Near-Infrared Upconversion Photoluminescence. *ACS Nano* **2010**, *4*, 3163–3168. [CrossRef] [PubMed]
59. Ramakrishna, P.V.; Pammi, S.V.N.; Samatha, K. UV-visible upconversion studies of Nd^{3+} ions in lead tellurite glass. *Solid State Commun.* **2013**, *155*, 21–24. [CrossRef]
60. Shang, Y.; Hao, S.; Liu, J.; Tan, M.; Wang, N.; Yang, C.; Chen, G. Synthesis of Upconversion β-$NaYF_4:Nd^{3+}/Yb^{3+}/Er^{3+}$ Particles with Enhanced Luminescent Intensity through Control of Morphology and Phase. *Nanomaterials* **2015**, *5*, 218–232. [CrossRef]
61. Wang, Y.-F.; Liu, G.-Y.; Sun, L.-D.; Xiao, J.-W.; Zhou, J.-C.; Yan, C.-H. Nd^{3+}-Sensitized Upconversion Nanophosphors: Efficient *in Vivo* Bioimaging Probes with Minimized Heating Effect. *ACS Nano* **2013**, *7*, 7200–7206. [CrossRef] [PubMed]
62. Li, X.; Wang, R.; Zhang, F.; Zhou, L.; Shen, D.; Yao, C.; Zhao, D. Nd^{3+} Sensitized Up/Down Converting Dual-Mode Nanomaterials for Efficient *in-vitro* and *in-vivo* Bioimaging Excited at 800 nm. *Sci. Rep.* **2013**, *3*. [CrossRef]

63. Shalav, A.; Richards, B.S.; Trupke, T. Application of $NaYF_4:Er^{3+}$ up-converting phosphors for enhanced near-infrared silicon solar cell response. *Appl. Phys. Lett.* **2005**, *86*, 013505. [CrossRef]
64. Tian, G.; Zheng, X.; Zhang, X.; Yin, W.; Yu, J.; Wang, D.; Zhang, Z.; Yang, X.; Gu, Z.; Zhao, Y. TPGS-stabilized $NaYbF_4$:Er upconversion nanoparticles for dual-modal fluorescent/CT imaging and anticancer drug delivery to overcome multi-drug resistance. *Biomaterials* **2015**, *40*, 107–116. [CrossRef] [PubMed]
65. Chen, G.Y.; Qiu, H.L.; Fan, R.W.; Hao, S.W.; Tan, S.; Yang, C.H.; Han, G. Lanthanide-doped ultrasmall yttrium fluoride nanoparticles with enhanced multicolor upconversion photoluminescence. *J. Mater. Chem.* **2012**, *22*, 20190. [CrossRef]
66. Jung, T.; Jo, H.L.; Nam, S.H.; Yoo, B.; Cho, Y.; Kim, J.; Kim, H.M.; Hyeon, T.; Suh, Y.D.; Lee, H.; *et al.* The preferred upconversion pathway for the red emission of lanthanide-doped upconverting nanoparticles, $NaYF_4:Yb^{3+},Er^{3+}$. *Phys. Chem. Chem. Phys.* **2015**, *17*, 13201. [CrossRef] [PubMed]
67. Tian, D.; Gao, D.; Chongb, B.; Liu, X. Upconversion improvement by the reduction of Na^+-vacancies in Mn^{2+} doped hexagonal $NaYbF_4:Er^{3+}$ nanoparticles. *Dalton Trans.* **2015**, *44*, 4133. [CrossRef] [PubMed]
68. Khan, A.F.; Yadav, R.; Mukhopadhya, P.K.; Singh, S.; Dwivedi, C.; Dutta, V.; Chawla, S. Core-shell nanophosphor with enhanced NIR–visible upconversion as spectrum modifier for enhancement of solar cell efficiency. *J. Nanopart. Res.* **2011**, *13*, 6837–6846. [CrossRef]
69. Chen, Y.; Liu, B.; Deng, X.; Huang, S.; Hou, Z.; Li, C.; Lin, J. Multifunctional Nd^{3+}-sensitized upconversion nanomaterials for synchronous tumor diagnosis and treatment. *Nanoscale* **2015**, *7*, 8574. [CrossRef] [PubMed]
70. Zhang, S.; Wang, J.; Xu, W.; Chen, B.; Yu, W.; Xu, L.; Song, H. Fluorescence Resonance energy transfer between $NaYF_4$:Yb,Tm upconversion nanoparticles and gold nanorods: Near-infrared responsive biosensor for streptavidin. *J. Lumin.* **2014**, *147*, 278–283. [CrossRef]
71. Luu, Q.; Hor, A.; Fisher, J.; Anderson, R.B.; Liu, S.; Luk, T.; Paudel, H.P.; Baroughi, M.F.; May, P.S.; Smith, S. Two-Color Surface Plasmon Polariton Enhanced Upconversion in $NaYF_4$:Yb:Tm Nanoparticles on Au Nanopillar Arrays. *J. Phys. Chem. C* **2014**, *118*, 3251–3257. [CrossRef]
72. Lin, H.; Xu, D.; Teng, D.; Yang, S.; Zhang, Y. Simultaneous size and luminescence control of $NaYF_4:Yb^{3+}/RE^{3+}$ (RE = Tm, Ho) microcrystals via Li^+ doping. *Opt. Mater.* **2015**, *45*, 229–234. [CrossRef]
73. Amemori, S.; Yanai, N.; Kimizuka, N. Metallonaphthalocyanines as triplet sensitizers for near-infrared photon upconversion beyond 850 nm. *Phys. Chem. Chem. Phys.* **2015**, *17*, 22557–22560. [CrossRef] [PubMed]
74. Huang, Z.; Li, X.; Mahboub, M.; Hanson, K.M.; Nichols, V.M.; Le, H.; Tang, M.L.; Bardeen, C.J. Hybrid Molecule—Nanocrystal Photon Upconversion across the Visible and Near-Infrared. *Nano Lett.* **2015**, *15*, 5552–5557. [CrossRef] [PubMed]
75. Monguzzi, A.; Braga, D.; Gandini, M.; Holmberg, V.C.; Kim, D.K.; Sahu, A.; Norris, D.J.; Meinardi, A.F. Broadband Up-Conversion at Subsolar Irradiance: Triplet-Triplet Annihilation Boosted by Fluorescent Semiconductor Nanocrystals. *Nano Lett.* **2014**, *14*, 6644–6650. [CrossRef]
76. Monguzzi, A.; Borisov, S.; Pedrini, J.; Klimant, I.; Salvalaggio, M.; Biagini, P.; Melchiorre, F.; Lelii, C.; Meinardi, F. Efficient Broadband Triplet-Triplet Annihilation-Assisted Photon Upconversion at Subsolar Irradiance in Fully Organic Systems. *Adv. Funct. Mater.* **2015**, *25*. [CrossRef]
77. Mahato, P.; Monguzzi, A.; Yanai1, N.; Yamada, T.; Kimizuka, N. Fast and long-range triplet exciton diffusion in metal-organic frameworks for photon upconversion at ultralow excitation power. *Nat. Mater.* **2015**, *14*, 924–930. [CrossRef] [PubMed]
78. Duan, P.; Yanai, N.; Kimizuka, N. Photon Upconverting Liquids: Matrix-Free Molecular Upconversion Systems Functioning in Air. *J. Am. Chem. Soc.* **2013**, *135*, 19056–19059. [CrossRef] [PubMed]
79. Duan, P.; Yanai, N.; Nagatomi, H.; Kimizuka, N. Photon Upconversion in Supramolecular Gel Matrixes: Spontaneous Accumulation of Light-Harvesting Donor—Acceptor Arrays in Nanofibers and Acquired Air Stability. *J. Am. Chem. Soc.* **2015**, *137*, 1887–1894. [CrossRef] [PubMed]
80. Ogawa1, T.; Yanai, N.; Monguzzi, A.; Kimizuka1, N. Highly Efficient Photon Upconversion in Self-Assembled Light-Harvesting Molecular Systems. *Sci. Rep.* **2015**, *5*. [CrossRef] [PubMed]
81. Hisamitsu, S.; Yanai, N.; Kimizuka, N. Photon-Upconverting Ionic Liquids: Effective Triplet Energy Migration in Contiguous Ionic Chromophore Arrays. *Angew. Chem. Int. Ed.* **2015**, *54*, 1–6. [CrossRef] [PubMed]
82. Vadrucci, R.; Weder, C.; Simon, Y.C. Low-power photon upconversion in organic glasses. *J. Mater. Chem. C* **2014**, *2*, 2837–2841. [CrossRef]
83. Vadrucci, R.; Weder, C.; Simon, Y.C. Organogels for low-power light upconversion. *Mater. Horiz.* **2015**, *2*, 120–124. [CrossRef]

84. Green, M.A.; Emery, K.; Hishikawa, Y.; Warta, W.; Dunlop, E.D. Solar cell efficiency tables. *Prog. Photovolt.* **2014**, *22*, 701–710. [CrossRef]
85. Lahoz, F. Ho^{3+}-doped nanophase glass ceramics for efficiency enhancement in silicon solar cells. *Opt. Lett.* **2008**, *33*, 2982–2984. [CrossRef] [PubMed]
86. Lahoz, F.; Perez-Rodriguez, C.; Hernandez, S.E.; Martin, I.R.; Lavin, V.; Rodriguez-Mendoza, U.R. Upconversion mechanisms in rare-earth doped glasses to improve the efficiency of silicon solar cells. *Sol. Energy Mater. Sol. Cells* **2011**, *95*, 1671–1677. [CrossRef]
87. Chen, D.; Lei, L.; Yang, A.; Wang, Z.; Wang, Y. Ultra-broadband near-infrared excitable upconversion core/shell nanocrystals. *Chem. Commun.* **2012**, *48*, 5898–5900. [CrossRef] [PubMed]
88. Shao, W.; Chen, G.; Ohulchanskyy, T.Y.; Kuzmin, A.; Damasco, J.; Qiu, H.; Yang, C.; Hans, Å.; Prasad, P.N. Lanthanide-Doped Fluoride Core/Multishell Nanoparticles for Broad-Band Upconversion of Infrared Light. *Adv. Opt. Mater.* **2015**, *3*, 575–582. [CrossRef]
89. Dobrovolsky, A.; Sukrittanon, S.; Kuang, Y.; Tu, C.W.; Chen, W.M.; Buyanova, I.A. Energy Upconversion in GaP/GaNP Core/Shell Nanowires for Enhanced Near-Infrared Light Harvesting. *Small* **2014**, *10*, 4403–4406. [PubMed]
90. Zhang, X.D.; Jin, X.; Wang, D.F.; Xiong, S.Z.; Geng, X.H.; Zhao, Y. Synthesis of $NaYF_4$:Yb, Er nanocrystals and its application in silicon thin film solar cells. *Phys. Status Solidi C* **2010**, *7*, 1128–1131.
91. De Wild, J.; Meijerink, A.; Rath, J.K.; van Sark, W.G.J.H.M.; Schropp, R.E.I. Towards upconversion for amorphous silicon solar cells. *Sol. Energy Mater. Sol. Cells* **2010**, *94*, 1919–1922. [CrossRef]
92. De Wild, J.; Rath, J.K.; Meijerink, A.; van Sark, W.G.J.H.M.; Schropp, R.E.I. Enhanced near-infrared response of a-Si:H solar cells with β-$NaYF_4$:Yb^{3+}(18%), Er^{3+}(2%) upconversion phosphors. *Sol. Energy Mater. Sol. Cells* **2010**, *94*, 2395–2398. [CrossRef]
93. Yang, F.; Liu, C.; Wei, D.; Chen, Y.; Lu, J.; Yang, S. Er^{3+}–Yb^{3+} co-doped TeO_2-PbF_2 oxyhalide tellurite glasses for amorphous silicon solar cells. *Opt. Mater.* **2014**, *36*, 1040–1043.
94. Kumar, P.; Gupta, B.K. New insight into rare-earth doped gadolinium molybdate nanophosphor assisted broad spectral converters from UV to NIR for silicon solar cells. *RSC Adv.* **2015**, *5*, 24729–24736. [CrossRef]
95. Kik, P.; Polman, A. Cooperative upconversion as the gain-limiting factor in Er doped miniature Al_2O_3 optical waveguide amplifiers. *J. Appl. Phys.* **2003**, *93*, 5008–5012. [CrossRef]
96. Pollack, S.; Chang, D. Ion-pair upconversion pumped laser emission in Er^{3+} ions in YAG, YLF, SrF_2, and CaF_2 crystals. *J. Appl. Phys.* **1988**, *64*, 2885–2893. [CrossRef]
97. Wang, X.F.; Yan, X.H.; Kan, C.X. Controlled synthesis and optical characterization of multifunctional ordered Y_2O_3:Er^{3+} porous pyramid arrays. *J. Mater. Chem.* **2011**, *21*, 4251–4256. [CrossRef]
98. Ivanova, S.; Pelle, F. Strong 1.53 μm to NIR-VIS-UV upconversion in Er-doped fluoride glass for high-efficiency solar cells. *J. Opt. Soc. Am. B* **2009**, *26*, 1930–1938. [CrossRef]
99. Ohwaki, J.; Wang, Y. New efficient upconversion phosphor $BaCl_2$:Er under 1.5 mm excitation. *Electron. Lett.* **1993**, *29*, 351–352. [CrossRef]
100. Chen, Y.; He, W.; Jiao, Y.; Wang, H.; Hao, X.; Lu, J.; Yang, S.E. β-NaYF4:Er^{3+} (10%) microprisms for the enhancement of a-Si:H solar cell near-infrared responses. *J. Lumin.* **2012**, *132*, 2247–2250. [CrossRef]
101. Cheng, Y.Y.; Fückel, B.; MacQueen, R.W.; Khoury, T.; Clady, R.G.; Schulze, T.F.; Ekins-Daukes, N.J.; Crossley, M.J.; Stannowski, B.; Lips, K.; *et al.* Improving the light-harvesting of amorphous silicon solar cells with photochemical upconversion. *Energy Environ. Sci.* **2012**, *5*, 6953–6959. [CrossRef]
102. Schulze, T.F.; Schmidt, T.W. Photochemical upconversion: Present status and prospects for its application to solar energy conversion. *Energy Environ. Sci.* **2015**, *8*, 103–125. [CrossRef]
103. Nattestad, A.; Cheng, Y.Y.; MacQueen, R.W.; Schulze, T.F.; Thompson, F.W.; Mozer, A.J.; Fückel, B.; Khoury, T.; Crossley, M.J.; Lips, K.; *et al.* Dye-Sensitized Solar Cell with Integrated Triplet–Triplet Annihilation Upconversion System. *J. Phys. Chem. Lett.* **2013**, *4*, 2073–2078. [CrossRef] [PubMed]
104. Schulze, T.F.; Cheng, Y.Y.; Fückel, B.; MacQueen, R.W.; Danos, A.; Davis, N.J.L.K.; Tayebjee, M.J.Y.; Khoury, T.; Clady, R.G.C.R.; Ekins-Daukes, N.J.; *et al.* Photochemical Upconversion Enhanced Solar Cells: Effect of a Back Reflector. *Aust. J. Chem.* **2012**, *65*, 480–485. [CrossRef]
105. Schulze, T.F.; Czolk, J.; Cheng, Y.; Fuckel, B.; MacQueen, R.W.; Khoury, T.; Crossley, M.J.; Stannowski, B.; Lips, K.; Lemmer, U.; *et al.* Efficiency Enhancement of Organic and Thin-Film Silicon Solar Cells with Photochemical Upconversion. *J. Phys. Chem. C* **2012**, *116*, 22794–22801. [CrossRef]

106. Schulze, T.F.; Cheng, Y.Y.; Khoury, T.; Crossley, M.J.; Stannowski, B.; Lips, K.; Schmidt, T.W. Micro-optical design of photochemical upconverters for thin-film solar cells. *J. Photonics Energy* **2013**, *3*, 034598. [CrossRef]
107. George, B.M.; Behrends, J.; Schnegg, A.; Schulze, T.F.; Fehr, M.; Korte, L.; Rech, B.; Lips, K.; Rohrmüller, M.; Rauls, E.; *et al.* Atomic Structure of Interface States in Silicon Heterojunction Solar Cells. *Phys. Rev. Lett.* **2013**, *110*, 136803. [CrossRef] [PubMed]
108. Gibart, P.; Auzel, F.; Guillaume, J.-C.; Zahraman, K. Below band gap IR response of substrate-free GaAs solar cells using two-photon upconversion. *Jpn. J. Appl. Phys.* **1996**, *351*, 4401–4402. [CrossRef]
109. Lin, H.Y.; Chen, H.N.; Wu, T.H.; Wu, C.S.; Su, Y.K.; Chu, S.Y. Investigation of Green Up-Conversion Behavior in $Y_6W_2O_{15}$:Yb^{3+},Er^{3+} Phosphor and Its Verification in 973 nm Laser-Driven GaAs Solar Cell. *J. Am. Ceram. Soc.* **2012**, *95*, 3172–3179. [CrossRef]
110. O'Regan, B.; Grätzel, M. A low-cost, high-efficiency solar cell based on dye-sensitized. *Nature* **1991**, *353*, 737–740. [CrossRef]
111. Grätzel, M. Photoelectrochemical cells. *Nature* **2001**, *414*, 338–344. [CrossRef] [PubMed]
112. Bisquert, J. Dilemmas of Dye-Sensitized Solar Cell. *Chem. Phys. Chem.* **2011**, *12*, 1633–1636. [CrossRef] [PubMed]
113. Basham, J.I.; Mor, G.K.; Grimes, C.A. Förster Resonance Energy Transfer in Dye-Sensitized Solar Cells. *ACS Nano* **2010**, *4*, 1253–1258. [CrossRef] [PubMed]
114. Chen, C.-Y.; Wang, M.; Li, J.-Y.; Pootrakulchote, N.; Alibabaei, L.; Ngoc-le, C.-H.; Decoppet, J.-D.; Tsai, J.-H.; Graetzel, C.; Wu, C.G.; *et al.* Highly Efficient Light-Harvesting Ruthenium Sensitizer for Thin-Film Dye-Sensitized Solar Cells. *ACS Nano* **2009**, *3*, 3103–3109. [CrossRef] [PubMed]
115. Huang, X.Y.; Wang, J.X.; Yu, D.C.; Ye, S.; Zhang, Q.Y.; Sun, X.W. Spectral conversion for solar cell efficiency enhancement using YVO_4:Bi^{3+}, Ln^{3+} (Ln = Dy, Er, Ho, Eu, Sm, and Yb) phosphors. *J. Appl. Phys.* **2011**, *109*, 113526. [CrossRef]
116. Bünzli, J.C.G.; Eliseeva, S.V. Lanthanide NIR luminescence for telecommunications, bioanalyses and solar energy conversion. *J. Rare Earths* **2010**, *28*, 824–842. [CrossRef]
117. Hafez, H.; Saif, M.; Abdel-Mottaleb, M.S.A. Down-converting lanthanide doped TiO_2 photoelectrodes for efficiency enhancement of dye-sensitized solar cells. *J. Power Sources* **2011**, *196*, 5792–5796. [CrossRef]
118. Levinson, R.; Berdahl, P.; Akbari, H. Solar spectral optical properties of pigments—Part I: Model for deriving scattering and absorption coefficients from transmittance and reflectance measurements. *Sol. Energy Mater. Sol. C* **2005**, *89*, 319–349. [CrossRef]
119. Hara, K.; Sato, T.; Katoh, R.; Furube, A.; Ohga, Y.; Shinpo, A.; Suga, S.; Sayama, K.; Sugihara, H.; Arakawa, H. Molecular Design of Coumarin Dyes for Efficient Dye-Sensitized Solar Cells. *J. Phys. Chem. B* **2003**, *107*, 597–606. [CrossRef]
120. Campbell, W.M.; Burrell, A.K.; Officer, D.L.; Jolley, K.W. Porphyrins as light harvesters in the dye-sensitised TiO_2 solar cell. *Coord. Chem. Rev.* **2004**, *248*, 1363–1379. [CrossRef]
121. Shan, G.-B.; Demopoulos, G.P. Near-Infrared Sunlight Harvesting in Dye-Sensitized Solar Cells via the Insertion of an Upconverter-TiO_2 Nanocomposite Layer. *Adv. Mater.* **2010**, *22*, 4373–4377. [CrossRef] [PubMed]
122. Shan, G.B.; Assaaoudi, H.; Demopoulos, G.P. Enhanced Performance of Dye-Sensitized Solar Cells by Utilization of an External, Bifunctional Layer Consisting of Uniform β-$NaYF_4$:Er^{3+}/Yb^{3+} Nanoplatelets. *ACS Appl. Mater. Interfaces* **2011**, *3*, 3239–3243. [CrossRef] [PubMed]
123. Yuan, C.Z.; Chen, G.Y.; Prasad, P.N.; Ohulchanskyy, T.Y.; Ning, Z.J.; Tian, H.; Sund, L.C.; Agren, H. Use of colloidal upconversion nanocrystals for energy relay solar cell light harvesting in the near-infrared region. *J. Mater. Chem.* **2012**, *22*, 16709–16713. [CrossRef]
124. Li, L.; Yang, Y.; Fan, R.; Jiang, Y.; Wei, L.; Shi, Y.; Yu, J.; Chen, S.; Wang, P.; Yang, B.; *et al.* A simple modification of near-infrared photon-to-electron response with fluorescence resonance energy transfer for dye-sensitized solar cells. *J. Power Sources* **2014**, *264*, 254–261. [CrossRef]
125. Wang, J.; Lin, J.; Wu, J.; Huang, M.; Lan, Z.; Chen, Y.; Tang, S.; Fan, L.; Huang, Y. Application of Yb^{3+}, Er^{3+}-doped yttrium oxyfluoride nanocrystals in dye-sensitized solar cells. *Electrochim. Acta* **2012**, *70*, 131–135. [CrossRef]
126. Zhang, J.; Shen, H.; Guo, W.; Wang, S.; Zhu, C.; Xue, F.; Hou, J.; Su, H.; Yuan, Z. An upconversion $NaYF_4$:Yb^{3+},Er^{3+}/TiO_2 coreeshell nanoparticle photoelectrode for improved efficiencies of dye-sensitized solar cells. *J. Power Sources* **2013**, *226*, 47–53. [CrossRef]

127. Liang, L.L.; Liu, Y.M.; Zhao, X.Z. Double-shell beta-NaYF4:Yb^{3+},Er^{3+}/SiO_2/TiO_2 submicroplates as a scattering and upconverting layer for efficient dye-sensitized solar cells. *Chem. Commun.* **2013**, *49*, 3958–3960.
128. Dyck, N.C.; Demopoulos, G.P. Integration of upconverting β-$NaYF_4$:Yb^{3+},Er^{3+}@TiO_2 composites as light harvesting layers in dye-sensitized solar cells. *RSC Adv.* **2014**, *4*, 52694–52701. [CrossRef]
129. Wu, J.H.; Wang, J.L.; Lin, J.M.; Lan, Z.; Tang, Q.W.; Huang, M.L.; Huang, Y.F.; Fan, L.Q.; Li, Q.B.; Tang, Z.Y. Enhancement of the Photovoltaic Performance of Dye-Sensitized Solar Cells by Doping $Y_{0.78}Yb_{0.20}Er_{0.02}F_3$ in the Photoanode. *Adv. Energy Mater.* **2012**, *2*, 78–81. [CrossRef]
130. Liu, M.; Lu, Y.; Xie, Z.B.; Chow, G.M. Enhancing near-infrared solar cell response using upconverting transparentceramics. *Sol. Energy Mater. Sol. Cells* **2011**, *95*, 800–803. [CrossRef]
131. Wang, J.; Wu, J.; Lin, J.; Huang, M.; Huang, Y.; Lan, Z.; Xiao, Y.; Yue, G.; Yin, S.; Sato, T. Application of Y_2O_3:Er^{3+} Nanorods in Dye-Sensitized Solar Cells. *ChemSusChem* **2012**, *5*, 1307–1312. [CrossRef] [PubMed]
132. Bai, J.; Zhao, R.; Han, G.; Li, Z.; Diao, G. Synthesis of 1D upconversion CeO_2:Er,Yb nanofibers via electrospinning and their performance in dye-sensitized solar cells. *RSC Adv.* **2015**, *5*, 43328–43333. [CrossRef]
133. Yuan, C.; Chen, G.; Li, L.; Damasco, J.A.; Ning, Z.; Xing, H.; Zhang, T.; Sun, L.; Zeng, H.; Cartwright, A.N.; *et al.* Simultaneous Multiple Wavelength Upconversion in a Core-Shell Nanoparticle for Enhanced Near Infrared Light Harvesting in a Dye-Sensitized Solar Cell. *ACS Appl. Mater. Interfaces* **2014**, *6*, 18018–18025. [CrossRef] [PubMed]
134. Zhao, P.; Zhu, Y.; Yang, X.; Jiang, X.; Shen, J.; Li, C. Plasmon-enhanced efficient dye-sensitized solar cells using core–shell-structured β-$NaYF_4$:Yb,Er@SiO_2@Au nanocomposites. *J. Mater. Chem. A* **2014**, *2*, 16523–16530. [CrossRef]
135. Li, L.; Yang, Y.; Fan, R.; Chen, S.; Wang, P.; Yang, B.; Cao, W. Conductive Upconversion Er, Yb-FTO Nanoparticle Coating To Replace Pt as a Low-Cost and High-Performance Counter Electrode for Dye-Sensitized Solar Cells. *ACS Appl. Mater. Interfaces* **2014**, *6*, 8223–8229. [CrossRef] [PubMed]
136. Yu, J.; Yang, Y.; Fan, R.; Liu, D.; Wei, L.; Chen, S.; Li, L.; Yang, B.; Cao, W. Enhanced Near-Infrared to Visible Upconversion Nanoparticles of Ho^{3+}-Yb^{3+}-F Tri-Doped TiO_2 and Its Application in Dye-Sensitized Solar Cells with 37% Improvement in Power Conversion Efficiency. *Inorg. Chem.* **2014**, *53*, 8045–8053. [CrossRef] [PubMed]
137. Boudreault, P.L.T.; Najari, A.; Leclerc, M. Processable Low-Bandgap Polymers for Photovoltaic Applications. *Chem. Mater.* **2011**, *23*, 456–469. [CrossRef]
138. Chen, H.Y.; Hou, J.H.; Zhang, S.Q.; Liang, Y.Y.; Yang, G.W.; Yang, Y.; Yu, L.P.; Wu, Y.; Li, G. Polymer solar cells with enhanced open-circuit voltage and efficiency. *Nat. Photonics* **2009**, *3*, 649–653. [CrossRef]
139. Hains, A.W.; Liang, Z.Q.; Woodhouse, M.A.; Gregg, B.A. Molecular Semiconductors in Organic Photovoltaic Cells. *Chem. Rev.* **2010**, *110*, 6689–6735. [CrossRef] [PubMed]
140. Chen, J.W.; Cao, Y. Development of Novel Conjugated Donor Polymers for High-Efficiency Bulk-Heterojunction Photovoltaic Devices. *Acc. Chem. Res.* **2009**, *42*, 1709–1718. [CrossRef] [PubMed]
141. He, Z.; Zhong, C.; Su, S.; Xu, M.; Wu, H.; Cao, Y. Enhanced power-conversion efficiency in polymer solar cells using an inverted device structure. *Nat. Photonics* **2012**, *6*, 591–595. [CrossRef]
142. You, J.; Dou, L.; Yoshimura, K.; Kato, T.; Ohya, K.; Moriarty, T.; Emery, K.; Chen, C.-C.; Gao, J.; Li, G.; *et al.* A polymer tandem solar cell with 10.6% power conversion efficiency. *Nat. Commun.* **2013**, *4*, 1446. [CrossRef] [PubMed]
143. Grätzel, M. Recent Advances in Sensitized Mesoscopic Solar Cells. *Acc. Chem. Res.* **2009**, *42*, 1788–1798. [PubMed]
144. Yella, A.; Mai, C.-L.; Zakeeruddin, S.M.; Chang, S.-N.; Hsieh, C.-H.; Yeh, C.-Y.; Grätzel, M. Molecular Engineering of Push—Pull Porphyrin Dyes for Highly Efficient Dye-Sensitized Solar Cells: The Role of Benzene Spacers. *Angew. Chem.* **2014**, *126*, 3017–3021. [CrossRef]
145. Günes, S.; Neugebauer, H.; Sariciftci, N.S. Conjugated Polymer-Based Organic Solar Cells. *Chem. Rev.* **2007**, *107*, 1324–1338. [CrossRef] [PubMed]
146. Wang, H.Q.; Batentschuk, M.; Osvet, A.; Pinna, L.; Brabec, C.J. Rare-Earth Ion Doped Up-Conversion Materials for Photovoltaic Applications. *Adv. Mater.* **2011**, *23*, 2675–2680. [CrossRef] [PubMed]
147. Wang, H.Q.; Stubhan, T.; Osvet, A.; Litzov, I.; Brabec, C.J. Up-conversion semiconducting MoO_3:Yb/Er nanocomposites as buffer layer in organic solar cells. *Sol. Energy Mater. Sol. Cells* **2012**, *105*, 196–201. [CrossRef]

148. Wu, J.L.; Chen, F.-C.; Chang, S.-H.; Tan, K.-S.; Tuan, H.-Y. Upconversion effects on the performance of near-infrared laser-driven polymer photovoltaic devices. *Org. Electron.* **2012**, *13*, 2104–2108. [CrossRef]
149. Adikaari, A.A.D.; Etchart, I.; Guering, P.H.; Berard, M.; Silva, S.R.P.; Cheetham, A.K.; Curry, R.J. Near infrared up-conversion in organic photovoltaic devices using an efficient Yb^{3+}:Ho^{3+} Co-doped Ln_2BaZnO_5 (Ln = Y, Gd) phosphor. *J. Appl. Phys.* **2012**, *111*. [CrossRef]
150. Guo, W.; Zheng, K.; Xie, W.; Sun, L.; Shen, L.; Liu, C.; He, Y.; Zhang, Z. Efficiency enhancement of inverted polymer solar cells by doping $NaYF_4$:Yb^{3+}, Er^{3+} nanocomposites in PCDTBT:PCBM active layer. *Sol. Energy Mater. Sol. Cells* **2014**, *124*, 126–132.
151. Yakutkin, V.; Aleshchenkov, S.; Chernov, S.; Miteva, T.; Nelles, G.; Cheprakov, A.; Baluschev, S. Towards the IR Limit of the Triplet–Triplet Annihilation-Supported Up-Conversion: Tetraanthraporphyrin. *Chem. Eur. J.* **2008**, *14*, 9846–9850. [CrossRef] [PubMed]
152. Boyer, J.C.; van veggel, F.C.J.M. Absolute quantum yield measurements of colloidal $NaYF_4$:Er^{3+},Yb^{3+} upconverting nanoparticles. *Nanoscale* **2010**, *2*, 1417–1419. [CrossRef] [PubMed]
153. Johnson, C.M.; Reece, P.J.; Conibeer, G.J. Slow-light-enhanced upconversion for photovoltaic applications in one-dimensional photonic crystals. *Opt. Lett.* **2011**, *36*, 3990–3992. [CrossRef] [PubMed]
154. Zhang, W.; Ding, F.; Chou, S.Y. Large Enhancement of Upconversion Luminescence of $NaYF_4$:Yb^{3+}/Er^{3+} Nanocrystal by 3D Plasmonic Nano-Antennas. *Adv. Mater.* **2012**, *24*, OP236–OP241. [CrossRef] [PubMed]
155. Xie, X.; Gao, N.; Deng, R.; Sun, Q.; Xu, Q.-H.; Liu, X. Mechanistic Investigation of Photon Upconversion in Nd^{3+}-Sensitized Core–Shell Nanoparticles. *J. Am. Chem. Soc.* **2013**, *135*, 12608–12611. [CrossRef] [PubMed]
156. Zou, W.; Visser, C.; Maduro, J.A.; Pshenichnikov, M.S.; Hummelen, J.C. Broadband dye-sensitized upconversion of near-infrared light. *Nat. Photonics* **2012**, *6*, 560–564. [CrossRef]
157. Pan, A.C.; del Canizo, C.; Canavos, E.; Santos, N.M.; Leitao, J.P.; Luque, A. Enhancement of up-conversion efficiency by combining rare earth-doped phosphors with PbS quantum dots. *Sol. Energy Mater. Sol. Cells* **2010**, *94*, 1923–1926.
158. Suyver, J.F.; Aebischer, A.; Biner, D.; Gerner, P.; Grimm, J.; Heer, S.; Kramer, K.W.; Reinhard, C.; Gudel, H.U. Novel materials doped with trivalent lanthanides and transition metal ions showing near-infrared to visible photon upconversion. *Opt. Mater.* **2005**, *27*, 1111–1130.
159. Simon, Y.C.; Weder, C. Low-Power Photon Upconversion through Triplet-Triplet Annihilation in Polymers. *J. Mater. Chem.* **2012**, *22*, 20817–20830. [CrossRef]
160. Singh-Rachford, T.N.; Nayak, A.; Muro-Small, M.L.; Goeb, S.; Therien, M.J.; Castellano, F.N. Supermolecular-Chromophore-Sensitized Near-Infrared-to-Visible Photon Upconversion. *J. Am. Chem. Soc.* **2010**, *132*, 14203–14211. [CrossRef] [PubMed]
161. Fückel, B.; Roberts, D.A.; Cheng, Y.Y.; Clady, R.; Piper, R.B.; Ekins-Daukes, N.J.; Crossley, M.J.; Schmidt, T.W. Singlet Oxygen Mediated Photochemical Upconversion of NIR Light. *J. Phys. Chem. Lett.* **2011**, *2*, 966–971. [CrossRef]
162. Singh-Rachford, T.N.; Castellano, F.N. Nonlinear photochemistry squared: Quartic light power dependence realized in photon upconversion. *J. Phys. Chem. A* **2009**, *113*, 9266–9269. [CrossRef] [PubMed]

12

Wavelength-Conversion-Material-Mediated Semiconductor Wafer Bonding for Smart Optoelectronic Interconnects

Kodai Kishibe, Soichiro Hirata, Ryoichi Inoue, Tatsushi Yamashita and Katsuaki Tanabe *

Department of Chemical Engineering, Kyoto University, Kyoto 615-8510, Japan;
k.kishibe@cheme.kyoto-u.ac.jp (K.K.); s.hirata@cheme.kyoto-u.ac.jp (S.H.); r.inoue.ku@gmail.com (R.I.);
t.yamashita@cheme.kyoto-u.ac.jp (T.Y.)
* Correspondence: tanabe@cheme.kyoto-u.ac.jp

Abstract: A new concept of semiconductor wafer bonding, mediated by optical wavelength conversion materials, is proposed and demonstrated. The fabrication scheme provides simultaneous bond formation and interfacial function generation, leading to efficient device production. Wavelength-converting functionalized semiconductor interfacial engineering is realized by utilizing an adhesive viscous organic matrix with embedded fluorescent particles. The bonding is carried out in ambient air at room temperature and therefore provides a cost advantage with regard to device manufacturing. Distinct wavelength conversion, from ultraviolet into visible, and high mechanical stabilities and electrical conductivities in the bonded interfaces are verified, demonstrating their versatility for practical applications. This bonding and interfacial scheme can improve the performance and structural flexibility of optoelectronic devices, such as solar cells, by allowing the spectral light incidence suitable for each photovoltaic material, and photonic integrated circuits, by delivering the respective preferred frequencies to the optical amplifier, modulator, waveguide, and detector materials.

Keywords: semiconductor; interface; wafer bonding; frequency conversion; optoelectronics; photonic device; solar cell; photonic integrated circuit

1. Introduction

The wafer bonding technique [1–3] is used to generate semiconductor heterostructures with low defect densities, which are difficult to obtain by the conventional epitaxial growth methods, owing to crystalline lattice mismatches. Therefore, semiconductor wafer bonding is promising for the realization of high-performance optoelectronic components and has been employed to fabricate various heterostructured devices, such as lasers [4–7], light-emitting diodes [8], photodetectors [9], and solar cells [6,10,11]. Here, we present optical wavelength-converting material (WCM)-mediated wafer bonding, as a means to simultaneously provide bond formation and interfacial function generation. This study is part of a series of developments regarding functional bonding interfaces; we have previously developed graphene-mediated semiconductor wafer bonding [12]. There exist various semiconductor wafer bonding schemes, such as semiconductor-to-semiconductor direct bonding [6,10,11] and oxide- [4,5], metal- [8], and polymer-mediated bonding [7]. However, there has been no WCM-mediated bonding reported so far, to the best of our knowledge. Our novel semiconductor-bonding scheme may improve, for example, multijunction photovoltaic cell efficiencies, by converting the wavelength of the light transmitted through the upper subcell to one that is highly absorbed by the lower subcell [13,14], as conceptually depicted in Figure 1. It also improves the performance of hybrid optical transceivers in photonic integrated circuits [15–18] by delivering suitable frequencies to each optical amplifier, modulator, waveguide, and detector material. Furthermore, our bonding is performed at room temperature and does not require heating, unlike the conventional wafer direct bonding processes, and therefore exhibits no risk of damaging the active material or retarding the production line.

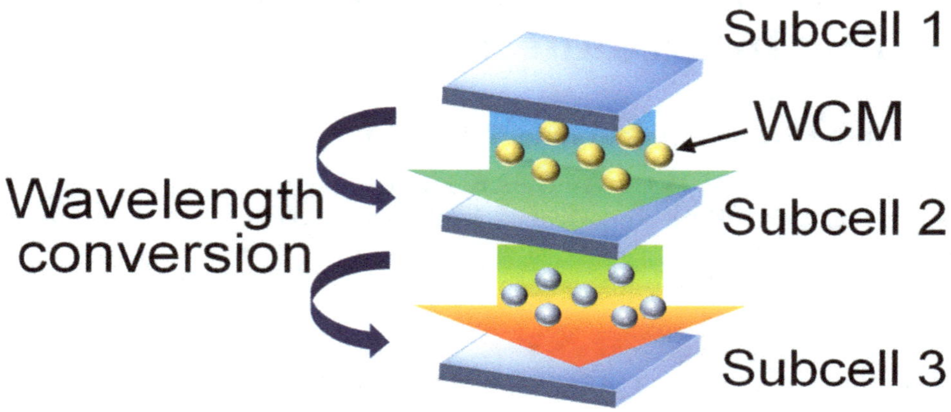

Figure 1. Conceptual drawing of the application of the wavelength-converting material (WCM)-mediated bonding for multijunction solar cells.

2. Experimental Methods

The polished surfaces of boron-doped *p*-type <100> Si wafers, with a doping concentration of ~1×10^{19} cm^{-3} and a thickness of 280 μm, were first coated using photoresist for the purpose of protecting the Si surfaces to be bonded in the dicing process into 64 mm^2. The diced wafers were dipped in acetone for 5 min to remove the photoresist coating and to degrease the Si surfaces to be bonded. They were then submerged in 9% hydrofluoric acid for 1 min to remove the SiO_2 native oxide layer formed on the Si wafers. We used a commercially available WCM of 4,7-bis(4-*tert*-butylphenyl)-2-octylbenzotriazole (RAYCREA, Nitto Denko Corp., Osaka, Japan) [19]—a fluorescent dye compound. However, the bare WCM was composed of solid-state particles and therefore could hardly be incorporated stably, in its original form, in semiconductor wafer bonding interfaces. Therefore, we used a hydrogel material as an adhesive and a viscous organic matrix to embed the fluorescent particles. A 2.5 w/v% polyacrylamide (PAM) aqueous solution (aq.) was prepared by mixing PAM powder with deionized water and stirring well to prevent the aggregation of the adhesive PAM particles. WCM, which was ground with a mortar to obtain uniform particle diameters, was mixed with the 2.5 w/v% PAM aq., and five types of mixtures (0, 1×10^{-4}, 1×10^{-3}, 5×10^{-3}, and 1×10^{-2} g/mL) were prepared. The prepared hydrogel containing WCM was then uniformly spin-coated onto the Si wafer. For a higher reproducibility of the experimental results, this spin-coating process was repeated three times. The Si piece coated with the hydrogel containing WCM was bonded to a bare Si piece at room temperature under a uniaxial pressure of 0.1 MPaG. Incidentally, the use of a WCM that can be molecularly dispersed in a polymer binder would have been technically more desirable, but such a WCM was not commercially available, and we were not equipped with the ability to synthesize it. Therefore, we employed the particulate WCM in this work.

The normal detachment stresses were measured as bonded interfacial strengths for the bonded samples. We connected the outer surface of the bonded sample to a digital spring weight scaler via a solid wire firmly attached to the sample surface using a household adhesive glue. Then, we pulled the scaler outward in the direction normal to the sample die until the bonded sample was debonded while the weight scaler recorded the maximum force at the point of delamination. For electrical measurements, an Au–Ge–Ni alloy (80:10:10 wt %) and pure Au layers with thicknesses of 30 and 150 nm, respectively, were sequentially deposited on the outer sides of the bonded samples as ohmic electrodes via thermal evaporation. For the interfacial observation, we used an infrared transmission analyzer (IRise-T, Moritex Corp., Asaka, Japan) with an optical wavelength of 1.2 μm. For the optical measurements of WCM, a sample containing 0.03 g/cm^2 of WCM between glass plates was prepared in the same manner as above but with no surface pretreatment. In this study, for the sake of simplicity, we used Si wafers as the representative semiconductor material, but our concept can be easily extended to other semiconductors, as numerous wafer bonding experimental demonstrations have been reported, thus far, between dissimilar semiconductor materials [4–11].

3. Results and Discussion

Figure 2 shows a cross-sectional scanning electron microscope image of the bonded interface (WCM concentration: 1×10^{-2} g/mL). As observed in this image, the wafers are firmly and uniformly in contact with each other, with sufficient mechanical strength to endure the cleavage of the bonded sample. The WCM-containing interlayer in this example may seem relatively thick, considering the field of thin-film optoelectronic devices. According to the required degree of the wavelength-converting function and interfacial mechanical stability, the interlayer thickness can be controlled in each application. As a general design direction, a thicker interlayer provides a sufficient wavelength conversion efficiency and higher bondability by mitigating the semiconductor surface roughness, whereas a thinner interlayer supports the advantages in the device weight, production cost, and throughput. The WCM uniformity in the interlayer may be improved by increasing the dispersion of the WCM particles, as discussed later. We measured the dependence of the bonding strength on the WCM concentration; the results are shown in Figure 3. At the present stage, the experimental reproducibility of the bonding strength is relatively low, and we therefore plotted the best measured values for each bonding condition. As observed in these results, the majority of the samples exhibited sufficient bonded interfacial strength to endure a series of optoelectronic device fabrication processes and user operations. The hydrogen bonds stemming from PAM presumably cause the adhesion to the semiconductor surfaces [20,21]. More specifically, hydrogen bonds may form between the $-NH_2$ groups of PAM and the Si surface terminated by $-OH$ groups, owing to the water contained in PAM [20]. In addition, the PAM matrix holds the WCM particles, suppressing their sedimentation. This is due to its viscosity, induced by the entanglement of PAM polymer chains. The results shown in Figure 3 indicate that, as the concentration of WCM increases, the interfacial bonding strength decreases. To further analyze the cause of this result, we observed the inner structure of the sample (WCM concentration: 1×10^{-2} g/mL) using an infrared transmission analyzer (Figure 4a). The black spots in Figure 4a identify the WCM via the accompanying emission image, and this image verifies that the WCM particles are not dispersed uniformly; rather, they appear aggregated. This random dispersion of WCM is due to our having no technical control of the spatial distribution of WCM at the present stage and also, presumably, to the energetic stability in the aggregated state of the WCM particles. Such clustered WCM particles may cause intense roughness that the PAM cannot mitigate. Therefore, the interfacial bonding strength significantly decreases at higher concentrations of WCM, as we observed larger aggregates for higher WCM concentrations (Figure 4). In addition, the wafers may experience serious damage, as evidenced by the cracks observed in Figure 4. This applies especially to higher WCM concentrations, because the bonding pressure is concentrated in particular areas of the agglomerated particles. Therefore, to improve the properties of our WCM-containing interfaces—such as the mechanical strength, electrical conductivity, and optical transparency—it may be effective to increase the dispersion of WCM by selecting the optimum preparation process conditions, such as the organic solvent species, the spin-coating rotation velocity, and the coating repetition. To account for the WCM concentration vs. the interfacial stability trade-off observed in Figure 3, it is important to adopt the optimal organic species and thickness of the matrix and the dispersion status of the WCM particles, according to the property requirements for each application. For instance, a design principle can be described as follows: a relatively thick interlayer with low-concentration WCM can fulfill the demands of both high mechanical strength and high optical conversion efficiency.

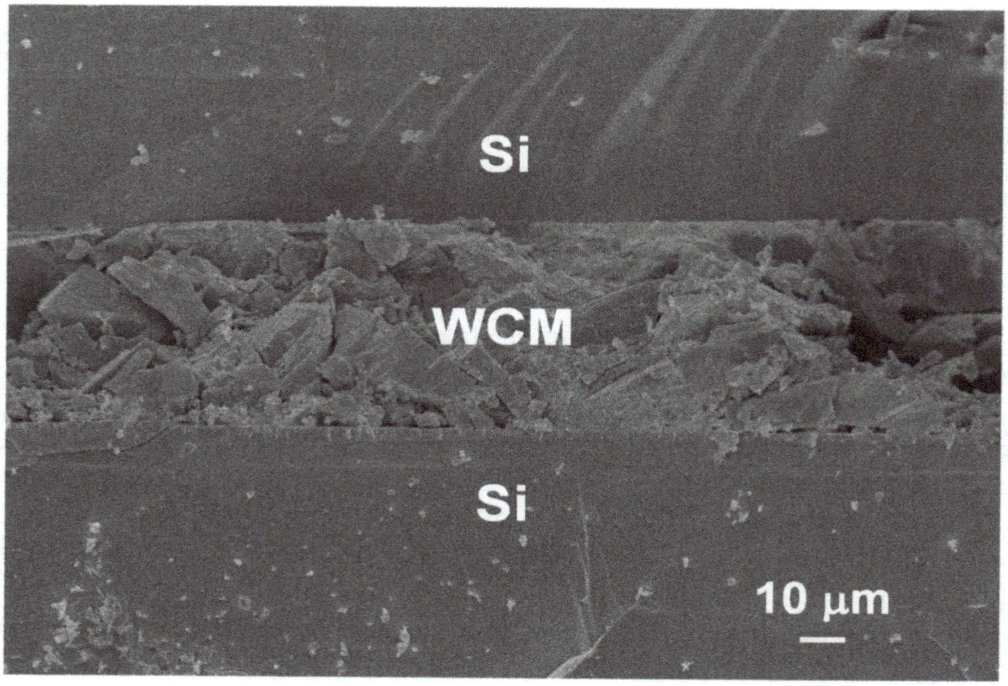

Figure 2. Cross-sectional scanning electron microscope image of a bonded interface (WCM concentration: 1×10^{-2} g/mL).

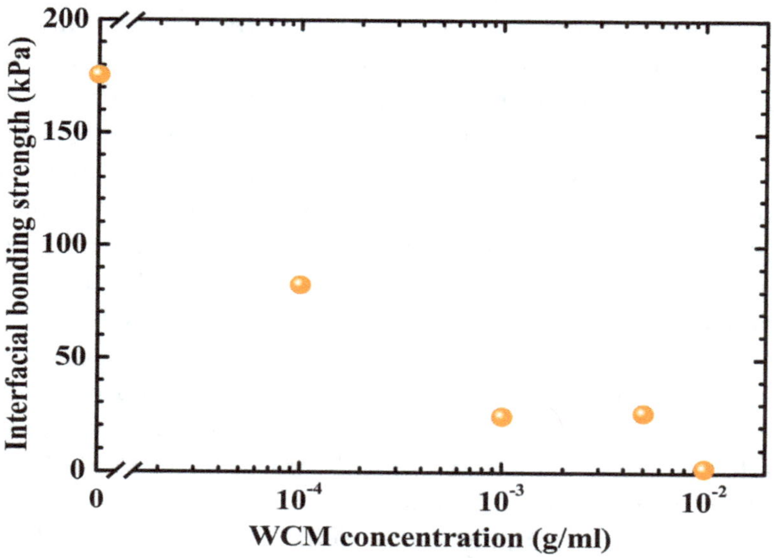

Figure 3. Interfacial bonding strength vs. WCM concentration.

Figure 5 shows the current–voltage (I–V) characteristics of the bonded interfaces for various concentrations of WCM. An ohmic interfacial electrical conductance with relatively low resistivity is observed when bonding with lower WCM concentrations. PAM, which is a hydrophilic polymer material, can contain water internally, and the water-originated ions are known to act as electrical carriers and provide electric conductance [21]. Intuitively, one might think that the presence of water has a negative effect on electronic component fabrication. However, in hydrogels, the water molecules are stably embedded in the matrices of organic polymer chains, and hydrogel-based devices have been successfully fabricated and operated [21], as in the field of organic electronics. As the concentration of WCM increases, the conductivity decreases, presumably because the highly agglomerated WCM for high WCM concentration conditions can induce voids between the wafers, as indicated in Figure 4. However, the strong decrease of the electrical conductivity with the increased concentration of WCM cannot be explained simply by a decreased width of the conductive joint, because the observed decrease is much larger than the loss of the joint area represented by Figure 4. The bonding interlayer thicknesses for the samples with WCM concentrations of 0 and 1×10^{-2} g/mL were measured at about 5 and 50 μm, respectively, by scanning electron microscopy. The interlayer thickness increase with WCM concentration can be attributed to the generation of larger WCM aggregates for higher WCM concentrations, as observed in Figure 4. Consequently, such large, and therefore tall, aggregated WCM clusters may mechanically resist the interlayer compression through the applied pressure in the bonding process. In this way, the bonding interlayer thickness becomes significantly larger for higher WCM concentrations. Overall, the unintentional combinative effect of such a decrease in the area and an increase in the length of the electrically conductive channel may have resulted, in our experiments, in the significant decrease of the interfacial electrical conductivity with WCM concentration. An increase in dispersion may be effective in obtaining, simultaneously, a high WCM concentration and a high electrical conductivity. A similar conclusion was reached in the previous case of WCM concentration vs. mechanical stability investigation, although the thickened dilute WCM interlayer strategy may also be applicable in this case. In addition, WCM particles with sizes much smaller than the total interlayer thickness may improve the density–conductivity trade-off. Moreover, the doping of electroconductive polymers, such as polyacetylene or polythiophene [22], in PAM may work effectively to produce a higher electrical conductivity.

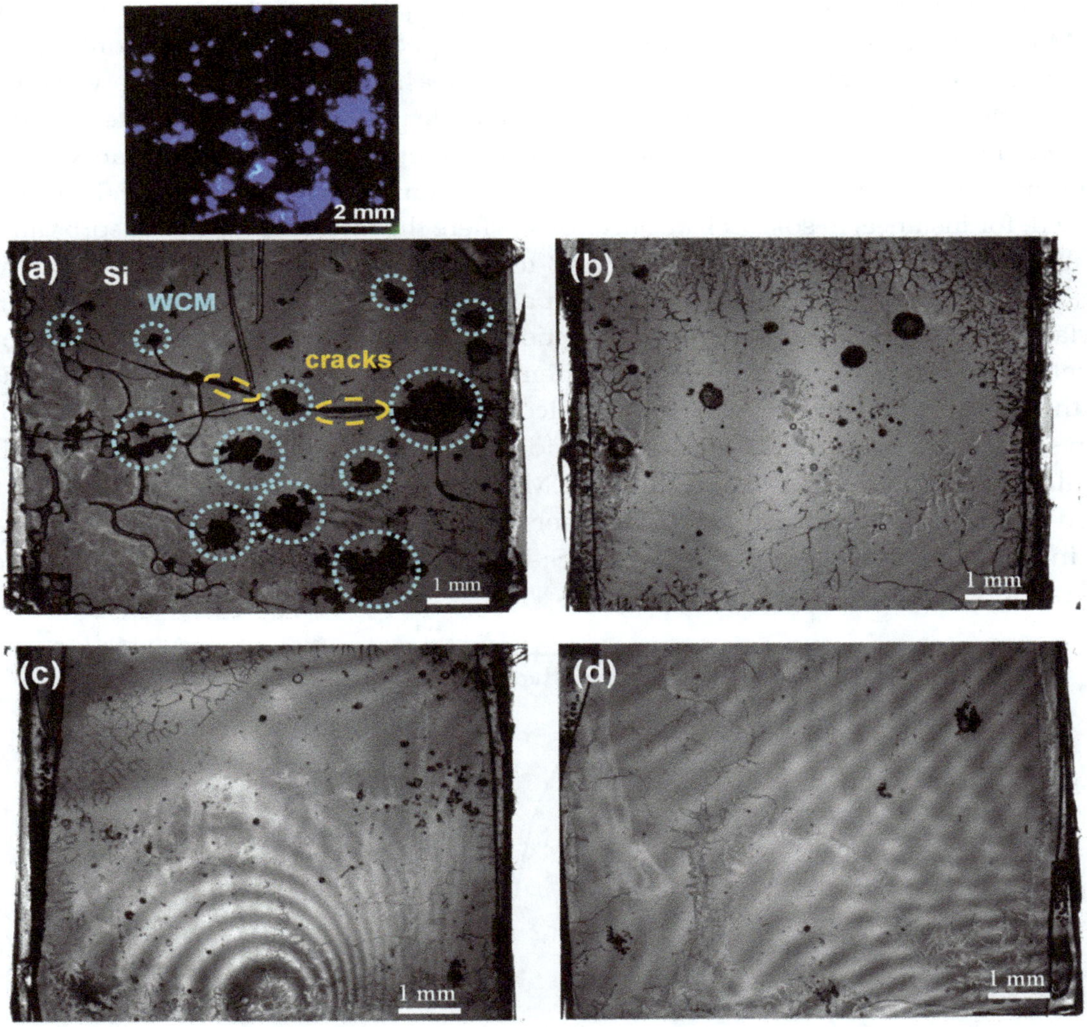

Figure 4. Infrared inner images of bonded interfaces (WCM concentration: (**a**) 1×10^{-2} g/mL, (**b**) 5×10^{-3} g/mL, (**c**) 1×10^{-3} g/mL, (**d**) 1×10^{-4} g/mL). The accompanying image (top) is an optical photograph of the debonded Si surface after the intentional debonding of the bonded sample of (**a**) under an ultraviolet lamp.

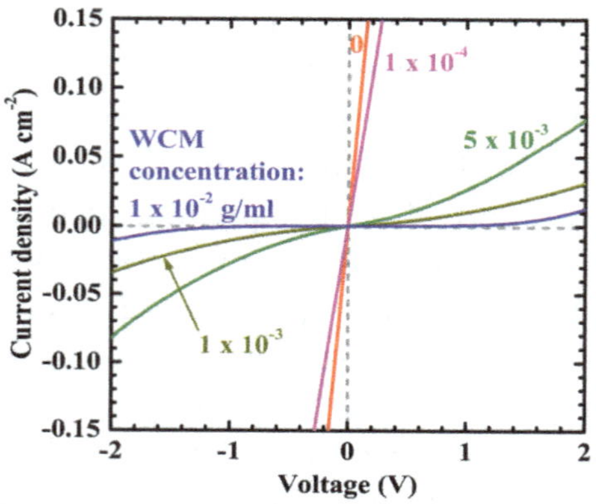

Figure 5. Interfacial current–voltage (*I–V*) characteristics on WCM concentration.

The inset of Figure 6 shows an optical photograph of the bonded glass plate sample, prepared for optical transmission measurements, under an ultraviolet lamp. In the snapshot of the inset of Figure 6, it is distinctly observed that the WCM at the bonded interface between the glass plates converts the incident ultraviolet light into visible blue light output. The transmittance spectrum of the bonded glass samples with WCM, along with the spectrum of the incident light, is shown in Figure 6. Note that, in our fiber-coupled optical transmission measurement setup, the sample-transmitted light scattered by the WCM particles is poorly coupled to the detection spectrometer, and therefore, the integration times of the spectrometer were adjusted so as to enable a reasonable comparison between the incident and the transmitted light. The change in the spectrum of the photon count in Figure 6 verifies that the WCM embedded in the bonded interface converted the incident light peaking at approximately 380 nm into another bundle of light peaking at approximately 460 nm. Such an optical wavelength conversion can be utilized, for instance, in solar cell applications, where the interfacial WCM absorbs an ultraviolet light (380 nm) and emits a visible light (460 nm) to fit the solar spectral irradiance to the spectral sensitivity of crystalline silicon absorption. This particular WCM can thus be utilized, for example, at the interfaces of (Al)(In)GaN/Si-based multijunction solar cells for the photocurrent improvement of the Si subcells. However, our WCM-bonding scheme can be used regardless of the material species, and thus, the most suitable combination of semiconductors can be selected, on demand, for each application situation. For example, stacked hybrid optical transmitters [4,7] in photonic integrated circuits could benefit from down-converting WCMs for longer wavelengths by enabling the use of GaAs-based or InP-based high-power diode lasers emitting at the red to the 900 nm region to be converted into the optimal wavelengths for the Si-based optical waveguides or silica-based optical fibers (1.3–1.6 μm).

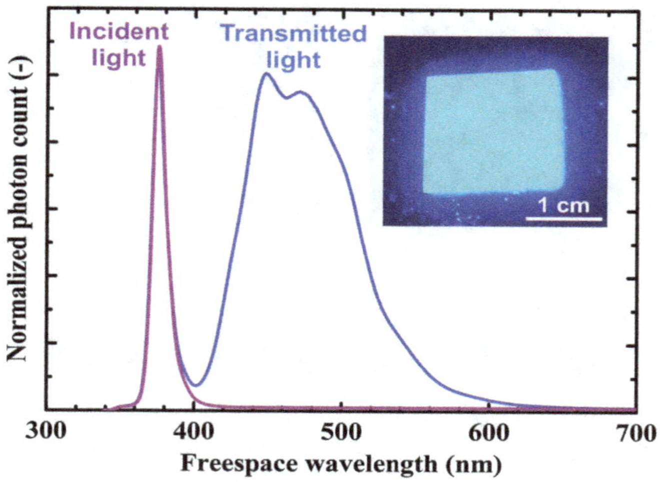

Figure 6. Transmission spectrum of the bonded samples with WCM and spectrum of the incident light. Inset: optical photograph of the bonded sample with WCM under an ultraviolet lamp.

4. Conclusions

In this study, we proposed and experimentally validated WCM-mediated semiconductor wafer bonding. This scheme could simultaneously support bond formation and interfacial function generation and thus potentially realize low-cost and high-throughput device production. The wavelength-converting heterointerfaces could improve, for example, multijunction solar cell efficiencies via photon management and current matching among the subcells and hybrid photonic-integrated circuitry performances by delivering the most suitable frequencies to each optical transceiver component. The bonding was carried out in ambient air at room temperature and therefore provides a cost advantage with regard to device manufacturing. We used Si wafers as the representative semiconductor material, for the sake of simplicity, but our concept can be extended to other semiconductors, as numerous wafer bonding experimental demonstrations between dissimilar semiconductor materials have been reported thus far [4–11]. We demonstrated the use of a down-conversion material in this study, but our bonding scheme can also be applied to up-conversion materials based, for instance, on harmonic generation [23,24] or triplet–triplet annihilation [25,26], for the efficient up-conversion of optical irradiance with a relatively low intensity, such as sunlight.

Author Contributions: K.T. conceived of the idea of the study and designed the experiments. K.T. set up the wafer bonding facility and carried out the initial bonding tests. K.K. and S.H. carried out the wafer bonding. K.K. fabricated the samples and carried out the measurements. R.I. supported the wafer bonding and the mechanical and electrical measurements. T.Y. supported the infrared imaging. K.K. and K.T. analyzed the data. K.T. supervised the study. K.K. and K.T. composed the manuscript. All the authors contributed to the discussion of the results.

References

1. Nakamura, T. Method of Making a Semiconductor Device. U.S. Patent 3,239,908, 15 March 1966.
2. Lasky, J.B. Wafer bonding for silicon-on-insulator technologies. *Appl. Phys. Lett.* **1986**, *48*, 78–80. [CrossRef]
3. Shimbo, M.; Furukawa, K.; Fukuda, K.; Tanzawa, K. Silicon-to-silicon direct bonding method. *J. Appl. Phys.* **1986**, *60*, 2987–2989. [CrossRef]
4. Van Campenhout, J.; Rojo-Romeo, P.; Regreny, P.; Seassal, C.; Van Thourhout, D.; Verstuyft, S.; Di Cioccio, L.; Fedeli, J.M.; Lagahe, C.; Baets, R. Electrically pumped InP-based microdisk lasers integrated with a nanophotonic silicon-on-insulator waveguide circuit. *Opt. Express* **2007**, *15*, 6744–6749. [CrossRef] [PubMed]
5. Tanabe, K.; Nomura, M.; Guimard, D.; Iwamoto, S.; Arakawa, Y. Room temperature continuous wave operation of InAs/GaAs quantum dot photonic crystal nanocavity laser on silicon substrate. *Opt. Express* **2009**, *17*, 7036–7042. [CrossRef] [PubMed]
6. Tanabe, K.; Watanabe, K.; Arakawa, Y. III-V/Si hybrid photonic devices by direct fusion bonding. *Sci. Rep.* **2012**, *2*, 349. [CrossRef]
7. Crosnier, G.; Sanchez, D.; Bouchoule, S.; Monnier, P.; Beaudoin, G.; Sagnes, I.; Raj, R.; Raineri, F. Hybrid indium phosphide-on-silicon nanolaser diode. *Nat. Photonics* **2017**, *11*, 297–300. [CrossRef]
8. Wong, W.S.; Sands, T.; Cheung, N.W.; Kneissl, M.; Bour, D.P.; Mei, P.; Romano, L.T.; Johnson, N.M. $In_xGa_{1-x}N$ light emitting diodes on Si substrates fabricated by Pd-In metal bonding and laser lift-off. *Appl. Phys. Lett.* **2000**, *77*, 2822–2824. [CrossRef]
9. Park, H.; Fang, A.W.; Jones, R.; Cohen, O.; Raday, O.; Sysak, M.N.; Paniccia, M.J.; Bowers, J.E. A hybrid AlGaInAs-silicon evanescent waveguide photodetector. *Opt. Express* **2007**, *15*, 6044–6052. [CrossRef]
10. Tanabe, K.; Fontcuberta i Morral, A.; Atwater, H.A.; Aiken, D.J.; Wanlass, M.W. Direct-bonded GaAs/InGaAs solar cell. *Appl. Phys. Lett.* **2006**, *89*, 102106. [CrossRef]
11. Dimroth, F.; Grave, M.; Beutel, P.; Fiedeler, U.; Karcher, C.; Tibbits, T.N.D.; Oliva, E.; Siefer, G.; Schachtner, M.; Wekkeli, A.; et al. Wafer bonded four-junction GaInP/GaAs//GaInAsP/GaInAs concentrator solar cells with 44.7% efficiency. *Prog. Photovolt.* **2014**, *22*, 277–282. [CrossRef]
12. Naito, T.; Tanabe, K. Fabrication of Si/graphene/Si Double Heterostructures by Semiconductor Wafer Bonding towards Future Applications in Optoelectronics. *Nanomaterials* **2018**, *8*, 1048. [CrossRef] [PubMed]
13. Gibart, P.; Auzel, F.; Guillaume, J.; Zahraman, K. Below band-gap IR response of substrate-free GaAs solar cells using two-photon up-conversion. *Jpn. J. Appl. Phys.* **1996**, *35*, 4401–4402. [CrossRef]

14. Fukuda, T.; Kato, S.; Kin, E.; Okaniwa, K.; Morikawa, H.; Honda, Z.; Kamata, N. Wavelength conversion film with glass coated Eu chelate for enhanced silicon-photovoltaic cell performance. *Opt. Mater.* **2009**, *32*, 22–25. [CrossRef]
15. Hill, M.T.; Dorren, H.J.S.; de Vries, T.; Leijtens, X.J.M.; den Besten, J.H.; Smalbrugge, B.; Oei, Y.S.; Binsma, H.; Khoe, G.D.; Smit, M.K. A fast low-power optical memory based on coupled micro-ring lasers. *Nature* **2004**, *432*, 206–209. [CrossRef]
16. Urino, Y.; Shimizu, T.; Okano, M.; Hatori, N.; Ishizaka, M.; Yamamoto, T.; Baba, T.; Akagawa, T.; Akiyama, S.; Usuki, T.; et al. First demonstration of high density optical interconnects integrated with lasers, optical modulators, and photodetectors on single silicon substrate. *Opt. Express* **2011**, *19*, B159–B165. [CrossRef]
17. Matsuda, N.; Le Jeannic, H.; Fukuda, H.; Tsuchizawa, T.; Munro, W.J.; Shimizu, K.; Yamada, K.; Tokura, Y.; Takesue, H. A monolithically integrated polarization entangled photon pair source on a silicon chip. *Sci. Rep.* **2012**, *2*, 817. [CrossRef]
18. Sun, C.; Wade, M.T.; Lee, Y.; Orcutt, J.S.; Alloatti, L.; Georgas, M.S.; Waterman, A.S.; Shainline, J.M.; Avizienis, R.R.; Lin, S.; et al. Single-chip microprocessor that communicates directly using light. *Nature* **2015**, *528*, 534–538. [CrossRef]
19. Kawamitsu, S.; Nakanishi, S.; Kurogi, M.; Onouchi, H. Fluorescent Dye Compound Having Benzotriazole Structure and Wavelength-Converting Encapsulant Composition Using Same. U.S. Patent US2017210903, 27 July 2017.
20. Nam, H.G.; Nam, M.G.; Yoo, P.J.; Kim, J. Hydrogen bonding-based strongly adhesive coacervate hydrogels synthesized using poly(*N*-vinylpyrrolidone) and tannic acid. *Soft Matter* **2019**, *15*, 785–791. [CrossRef]
21. Zhang, Q.; Liu, X.; Duan, L.; Gao, G. Ultra-stretchable wearable strain sensors based on skin-inspired adhesive, tough and conductive hydrogels. *Chem. Eng. J.* **2019**, *365*, 10–19. [CrossRef]
22. Mengistie, D.A.; Ibrahem, M.A.; Wang, P.C.; Chu, C.W. Highly conductive PEDOT: PSS treated with formic acid for ITO-free polymer solar cells. *ACS Appl. Mater. Interfaces* **2014**, *6*, 2292–2299. [CrossRef]
23. Miller, G.D.; Batchko, R.G.; Tulloch, W.M.; Weise, D.R.; Fejer, M.M.; Byer, R.L. 42%-efficient single-pass cw second-harmonic generation in periodically poled lithium niobate. *Opt. Lett.* **1997**, *22*, 1834–1836. [CrossRef] [PubMed]
24. Carmon, T.; Vahala, K.J. Visible continuous emission from a silica microphotonic device by third-harmonic generation. *Nat. Phys.* **2007**, *3*, 430–435. [CrossRef]
25. Murakami, Y.; Himuro, Y.; Ito, T.; Morita, R.; Niimi, K.; Kiyoyanagi, N. Transparent and nonflammable ionogel photon upconverters and their solute transport properties. *J. Phys. Chem. B* **2016**, *120*, 748–755. [CrossRef] [PubMed]
26. Ishii, A.; Hasegawa, M. Solar-pumping upconversion of interfacial coordination nanoparticles. *Sci. Rep.* **2017**, *7*, 41446. [CrossRef] [PubMed]

13

Investigating the Effect of Carbon Nanotube Diameter and Wall Number in Carbon Nanotube/Silicon Heterojunction Solar Cells

Tom Grace [1], LePing Yu [1], Christopher Gibson [1], Daniel Tune [1,2], Huda Alturaif [3], Zeid Al Othman [3] and Joseph Shapter [1,*]

[1] Centre for Nanoscale Science & Technology (CNST), Flinders University of South Australia, South Australia 5042, Australia; tom.grace@flinders.edu.au (T.G.); leping.yu@flinders.edu.au (L.Y.); christopher.gibson@flinders.edu.au (C.G.); daniel.tune@flinders.edu.au (D.T.)

[2] Institute of Nanotechnology, Karlsruhe Institute of Technology, 76021 Karlsruhe, Germany

[3] Advanced Material Research Chair, Chemistry Department, College of Science, King Saud University, Riyadh 11451, Saudi Arabia; h.a.alturaif@hotmail.com (H.A.); zaothman@ksu.edu.sa (Z.A.O.)

* Correspondence: joe.shapter@flinders.ed.au

Academic Editors: Guanying Chen, Zhijun Ning and Hans Agren

Abstract: Suspensions of single-walled, double-walled and multi-walled carbon nanotubes (CNTs) were generated in the same solvent at similar concentrations. Films were fabricated from these suspensions and used in carbon nanotube/silicon heterojunction solar cells and their properties were compared with reference to the number of walls in the nanotube samples. It was found that single-walled nanotubes generally produced more favorable results; however, the double and multi-walled nanotube films used in this study yielded cells with higher open circuit voltages. It was also determined that post fabrication treatments applied to the nanotube films have a lesser effect on multi-walled nanotubes than on the other two types.

Keywords: carbon nanotubes; solar cells; carbon nanotube (CNT)/Si heterojunction solar cells; double-walled carbon nanotube (DWCNT); multi-walled carbon nanotubes (MWCNT)

---

1. Introduction

The search for efficient, low-cost renewable energy sources is one of great importance in the modern world. As the conventional fossil fuel sources of electricity become scarcer, and are discovered to cause severe problems for our planet's climate, it becomes all the more imperative to seek out new and innovative ways of exploiting sustainable resources.

Solar energy is the quintessential sustainable resource, however there are some significant disadvantages of current commercial solar cells. Firstly, silicon solar panels have a high manufacturing cost due to the need for high purity, processed silicon to produce high efficiency solar panels. While other semiconducting materials may be used in photovoltaic (PV) devices, they will generally consist of alloys of rare and/or toxic elements such as arsenic, cadmium, indium, gallium, germanium and ruthenium [1]. Innovations in PV technologies are making solar cells more and more competitive. Some methods of improving the economic viability of solar capture technology involve investigating solar cells containing organic molecules, quantum dots, or dye-sensitized solar cells [2]. However, this paper investigates a cell design that combines the already established good solar properties of silicon with an inexpensive material, namely carbon nanotubes (CNTs). Carbon nanotubes are considered to be an exciting prospect for solar cell integration due to a wide range of unique and interesting properties, such as; high charge carrier mobilities, ballistic transport properties, high optical transmittance, low light reflectance [3–5], and photoelectrochemical effects under light irradiation [3,5–9]. In addition, there are many different types of nanotubes to choose from with a wide array of possible bandgaps, thus the bandgap can be "tuned" for the situation. CNTs are also highly resistant to damage, whether it be mechanical, chemical or radiation induced [7]. Coupled with their potential low cost due to the abundance of carbon [4] and the miniscule quantities of material required in applications, it is clear why they are an exciting material for research. Solar cells consisting of a heterogeneous junction (heterojunction) between a (generally n-doped) silicon substrate and a CNT film have been shown to be an appealing prospect for the future of solar energy capture technologies.

In 2007, Wei *et al.* designed a new kind of cell in which CNTs function not only as transport charge carriers, but also assist in exciton separation [10–13], using double-walled CNTs (DWCNTs) deposited via water expansion and aqueous film transfer of an as grown chemical vapor deposition (CVD) film [11,12]. While these cells had a power conversion efficiency of only 1.3% (compared to commercial silicon cells at generally 13%–25%) [14,15] many improvements have since been made to the cell design and doping methods, with 15% efficiencies reported in 2012 [2] and 17% efficiencies reported in 2015 [6,16]. Although, for the most part, these efficiencies could only be achieved via the use of an anti-reflection coating on the surface of the solar cell, with the 15% value achieved by depositing a TiO_2 layer over a solar cell which previously achieve efficiencies of around 8% and the 17% efficiency was achieved through the use of a molybdenum oxide layer on a solar cell design with an intrinsic efficiency of 11.1%. In terms of solar cell intrinsic efficiencies, without an anti-reflection layer, the highest in 2013 was 11.5% [17], which was improved to a record high intrinsic efficiency of 13.85% in 2015 [18]. Thus, in less than a decade the cell efficiency has been improved by a factor of 10. This rapid improvement is one of the reasons for much excitement around this solar cell design. In addition, these cells are interesting for future research as their manufacturing process is both simple and scalable [10]. The typical architecture for these solar cells is much like that of a single junction crystalline silicon solar cell with the emitter layer replaced by a film of p-doped CNTs [11]. While Wei's initial design used DWCNTs, most studies have since used single-walled carbon nanotubes (SWCNTs), and multi-walled carbon nanotubes (MWCNTs) may also be used [10,19]. In all of these cases, the CNT film acts as a component of the heterojunction to enable charge separation, as a highly conductive network for charge transport and as a transparent electrode to allow good light illumination and charge collection [10].

Li *et al.* [20] found that post-fabrication treatment of a SWCNT layer with the *p*-type dopant thionyl chloride ($SOCl_2$) increased the power conversion efficiency of the cells by over 45% by lowering the sheet resistance, and increasing the short circuit current density (J_{SC}) and open circuit voltage. Hall Effect measurements showed that the $SOCl_2$ treatment led to an increase of carrier density from 3.1×10^{15} to 4.6×10^{17} cm^{-2} and an improvement from 0.23 to 1.02 cm^2 V^{-1} s^{-1} of the effective mobilities [11,20]. In addition, it was found that $SOCl_2$ treatment adjusted the Fermi level and shifted the major conduction mechanism in the SWCNT layer from hopping towards tunneling [21,22].

Jia *et al.* [23] performed the first comparison between SWCNTs, DWCNTs and MWCNTs for use in CNT/Si cells in terms of area density of the films. It was found that SWCNTs are superior to MWCNTs at low densities and that the density (and thus optical transmittance) is vitally important in the performance of these cells. Increasing film transparency (lowering CNT density) increases the efficiency of the cells by allowing more light to reach the silicon, while decreasing the transparency (increasing the CNT density) increases the efficiency by lowering the sheet resistance across the film [11,23]. Thus, some optimal thickness must exist to achieve maximum efficiency. This research team also found their DWCNT cells to be significantly superior to both the SWCNT and MWCNT cells. However, their DWCNT films were produced using a different method to the SWCNT and MWCNT films [23] and were significantly longer and more pristine. This makes it difficult to draw a good comparison, as the nanotube film properties are highly dependent on film morphology [11] and fabrication route.

The aim of the research reported in this paper was to provide a more reliable comparison by creating suspensions of single, double and multi-walled carbon nanotubes under the same solvent conditions and examining the differences in solar cell properties between the different types. This study further improves on previous nanotube comparisons by using nanotubes that were specifically chosen to be of similar length and the films were all produced using the exact same procedure.

Nanotube/Silicon

2. Results

Films were produced for each nanotube sample and were deposited on silicon substrates and imaged with scanning electron microscopy (SEM) to determine the film morphology. The images are shown in Figure 1. It is immediately noticeable that the DWCNT-2 suspension did not form a homogeneous film as the SEM image shows that the nanotubes clump together rather than spread across the membrane during film formation. The poor film morphology is due to issues dispersing DWCNTs in suspension. There are several possible reasons for this poor dispersion including tube length, contaminants and the surface properties of the nanotubes in the sample. Due to the poor film morphologies and coverage obtained for the DWCNT-2 sample, further work with this sample proved unfruitful and thus it will not be discussed in the rest of this study. The DWCNT-1 sample also did not form a fully homogenous surface covering film, however the film coverage was much closer to the SWCNT and MWCNT samples than DWCNT-2. Due to difficulties faced in suspending the DWCNTs, those suspensions were more dilute than the SWCNT or MWCNT suspensions, despite the addition of the same volume of dry nanotubes. It may be that the DWCNTs tend to remain bundled together more in suspension than the other nanotube types and thus form less homogeneous, more clustered films. There does not appear to be a large difference in the film morphology between the SWCNTs and MWCNTs: all three of these samples appear to form good, homogeneous coverings. There are noticible holes in the SWCNT-1 films, which were formed during the vaccuum filtration process when the vacuum was allowed to run for a time after all liquid has passed through. These are likely more prominent in this example as the film formed was thicker than the others. Overall, the SEM images show that all the types of nanotubes form suspensions well enough to produce homogenous films on a substrate and are thus usable for PV solar cell projects.

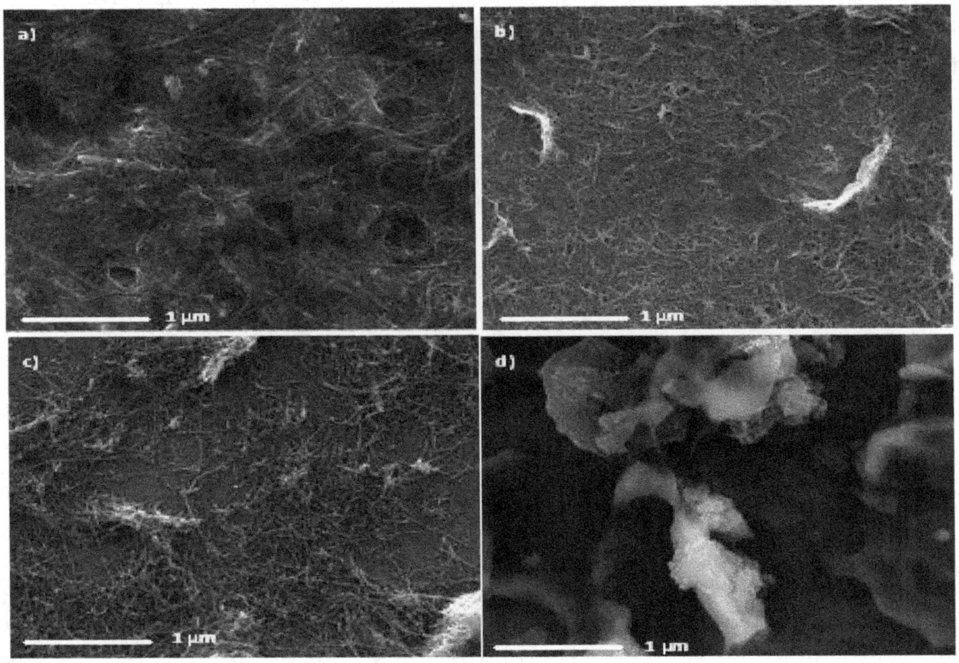

Figure 1. *Cont.*

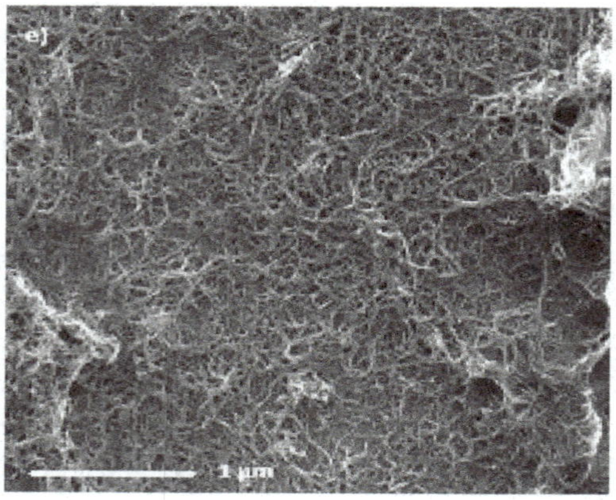

Figure 1. Scanning electron microscopy (SEM) images of various types of carbon nanotube (CNT) samples on Si: (**a**) single-walled carbon nanotube sample 1 (SWCNT-1); (**b**) single-walled carbon nanotube sample 2 (SWCNT-2); (**c**) double-walled carbon nanotube sample 1 (DWCNT-1); (**d**) DWCNT-2; and (**e**) multi-walled carbon nanotube (MWCNT).

Optical absorption spectroscopy was performed on the nanotube films to determine the thickness and was also performed on the nanotube solutions to help confirm the nanotube types. It is expected that the ultra-violet/visible (UV/Vis) spectrum for MWCNT and DWCNT nanotubes will be featureless with an increase in absorption as the light wavelength decreases. A different spectrum is expected for SWCNT nanotubes, for which the van Hove singularities present in the SWCNT density of states lead to characteristic peaks in the optical spectra.

It can be observed from Figure 2 that the SWCNT spectra display peaks in the regions of 600–800 nm due to the S_{11} transition and 350–500 nm regions due to the M_{11} transition, and thus show the presence of SWCNTs in these solutions. This is the expected shape for UV/Vis spectra of single-walled nanotubes, due to their unique density of states allowing electronic transitions in the UV/Vis range. The differences in peak position in the two SWCNT samples are likely caused by species of different chirality being present in each sample. It is also noticeable that the SWCNT-1 sample had much less distinct peaks than the SWCNT-2 sample. This is likely due to the presence of a range of closely related nanotube chiralities in sample SWCNT-1. The spectra of the DWCNT and MWCNT samples are relatively featureless with an increase in absorbance at the lower wavelength end of the spectra as expected.

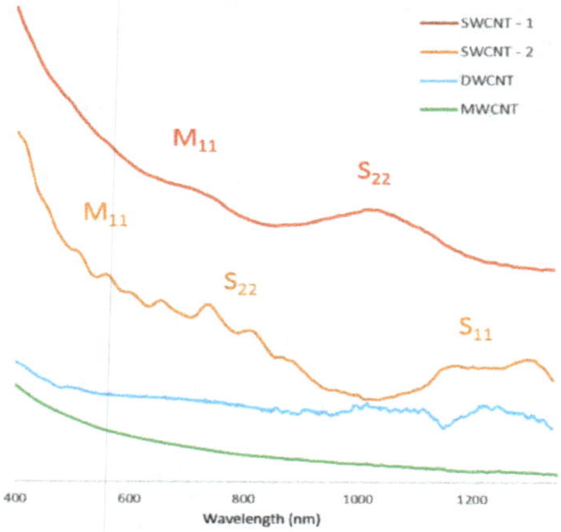

Figure 2. Ultra-violet/visible (UV/Vis) absorption spectra for all as prepared CNT suspensions. The absorption values have been offset to allow easier viewing. Included in the figure are reference labels for semiconducting tube Van Hove singularity transitions S_{11} and S_{22} and metallic tube Van Hove singularity transition M_{11}.

In Raman spectroscopy, carbon nanotubes display characteristic peaks at around 1580 cm^{-1} and at around 1350 cm^{-1} known as the G and D bands, respectively. The G band is characteristic of highly ordered carbon species such as graphite or CNTs while the D band is characteristic of disordered carbon species. Thus the intensity of the G band relative to that of the D band is a strong indicator of crystallinity [24]. A radial breathing mode (RBM) between 100 and 500 cm^{-1} can also be uniquely observed in Raman spectra of carbon nanotubes and is seen as direct evidence for the presence of SWCNTs [25] or DWCNTs [24] in the sample. An overtone of the D mode, known as the G′ band (an overtone of the D band) is from a two-phonon, second-order Raman scattering process and expected to be seen at around 2700 cm^{-1} while a signal at 1550 cm^{-1} is known as a Breit-Wigner-Fano (BWF) band and is attributed to metallic carbons (metallic nanotubes in this case) [25]. Figure 3 displays the spectra obtained for each nanotube sample. As well as the characteristic D and G bands for CNTs the spectra also display the G′ band at 2700 cm^{-1} and while no RBM can be observed in the MWCNT sample. Peaks between 100 and 500 cm^{-1} can be observed in all other samples in Figure 3a. The ratio of the intensities of the D and G bands (D/G ratio) is often used to measure the disorder in carbon nanotube samples. The D/G ratio in the MWCNT sample is significantly different than that of the other four samples, as expected for MWCNTs [24]. The D/G peak height ratios for all nanotube types are shown in Table 1. It can be seen from the values in Table 1 that the SWCNT-1 and DWCNT-1 nanotubes exhibit the lowest ratios, and are thus the most highly ordered nanotube or purest species in this study. The SWCNT-2 sample exhibited a higher D/G ratio, indicating a higher level of disorder or impurity, where as the MWCNT sample gave a ratio above 1, due to the D peak being higher than the G peak, which is expected for MWCNTs [24,25]. The amount of CNT disorder is important to investigate as chemical reactions with nanotubes are more likely to occur at disordered regions than ordered regions.

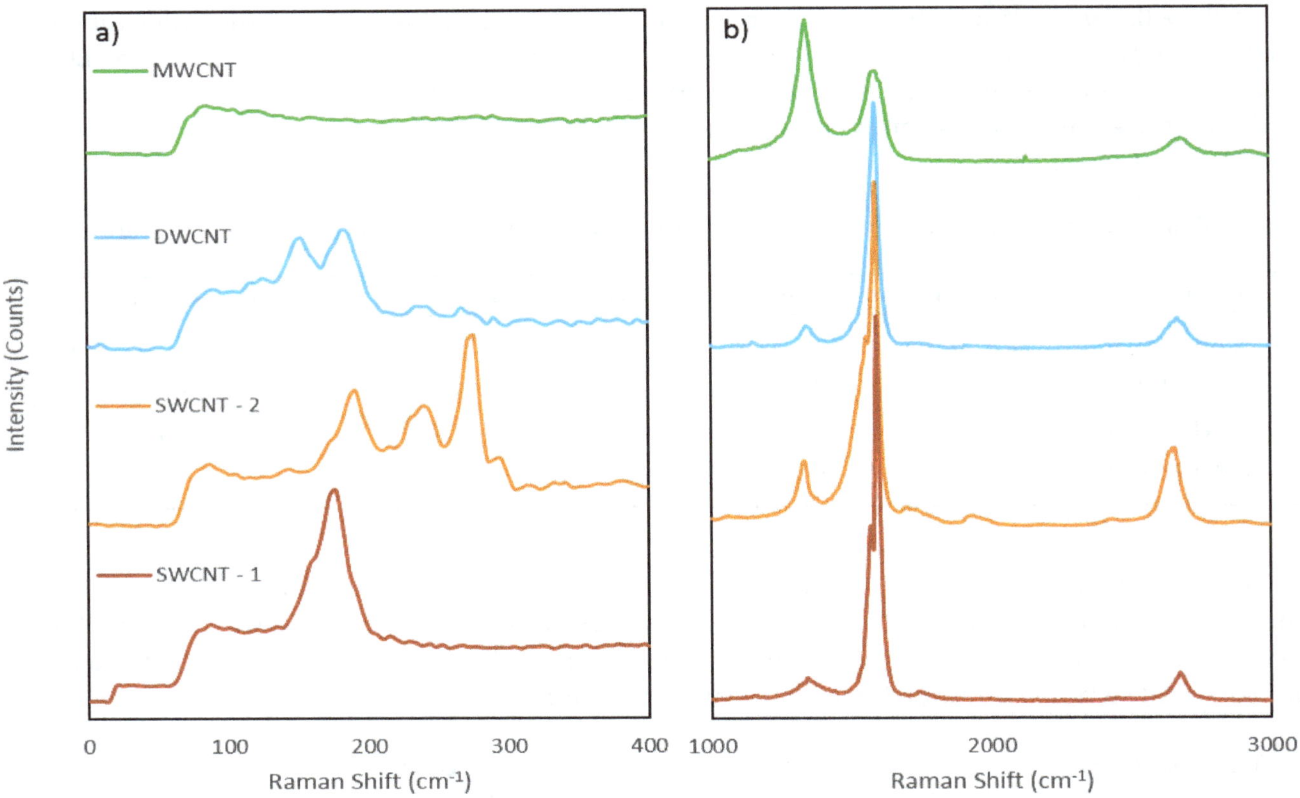

Figure 3. (**a**) Low wavenumber region of the Raman spectra for all CNT samples showing the D, G, G′ and Breit-Wigner-Fano (BWF) bands, the intensity values have been offset to allow for easier viewing. (**b**) High wavenumber region of the Raman spectra for all CNT samples, showing the radial breathing mode (RBM) region. The intensity values have been offset to allow for easier viewing.

Table 1. D band to G band ratios for each nanotube type (two single-walled carbon nanotube samples (SWCNT-1 and SWCNT-2), one double-walled carbon nanotube sample (DWCNT) and one multi-walled carbon nanotube sample (MWCNT)).

Nanotube Sample	SWCNT-1	SWCNT-2	DWCNT	MWCNT
D/G Ratio	0.049	0.171	0.078	1.55

The Raman shift at which the RBM occurs can be used to calculate the nanotube diameter for SWCNTs as the RBM frequency is inversly proportional to the diameter of the tubes [26]. The equation relating the RBM shift to the diameter is: RBM shift $(cm^{-1}) = \frac{\mathrm{A}}{d_t\ (nm)} + \mathrm{B}$ where A has an approximate value of 234 nm$\cdot$ cm^{-1} as determined from first principles (*ab initio*) calculations and B is approximately 10 cm^{-1}, which corrects the intertube interaction frequencies in SWCNT bundles [26,27]. This equation is applied to the RBM data shown in Figure 3a to give a calculated diameter for the SWCNT-1, SWCNT-2 and DWCNT nantube samples and is compared with the manufacturer supplied diameters in Table 2 (note that the value of B was 0 in the DWCNT case as B is a correction for SWCNT bundles). Firsty, the calculation shows a very good agreement with the manufacturer supplied diameters for both of the SWCNT samples, with the calculated value(s) falling within the range given by the manufacturers. The DWCNT calculation does not agree, with the large peaks giving diameters around 0.75 nm smaller than the lowest diameter given by the manufacturer. This discrepancy can be explained by assigning the visible RBM peaks to the inner tubes of the DWCNTs since the interwall difference is known to be around 0.33–0.41 nm [28,29]. To achieve a diameter of >2 nm the RBM peak for the outer tube would have to be at or below 117 cm^{-1}. The two small peaks to the left of the larger peaks on the DWCNT spectrum are closer to this range, with the left most peak occuring at around the expected 117 cm^{-1} and thus these smaller peaks could be due to the outer tubes.

Table 2. Calculation of nanotube diameter from radial breathing mode (RBM) Raman shift.

Nanotube Sample	RBM Raman Shift (cm^{-1})	Calculated Diameter (nm)	Supplied Diameter (nm)
SWCNT-1	177	1.40	1.4–1.5
SWCNT-2	191, 239, 276	1.29, 1.02, 0.88	0.8–1.2
DWCNT Small Peaks	115, 125	2.03, 1.87	2–4
DWCNT Large Peaks	154, 186	1.52, 1.25	2–4

A series of films of different transparencies were prepared to determine suspension volume required to give similar transmittances. The data from these experiments are provided in Table S1. Films of each nanotube sample were produced at approximately the same transmittance for use in the solar cells. All samples displayed good transmittance values between 58% and 65%.

The sheet resistances of the nanotube films at each stage of treatment are shown in Table 3. The data show that the SWCNT films had a lower sheet resistance than the DWCNT and MWCNT films. Since the amount of material in the films was the same, as indicated by the optical measurements, this suggests improved charge carrier transport in the SWCNT films. The changes in sheet resistance with treatment are expected and are due to carrier doping and reductions in tube-tube contact resistance.

Table 3. Average sheet resistance for each carbon nanotube (CNT) sample with treatment ($\Omega \cdot sq^{-1}$).

Film Type with Transmittance Percentage	As Prepared ($\Omega \cdot sq^{-1}$)	HCl Treatment 1 ($\Omega \cdot sq^{-1}$)	Thionyl Chloride Treatment ($\Omega \cdot sq^{-1}$)	HCl Treatment 2 ($\Omega \cdot sq^{-1}$)
SWCNT-1 60%	1440 ± 8.2%	951 ± 1.9%	693 ± 42%	543 ± 2.9%
SWCNT-2 65%	4070 ± 7.0%	3650 ± 2.6%	1880 ± 4.0%	2410 ± 12%
DWCNT 58%	138,000 ± 84%	4190 ± 39%	19,600 ± 138%	2550 ± 38%
MWCNT 60%	3340 ± 6.0%	3520 ± 6.0%	3020 ± 23%	2890 ± 27%

2.1. Solar Cell Performance

Films of all nanotube samples were deposited on the standard solar cell substrates described in the experimental details and their solar cell performance was measured four times: once directly after fabrication and then once after each treatment (HF etch, $SOCl_2$ treatment, and 2nd HF etch).

Figure 4 shows the current-voltage characteristics of the best performing cells for each nanotube type after the complete treatment sequence, and the corresponding dark curves are shown in Figure S1. Both SWCNT samples show similar behavior under illumination, the only difference being a higher short-circuit current density (J_{SC}) for SWCNT-1 and a slightly higher open circuit voltage (V_{OC}) for SWCNT-2. Both the DWCNT and MWCNT samples display lower J_{SC} values. However, it can be seen that the DWCNT and the MWCNT cells produced a significantly higher V_{OC} than all the other cells. This is unexpected as the V_{OC} is determined by the energy levels of the system and should not be greatly affected by a change in nanotube type in this regard. It is likely, however, that this higher V_{OC} is caused by a lower rate of recombination in these nanotube types.

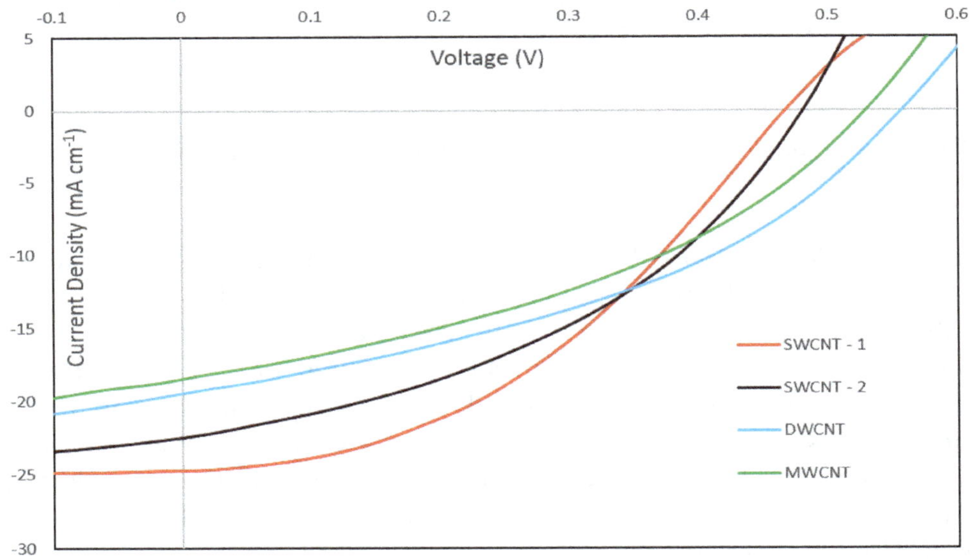

Figure 4. Current density *vs* voltage (J/V) curves for best performing cells for each CNT sample after the second HF etch. Curves for two single-walled carbon nanotube samples (SWCNT-1 and SWCNT-2), one double-walled carbon nanotube sample (DWCNT) and one multi-walled carbon nanotube sample (MWCNT) are shown.

Figure 5 shows how the power conversion efficiency (PCE) of the best performing cell of each nanotube type varies with treatment. By the second HF etch the best cell PCEs ranged from 3.78% for the MWCNT sample to 4.79% for the SWCNT-1 sample. This difference of 1% absolute, or over 25% relative (based on the MWCNT device), between the two nanotube types is quite significant considering the care taken to control other differences between devices and the fact that the average relative error was around 15%. The ratio of the direct current (DC) electrical to optical conductivity, σ_{DC}/σ_{OP} for all films is very similar (see Table S1). The value is slightly higher for the SWCNT-1 film, which likely explains its higher PCE value. The origin of the large increase in performance of the DWCNT devices after the 2nd HF treatment is unknown. It is not consistent with a model in which the $SOCl_2$ is the *p*-type dopant increasing conductivity and lowering Fermi energy relative to the silicon, and thus improving device performance, as this effect should be observed after the $SOCl_2$ treatment. It could possibly be due to acid digestion of the non-nanotube carbonaceous impurities (up to 40% by weight in the starting material) by the sequence of aggressive chemical treatments, resulting in better contact between the (purer) nanotube film and the underlying silicon, but further investigations are required to shed light on this.

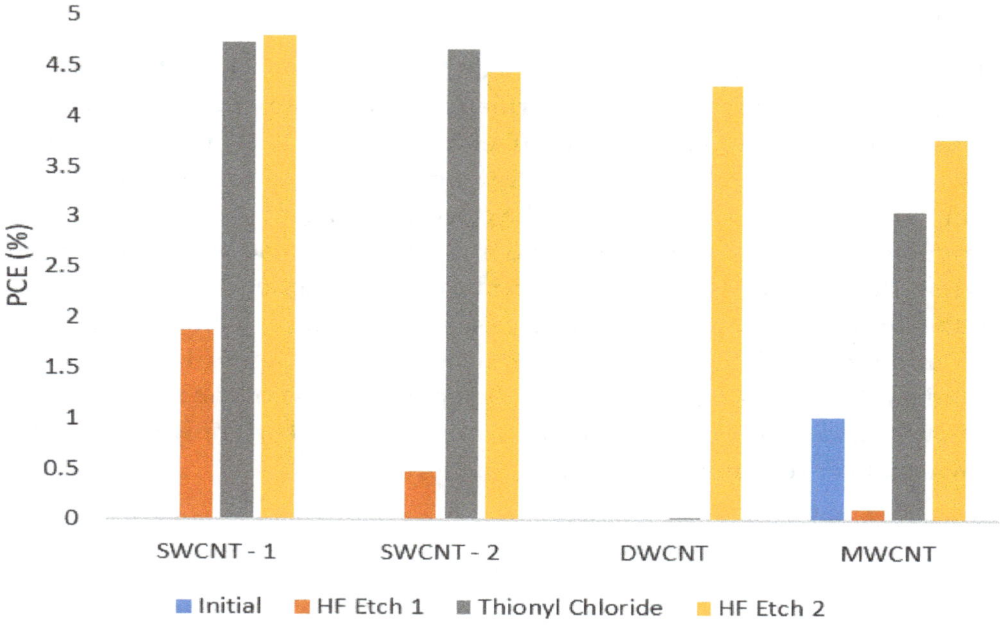

Figure 5. Solar cell efficiencies (%) for all nanotube types with treatment. PCE: power conversion efficiency.

Cells containing MWCNTs generally performed better at the initial stage of testing, but did not improve to the same extent as the SWCNT and DWCNT cells with doping. This can be explained by considering the surface area to volume ratio of MWCNTs compared to SWCNTs and DWCNT. Multi-walled nanotubes have a larger bulk than the other types for the same surface area, and thus surface acting treatments such as HF and $SOCl_2$ will not have the same effect on bulk MWCNT films. Additionally, unlike SWCNTs, the large diameter and complex mixing of wall types and energy states in MWCNTs means they only have metallic character and therefore there are no tube-tube energy barriers to be lowered by doping. Thus, the MWCNTs are less affected by the acid and the thionyl chloride treatments used in this study.

3. Discussion

Table 4 shows a breakdown of the solar cell data. It can be seen that the highest J_{SC} was achieved by the cell that gives the highest efficiency. This is expected, as the efficiency is proportional to the current density. The efficiency is also proportional to the open circuit voltage (V_{OC}) and the fill factor (FF). Thus the fact that the SWCNT-1 sample also achieved the equal highest fill factor is expected. It is surprising that the open circuit voltage is lower for both SWCNT samples than for the DWCNT and MWCNT samples. The higher voltages for these samples can be attributed to the smaller saturation current (J_{SAT}) values for these cells. A smaller J_{SAT} indicates less current flowing in the undesired direction in the cell. The relation between J_{SAT} and V_{OC} for the best performing cells for each sample can be seen in Figure S2. The huge error values attached to the J_{SAT} measurements are due to the order of magnitude differences between the J_{SAT} values in each cell for each sample. The improved efficiencies seen in the SWCNT samples over the DWCNT and MWCNT samples can be partially understood by observing the difference between the estimated shunt resistance (R_{shunt}) and series resistance (R_{series}). The series resistance value represents the amount of resistance opposing current flow in the desired direction and thus the ideal situation for a solar cell is to have a low R_{series}. Conversely, the shunt resistance value represents the amount of resistance opposing current flow over "short-cuts" in the cell circuit, thus the ideal value for R_{shunt} is high. Table 4 shows that the difference between R_{shunt} and R_{series} is an order of magnitude in the SWCNT cases, but is less than this in the DWCNT and MWCNT cases. This is a likely explanation for the slightly improved PCE for the SWCNT samples. The SWCNT-1 sample produced a higher diode ideality than the other three samples, given that the ideal value is 1 it is apparent that this cell was not as ideal as the other samples. This is likely indicative of a poorer contact with the silicon substrate in the solar cell, when compared with the other samples.

Table 4. Solar cell properties for best performing cells for each CNT sample in bold text, average properties and error values for sets of three (two for the Carbon Allotropes DWCNT) cells in regular text. J_{SC}: Short circuit current density. V_{OC}: Open circuit voltage. PCE: Power conversion efficiency. FF: fill factor.

	SWCNT-1	SWCNT-2	DWCNT	MWCNT
J_{SC} (mA cm^{-2})	**24.7;** 24 ± 4.3%	**22.5;** 22.2 ± 1.5%	**19.5;** 18.9 ± 4.5%	**18.4;** 18.1 ± 2.5%
V_{OC} (V)	**0.468;** 0.427 ± 9.17%	**0.483;** 0.449 ± 7.8%	**0.58;** 0.553 ± 1.3%	**0.533;** 0.465 ± 17.6%
FF	**0.41;** 0.41 ± 8.64%	**0.41;** 0.39 ± 5.13%	**0.40;** 0.40 ± 0.0%	**0.38;** 0.35 ± 8.7%
PCE%	**4.79;** 4.21 ± 18.74%	**4.45;** 3.90 ± 13.8%	**4.31;** 4.14 ± 6.0%	**3.78;** 3.03 ± 28.2%
R_{shunt} (Ohm)	**6.42 × 10^3;** 2.81 × 10^3 ± 112%	**1.06 × 10^3;** 7.91 × 10^2 ± 30.2%	**8.42 × 10^2;** 9.61 × 10^2 ± 17.5%	**8.07 × 10^2;** 6.65 × 10^2 ± 24.8%
R_{series} (Ohm)	**1.21 × 10^2;** 1.08 × 10^2 ± 16.3%	**1.11 × 10^2;** 1.12 × 10^2 ± 9.4%	**1.44 × 10^2;** 1.57 × 10^2 ± 11.3%	**1.64 × 10^2;** 1.68 × 10^2 ± 2.1%
Diode Ideality	**2.37;** 3.27 ± 23.96%	**1.98;** 1.95 ± 16.4%	**1.67;** 1.93 ± 19.3%	**1.94;** 2.75 ± 33.2%
J_{SAT} (mA cm^{-2})	**2.20 × 10^{-3};** 2.45 × 10^{-1} ± 106%	**4.64 × 10^{-4};** 1.81 × 10^{-3} ± 145%	**1.35 × 10^{-5};** 1.26 × 10^{-4} ± 126%	**1.31 × 10^{-4};** 1.39 × 10^{-1} ± 171%
Film Transmittance (%) Final Sheet Resistance of Film (Ω· sq^{-1})	60 543 ± 2.9%	65 2410 ± 12%	58 2550 ± 38%	60 2890 ± 27%

Overall the SWCNT samples performed better than the other nanotube types. This is expected from the sheet resistance measurements, as the SWCNT samples exhibited lower resistance after the final treatment. This improved resistance is likely due to the better morphology of the SWCNT films compared to the DWCNT films. There has been some work to suggest that DWCNTs should conduct charge better [30]. The fact that the DWCNT sample produces a poorer film is likely traceable to the fact that the suspension is more difficult to make and the lower purity of most DWCNT samples. The MWCNT sample showed a good film morphology and overall good sheet resistance. However the resistance of these films did not decrease with treatment at the same rate as for the SWCNT films.

4. Materials and Methods

Five nanotube samples were used in this study, as shown in Table 5.

Table 5. Types of nanotube samples used.

Type	Company	Diameter (nm)	Length (nm)	Purity (%)
SWCNT-1	Carbon Solutions (Riverside, CA, USA)	1.4–1.5	500–1500	90
SWCNT-2	NanoIntegris (Boisbriand, QC, Canada)	0.8–1.2	100–1000	95
DWCNT-1	Carbon Allotropes ((Kensington, NSW, Australia)	2–4	<1500	>>60
DWCNT-2	Sigma-Aldrich (St Louis, MI, USA)	3.5	3000	90
MWCNT	Sigma-Aldrich	9.5	1500	95

Each nanotube suspension was produced in the same manner. The nanotubes (95 mg) were added to an aqueous TritonX-100 solution (1% v/v, 50 mL) (Sigma-Aldrich, St. Louis, MI, USA) and bath sonicated ($\approx$50 W_{RMS} (root mean squared Watts), 3 $\times$ 20 min intervals, Elmasonic S 30H (Singen, Germany). In between each sonication the sonicating bath water was changed to prevent the suspension heating too much during sonication. The resulting sonicated suspension was centrifuged (1 h, 17,500 g, Beckmann-Coulter Allegra X-22 (Brea, CA, USA) then the upper two thirds of the liquid in each tube was carefully extracted, combined, and centrifuged again with the same parameters. The upper two thirds of the liquid in each tube was carefully extracted and combined. The remaining third contained black lumps of unsuspended carbon and was discarded.

The photovoltaic cell substrates were produced by cutting silicon pieces of approximately 1.5 cm^2 from a wafer of n-doped (phosphorous) silicon with a 100 nm oxide layer on one side. Each piece was patterned using a positive photoresist (Methoxypropyl acetate) applied via spin coating (30 s, 3000 rpm) and soft baked (100 °C, 1 min). The resist was exposed to UV light through a mask and the exposed resist was dissolved in a developer solution (trimethyl ammonium hydroxide) to leave an active area of 0.079 cm^2 still covered. The substrate was sputter coated with a 5 nm layer of chromium and a 145 nm layer of gold. The thickness is controlled by a quartz crystal microbalance. The gold coated substrates were submerged in acetone (30 min) before sonicating briefly to dissolve the unexposed resist. One drop of buffered oxide etch (BOE, 6:1 ratio of 40% NH_4F and 49% hydrofluoric acid (HF)) was placed on the active area for 90 s in order to remove the 100 nm thermal oxide and allow the nanotubes to contact the silicon [31]. A schematic of the photovoltaic cell substrate is shown in Figure 6.

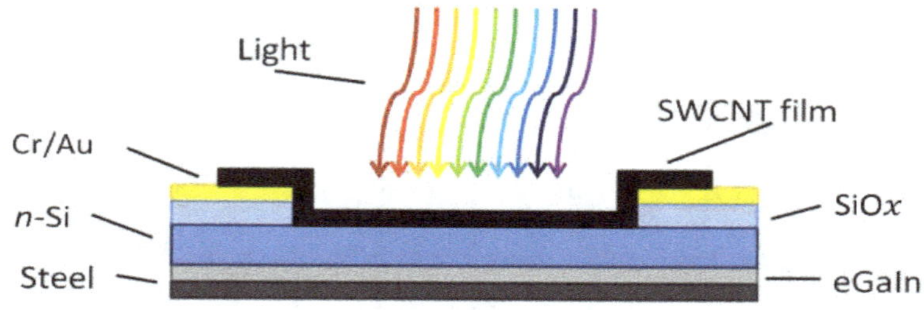

Figure 6. Simple schematic of the cells used in this experiment (Not to scale). Gallium indium eutectic (eGaIn).

The nanotube films were produced via vacuum filtration. This was accomplished by first mixing the required amount of nanotube suspension (dependent on the concentration of the suspension in question, can vary from tens of microliters to up to 15 mL) with MilliQ water (Kansas City, MO, USA) to produce a solution of 250 mL. This solution was filtered using a vacuum produced with a water aspirator through a series of two microporous filter papers. The bottom filter paper (VSWP, Millipore (Billerica, MA, USA), 0.025 μm pore size) was patterned with holes the size of the desired nanotube films (0.5 cm^2 in this case), while the top filter paper (HAWP, Millipore (Billerica, MA, USA), 0.45 μm pore size) was unpatterned. The difference in flow rate through the filter papers causes the solution to pass preferentially through the top film where the bottom film is patterned. The nanotubes are thus caught by the top film in the same shape as the template film. The solution that had passed through both films was passed through the filtration apparatus two more times, to ensure enough nanotubes were retained on the top film. Pure MilliQ water was then passed through to wash out Triton X-100 surfactant remaining in the nanotube film. In most cases in these experiments the template used in each filtration produced four 0.5 cm^2 films, one for attachment to glass to measure the optical transmittance, sheet resistance and Raman spectra and the other three for attachment to solar cells.

Nanotube films were attached to either glass (prewashed in ethanol) or the silicon substrate in the same manner. The films were cut from the filter paper and placed nanotube side down on the substrate. They were wetted with a small drop of water and sandwiched with a piece of Teflon and a glass piece and clamped together. The clamped substrate was heated at around 80 °C for 15 min then left to cool in the dark for 1 h. The substrates were washed in acetone (first wash 30 min, second and third wash 30 min with stirring) to remove the attached filter paper. To complete cell preparation, the reverse side of each piece of silicon was manually scratched to remove the oxide layer. A gallium indium eutectic (eGaln) was applied to the scratched surface and then a piece of stainless steel was attached as the back contact.

The cells were tested after the steel attachment and then a further 3 times after different post fabrication treatment steps. Firstly a drop of 2% HF was applied to the active area to etch silicon oxide formed between the silicon and the nanotube film produced during the film attachment step. Secondly, the nanotube film was treated with a few drops of thionyl chloride ($SOCl_2$) and left until it evaporated, the residue was washed with ethanol prior to testing. The last step was another 2% HF treatment, which is observed to significantly improve performance, though the mechanism for this is still unclear [32].

At each stage of treatment the solar cells were tested by applying contacts to the back and front electrodes and reading the current density as the voltage is ramped from 1 V to −1 V. This is performed both in the absence of light and under illumination provided by a solar spectrum simulator at 100 $mW \cdot cm^{-1}$ to measure both a "dark" and "light" curve the light is filtered through an AM1.5G filter (Irvine, CA, USA). The irradiance at the sample was kept constant by measuring with a silicon reference cell (PV Measurements, National Institute of Standards and Technology (NIST)-traceable calibration). The information was collected using a Keithley 2400 SourceMeter (Solon, OH, USA) attached to a PC running a program written in LabView (Austin, TX, USA).

The amount of light that passes through the CNT film to reach the CNT/Si junction directly affects the amount of energy that the cell can produce. Thus, it was important to perform nanotube type comparisons between films that allowed a similar amount of light to pass through. A series of films were produced from each sample using different amounts of nanotube suspension. The average light transmittance over a wavelength range of 300–1100 nm was determined for each sample and samples with the same or similar percentage transmittances were used on solar cells for comparative experiments.

Sheet resistance measurements were performed on nanotube films using a four point probe attached to Keithlink software (Solon, OH, USA), three sets of five measurements were performed on each film at different orientations and the results were averaged. The resistance was measured at each stage of the solar cell treatment process. However, as HF etches glass, a 2% HCl treatment was performed for 10 s as per the HF treatment. Scanning electron microscopy (SEM) was performed using an FEI Inspect F50 (Hillsboro, OR, USA) on nanotube films mounted on silicon wafers.

Raman spectra of the various nanotube types were obtained using a Witec Alpha R confocal Raman microscope (Ulm, Germany). The laser used was a Nd:YAG 532 nm (2.33 eV) laser and the power used at the sample was less than 10 mW. A 40× magnification objective with a Numerical Aperture (NA) of 0.6 was used. The data collected were single spectra with 10 spectra collected per sample at 5 different locations on each sample. The integration time was between 5 and 10 s with 3 accumulations per spectrum.

5. Conclusions

Suspensions of each nanotube type were produced under the exact same solvent conditions and procedures. This allowed an unambiguous comparison to be made between different nanotube types in a silicon/CNT heterojunction solar cell in which the suspension and film preparation were identical.

It was observed that although one SWCNT sample produced a higher efficiency than the other samples, both single-walled samples and the double-walled sample produced similar power conversion efficiencies as each other, with a difference of less than 0.5% between the three samples. Single-walled films produced better short circuit current densities, whereas double-walled and multi-walled carbon nanotube films displayed higher open circuit voltages. Fill factors were found to be similar across the board, with SWCNT films producing better R_{Shunt} to R_{Series} ratios. It was discovered that cells made with MWCNT films improve to a lesser extent with doping treatment. Overall, the results of this study indicate that when variables other than nanotube type are controlled, large diameter single-walled carbon nanotube films provide the best performance in silicon/CNT solar cells. This conclusion differs from some previous work in which other variables were not well controlled, and will inform future research and development in this field.

Supplementary Materials
Figure S1: Dark J/V curves for cells for each type of sample after the second HF etch: (**a**) SWCNT-1; (**b**) SWCNT-2; (**c**) DWCNT; and (**d**) Sigma Aldrich MWCNT. Figure S2: A plot of the relation between J_{SAT} and V_{OC} for the best performing cells for each sample. Table S1: Sheet resistance as a function of thickness. The values marked with an asterisk were the volumes used to produce films for solar cells in this study.

Acknowledgments: This work is supported by the Australian Microscopy and Microanalysis Research Facility (AMMRF), and The Flinders University Centre for Nanoscale Science & Technology (CNST) and was financially supported by Vice Deanship of Research Chairs, King Saud University.

Author Contributions: Daniel Tune and Joseph Shapter conceived and designed the experiments; Tom Grace, LePing Yu, Christopher Gibson and Huda Alturaif performed the experiments; Tom Grace, Christopher Gibson, Zeid Al Othman and Daniel Tune analyzed the data; Joseph Shapter and Zeid Al Othman contributed reagents/materials/analysis tools; and Ton Grace wrote the paper.

Abbreviations

The following abbreviations are used in this manuscript:

SWCNT:	Single-walled Carbon Nanotube
DWCNT:	Double-walled Carbon Nanotube
MWCNT:	Multi-walled Carbon Nanotube
CNT:	Carbon Nanotube
PCE:	power conversion efficiency
CVD:	chemical vapor deposition
PV:	photovoltaics
RBM:	radial breathing mode

References

1. Colinge, J.-P.C.; Cynthia, A. *Physics of Semiconductor Devices*; Kluwer Academic: Norwell, MA, USA, 2002; p. 436.
2. Shi, E.; Zhang, L.; Li, Z.; Li, P.; Shang, Y.; Jia, Y.; Wei, J.; Wang, K.; Zhu, H.; Wu, D. TiO_2-coated carbon nanotube-silicon solar cells with efficiency of 15%. *Sci. Rep.* **2012**, *2*. [CrossRef] [PubMed]
3. Durkop, T.G.; Cobas, S.A.; Fuhrer, M.S. Extraordinary mobility in semiconducting carbon nanotubes. *Nano Lett.* **2004**, *4*, 35–39. [CrossRef]
4. Barnes, T.M.; Blacknurn, J.L.; van de Lagemaat, J.; Coutts, T.J.; Heben, M.J. Reversibility, dopand desorption, and tunneling in the temperature-dependent conductivity of type-separated, conductive carbon nanotube networks. *ACS Nano* **2008**, *2*, 1968–1976. [CrossRef] [PubMed]
5. Baughman, R.H.; Zakhidov, A.A.; de Heer, W.A. Carbon nanotubes—The route toward applications. *Science* **2002**, *297*, 787–792. [CrossRef] [PubMed]
6. Wang, F.; Kozawa, D.; Miyauchi, Y.; Hiraoka, K.; Mouri, S.; Ohno, Y.; Matsuda, K. Considerably improved photovoltaic performance of carbon nanotube-based solar cells using metal oxide layers. *Nat. Commun.* **2015**, *6*. [CrossRef] [PubMed]
7. Tzolov, M.B.; Kuo, T.-F.; Straus, D.A.; Yin, A.; Xu, J. Carbon nanotube-silicon heterojunction arrays and infrared photocurrent responses. *J. Phys. Chem. C* **2007**, *111*, 5800–5804. [CrossRef]
8. Javey, A.; Guo, J.; Farmer, D.B.; Wang, Q.; Yenilmez, E.; Gordon, R.G.; Lundstrom, M.; Dai, H. Self-aligned ballistic molecular transistors and electrically parallel nanotube arrays. *Nano Lett.* **2004**, *4*, 1319–1322. [CrossRef]
9. Kongkanand, A.; Domínguez, R.M.; Kamat, P.V. Single wall carbon nanotube scaffolds for photoelectrochemical solar cells. Capture and transport of photogenerated electrons. *Nano Lett.* **2007**, *7*, 676–680. [CrossRef] [PubMed]

10. Jia, Y.; Wei, J.; Wang, K.; Cao, A.; Shu, Q.; Gui, X.; Zhu, Y.; Zhuang, D.; Zhang, G.; Ma, B. Nanotube-silicon heterojunction solar cells. *Adv. Mater.* **2008**, *20*, 4594–4598. [CrossRef]
11. Tune, D.D.; Flavel, B.S.; Krupke, R.; Shapter, J.G. Carbon nanotube-silicon solar cells. *Adv. Energy Mater.* **2012**, *2*, 1043–1055. [CrossRef]
12. Wei, J.; Jia, Y.; Shu, Q.; Gu, Z.; Wang, K.; Zhuang, D.; Zhang, G.; Wang, Z.; Luo, J.; Cao, A. Double-walled carbon nanotube solar cells. *Nano Lett.* **2007**, *7*, 2317–2321. [CrossRef] [PubMed]
13. Castrucci, P. Carbon nanotube/silicon hybrid heterojunctions for photovoltaic devices. *Adv. Nano Res.* **2014**, *2*, 23–56. [CrossRef]
14. Honsberg, C.; Bowden, S. PVCDROM. Available online: http://www.pveducation.org/pvcdrom (accessed on 22 September 2014).
15. Green, M.A.; Emery, K.; Hishikawa, Y.; Warta, W.; Dunlop, E.D. Solar cell efficiency tables (Version 45). *Prog. Photovolt. Res. Appl.* **2015**, *23*, 1–9. [CrossRef]
16. Yu, L.; Tune, D.D.; Shearer, C.J.; Shapter, J.G. Implementation of antireflection layers for improved efficiency of carbon nanotube–silicon heterojunction solar cells. *Sol. Energy* **2015**, *118*, 592–599. [CrossRef]
17. Jung, Y.; Li, X.; Rajan, N.K.; Taylor, A.D.; Reed, M.A. Record high efficiency single-walled carbon nanotube/silicon p–n junction solar cells. *Nano Lett.* **2012**, *13*, 95–99. [CrossRef] [PubMed]
18. Harris, J.M.; Semler, M.; May, S.; Fagan, J.A.; Hobbie, E.K. Nature of record efficiency fluid-processed nanotube–silicon heterojunctions. *J. Phys. Chem. C* **2015**, *119*, 10295–10303. [CrossRef]
19. Castrucci, P.; Scilletta, C.; Del Gobbo, S.; Scarselli, M.; Camilli, L.; Simeoni, M.; Delley, B.; Continenza, A.; De Crescenzi, M. Light harvesting with multiwall carbon nanotube/silicon heterojunctions. *Nanotechnology* **2011**, *22*. [CrossRef] [PubMed]
20. Li, Z.; Kunets, V.P.; Saini, V.; Xu, Y.; Dervishi, E.; Salamo, G.J.; Biris, A.R.; Biris, A.S. $SOCl_2$ enhanced photovoltaic conversion of single wall carbon nanotube/n-silicon heterojunctions. *Appl. Phys. Lett.* **2008**, *93*. [CrossRef]
21. Li, Z.; Kunets, V.P.; Saini, V.; Xu, Y.; Dervishi, E.; Salamo, G.J.; Biris, A.R.; Biris, A.S. Light-harvesting using high density *p*-type single wall carbon nanotube/*n*-type silicon heterojunctions. *ACS Nano* **2009**, *3*, 1407–1414. [CrossRef] [PubMed]
22. Tune, D.D.; Blanch, A.J.; Krupke, R.; Flavel, B.S.; Shapter, J.G. Nanotube film metallicity and its effect on the performance of carbon nanotube–silicon solar cells. *Phys. Status Solidi* **2014**, *211*, 1479–1487. [CrossRef]
23. Jia, Y.; Li, P.; Wei, J.; Cao, A.; Wang, K.; Li, C.; Zhuang, D.; Zhu, H.; Wu, D. Carbon nanotube films by filtration for nanotube-silicon heterojunction solar cells. *Mater. Res. Bull.* **2010**, *45*, 1401–1405. [CrossRef]
24. Liu, L.; Kong, L.B.; Yin, W.Y.; Chen, Y.; Matisine, S. Carbon Nanotubes. In *Chapter 5—Microwave Dielectric Properties of Carbon Nanotube Composites*; Marulanda, J.M., Ed.; Intech: Vukovar, Croatia, 2010.
25. Dresselhaus, M.S.; Dresselhaus, G.; Saito, R.; Jorio, A. Raman spectroscopy of carbon nanotubes. *Phys. Rep.* **2005**, *409*, 47–99. [CrossRef]
26. Arvanitidis, J.; Christofilos, D.; Papagelis, K.; Andrikopoulos, K.; Takenobu, T.; Iwasa, Y.; Kataura, H.; Ves, S.; Kourouklis, G. Pressure screening in the interior of primary shells in double-wall carbon nanotubes. *Phys. Rev. B* **2005**, *71*. [CrossRef]
27. Pfeiffer, R.; Kuzmany, H.; Kramberger, C.; Schaman, C.; Pichler, T.; Kataura, H.; Achiba, Y.; Kürti, J.; Zólyomi, V. Unusual high degree of unperturbed environment in the interior of single-wall carbon nanotubes. *Phys. Rev. Lett.* **2003**, *90*. [CrossRef] [PubMed]
28. Dresselhaus, M.; Dresselhaus, G.; Saito, R. Physics of carbon nanotubes. *Carbon* **1995**, *33*, 883–891. [CrossRef]
29. Ren, W.; Li, F.; Chen, J.; Bai, S.; Cheng, H.-M. Morphology, diameter distribution and Raman scattering measurements of double-walled carbon nanotubes synthesized by catalytic decomposition of methane. *Chem. Phys. Lett.* **2002**, *359*, 196–202. [CrossRef]
30. Moore, K.E.; Flavel, B.S.; Ellis, A.V.; Shapter, J.G. Comparison of double-walled with single-walled carbon nanotube electrodes by electrochemistry. *Carbon* **2011**, *49*, 2639–2647. [CrossRef]
31. Tune, D.D.; Flavel, B.S.; Quinton, J.S.; Ellis, A.V.; Shapter, J.G. Single-walled carbon nanotube/polyaniline/n-silicon solar cells: Fabrication, characterization, and performance measurements. *ChemSusChem* **2013**, *6*, 320–327. [CrossRef] [PubMed]
32. Li, X.; Hung, J.-S.; Nejeti, S.; McMillon, L.; Huang, S.; Osuji, C.O.; Hazari, N.; Taylor, A.D. The role of HF in oxygen removal from carbon nanotubes: Implications for high performance carbon electronics. *Nano Lett.* **2014**, *14*, 6179–6184. [CrossRef] [PubMed]

14

Graphene and Carbon Quantum Dot-Based Materials in Photovoltaic Devices: From Synthesis to Applications

Sofia Paulo [1,2], Emilio Palomares [1,3,*] and Eugenia Martinez-Ferrero [2,*]

1 Institut Català d'Investigació Química (ICIQ), The Barcelona Institute of Science and Technology, Avda. Països Catalans 16, Tarragona 43007, Spain; spaulo@iciq.es
2 Fundació Eurecat, Avda. Ernest Lluch 36, Mataró 08302, Spain
3 ICREA, Passeig Lluís Companys 23, Barcelona 08010, Spain
* Correspondence: epalomares@iciq.es (E.P.); emartinez@eurecat.org (E.M.-F.)

Academic Editor: Guanying Chen

Abstract: Graphene and carbon quantum dots have extraordinary optical and electrical features because of their quantum confinement properties. This makes them attractive materials for applications in photovoltaic devices (PV). Their versatility has led to their being used as light harvesting materials or selective contacts, either for holes or electrons, in silicon quantum dot, polymer or dye-sensitized solar cells. In this review, we summarize the most common uses of both types of semiconducting materials and highlight the significant advances made in recent years due to the influence that synthetic materials have on final performance.

Keywords: quantum dots; graphene; carbon; photovoltaics; solar cells

1. Introduction

Converting solar energy efficiently into either electrical or fuel sources remains one of mankind's biggest challenges [1]. Despite the rapid progress that has been made in recent years in research into third generation solar cells, silicon is still the biggest and most important player in the PV industry. Even so, such new technologies as mixed halide perovskite solar cells are quickly catching-up in efficiency (the current record of efficiency is above 22% at 1 sun) [2,3]. Dye-sensitized and organic solar cells (which include polymer- and small organic molecule-based solar cells) have already shown their potential for applications like building integrated photovoltaics. Whereas semiconductor-based quantum dots, typically composed of cadmium or lead derivatives, have such excellent optical properties that they have been used in a wide array of optoelectronic devices such as solar cells, light emitting diodes, bioimaging or optical sensors [4–6].

In this context, carbon-based quantum dots have emerged as potential candidates for application in such devices. Since their discovery early 2000s, carbon-based quantum dots have been the focus of intensive research because of their excellent luminescent properties, good solubility and biocompatibility [7,8]. This research effort increased exponentially after the Nobel Prize awarded to Novoselov and Geim for discovering graphene and describing its properties [9].

These carbon-based nanostructures are in fact two different allotropes (Figure 1). On the one hand, carbon quantum dots (CDs) are quasi-spherical nanoparticles less than 10 nm in diameter, formed by crystalline sp2 graphite cores, or amorphous aggregations, which have a quantum confinement effect. On the other hand, there are dots—the so-called graphene quantum dots (GDs)—made up of single or very few graphene lattices (<10) that have quantum confinement effect and edge effects. GDs are usually more crystalline than CDs because their conjugated domains are larger and their structure regular. Both allotropes are functionalized with complex surface groups, specially oxygen-related molecules such as carboxylates or hydroxylate derivatives that remain after the synthetic procedure and enhance the optical properties and the solubility of the particles [10,11]. It should be pointed out that variability in the fabrication of these materials results in different surface functionalization and the addition of complexity to the hybridization of the carbon atoms.

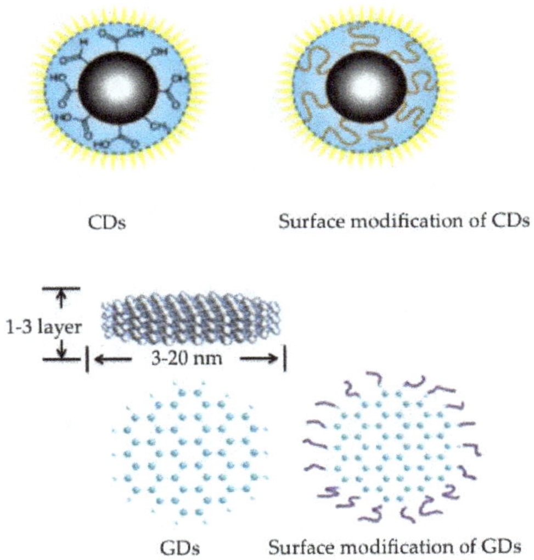

Figure 1. Illustration of CD (top) and GD (bottom) structures. Reproduced with permission of [12–14].

Carbon-based dots have many advantages over non-carbon dots because of their chemical inertness and lower citotoxicity photobleaching and cost. For instance, they can be produced from biomass. In recent years, carbon-based dots have been tested as fluorescent probes, in light emitting diodes, solar cells, biosensors, supercapacitors, lithium ion batteries and catalysts [15–20] and have even been combined with non-carbon dots in optoelectronic applications [21]. Despite their excellent optical properties, they have not performed in photovoltaics as well as non-carbon based quantum dot solar cells. As far as we know, no exhaustive review has been made of carbon-based dots used in photovoltaics. Therefore, in an attempt to understand why these nanostructures have so far failed to realize their potential, in this review we analyze the main achievements in the link between functionality and the synthesis of the material. We aim to give a general overview of how these promising carbon nanostructures can be applied in PV dividing this feature article into the following parts:

1. General synthetic approaches.
2. Photonic properties.
3. Graphene quantum dots in photovoltaic devices.
4. Carbon quantum dots in photovoltaic devices.
5. Outlook and perspectives.

At the end of the manuscript, we have included a list of the abbreviations used throughout the text and Tables 1 and 2 summarize the research done on graphene and carbon quantum dots in photovoltaics, respectively.

Table 1. Summary of the synthetic techniques of CDs included in this article, the resulting size and functional groups and the performance of the photovoltaic cells in which they are used.

Synthesis [1]	Carbon Source	Average Size (nm)	Surface Groups	Solar Cell [2]	Jsc (mA/cm^2)	Voc (V)	FF (%)	η (%)	Effect	R [3]
H	γ-butyrolactone	9 ± 6	Sulfonate, carboxyl, hydroxyl, alkyl	DSSC	0.53	0.38	64	0.13	Emissive traps on the dot surface and enhancement of recombination	[22]
H	Citric acid	1–2	carboxyl	SMOPV PSC	13.32 9.98	0.904 0.609	63.7 54.8	7.67 3.42	Increment in exciton separation and charge collection	[23]
H	CCl_4	1.5–3.3	Amino, carboxylic	DSSC	0.33	0.370	43	0.13	Contribution to light absorption	[24,25]
H	Polystyrene-co-maleic anhydride	---	---	PSC	13.61	0.870	59.5	7.05	Improvement of absorption in the UV and charge transport	[26]
M	Citric acid	200 [4] 1.2 [5]	Carboxylic, primary amines	QDSC	16.6 [4] 2.0 [5]	0.708 [4] 0.550 [5]	46 [4] 16 [5]	5.4 [4] 0.18 [5]	Improved charge extraction	[27]
E	Graphite rods	<4	Hydroxyl, carboxyl, aromatic groups, epoxide/ether	DSSC	0.02	0.580	35	0.0041	Non-optimized electrolyte and electrode	[28]
H	Biomass (chitin, chitosan, glucose)	14.1 ± 2.4 chitin 8.1 ± 0.3 chitosan 2.57 ± 0.04 glucose	Amine, amide, hydroxyl	DSSC	0.674 [6]	0.265 [6]	43 [6]	0.077 [6]	Influence of surface groups	[29]
E	Graphite rod	4.5	---	Si	30.09	0.510	59.3	9.1	Improvement of absorption in the UV and decrease of recombination	[30]
H	Ascorbic acid	3–4	Carboxylic, hydroxyl	DSSC	8.40	0.610	62	3.18	Improvement of light absorption	[31]
S	Citric acid	1.5	Aldehyde, carboxylic	PSC	0.288	1.588	48.5	0.23	Insulating character of oleylamine ligand	[32]
H	Glucose	16	---	QDSC	1.88	0.605	31	0.35	Increment of charge transfer and decrease of recombination	[33]
E	Graphite rods	<10	---	DSSC	0.64	0.500	--	0.147	Improvement of absorption in the UV and decrease of recombination	[34]
H	Citric acid	2–3	---	PerSC	7.83	0.515	74	3.00	Non-optimized device	[35]

[1] H: hydrothermal, M: microwave, E: electrochemical, S: soft template synthesis; [2] DSSC: dye-sensitized solar cell, SMOPV: small molecule organic solar cell, PSC: polymer-based organic solar cells, Si: silicon-based solar cell, QDSC: quantum dot sensitized solar cell, PerSC: perovskite-based solar cell; [3] R: reference; [4]: CDs-Au particles; [5]: CDs; [6] best results obtained from the combination of chitosan- and chitin-derived CDs.

Table 2. Summary of the synthetic techniques of GDs included in this article, the resulting size and functional groups and the performance of the photovoltaic cells in which they are used.

Synthesis [1]	Carbon Source	Size (nm)	Surface Groups	Solar Cell [2]	Jsc (mA/cm^2)	Voc (V)	FF (%)	η (%)	Effect	R^3
H	Bromobenzoic acid	13.5	1,3,5 trialkyl phenyl	DSSC	0.2	0.48	58	0.055	Poor charge injection due to low affinity of GDs to titania	[36]
M	Glucose	3.4	---	Si	37.47	0.61	72.51	16.55	Improvement of absorption in the UV	[37]
E	Graphite rod	5–10	Hydroxyl, epoxy, carboxylic, carbonyl	PerSC	17.06	0.937	63.5	10.15	Improvement of charge extraction	[38]
E	Graphene film	3–5	Hydroxyl, carbonyl	PSC	6.33	0.67	30	1.28	Increment of exciton separation and charge transport. Non-optimized morphology	[39]
A	Graphite	8.5	---	DSSC	0.45	0.8	50	0.2	Inefficient hole collection due to non-optimized thickness of GD layer	[40]
M	Glucose	2.9	---	Si	36.26	0.57	63.87	13.22	Improvement of absorption in the UV and conductivity	[41]
A+H	Graphene oxide	2–6	Epoxy, carboxyl	Si	23.38	0.51	55	6.63	Reduction in current leakage	[42,43]
A	Carbon black	10	Hydroxyl, carboxyl	DSSC	14.36	0.723	50.8	5.27	Reduction in internal resistance and increment of charge transfer	[44]
A+H	Graphene oxide	<1	Epoxy, carbonyl, hydroxyl	PSC	15.2	0.74	67.6	7.6	Increment in conductivity	[45]
A+H	Graphene sheets	9	Carboxyl [4]	PSC	3.51	0.61	53	1.14	Increase in exciton separation and charge transport	[46]
A+H	Graphene oxide	50	PEG	DSSC	14.07	0.66	59	6.1	Increase in light absorption	[47]
M	Glucosamine hydrochloride	4.3	amine	DSSC	5.58	0.583	66	2.15	Increase in light absorption and decrease of recombination	[48]
A	Carbon fibers	20–30	---	PSC SMOPV	10.2 11.36	0.52 0.92	66.3 65.2	3.5 6.82	Increase in conductivity	[49,50]

[1] H: hydrothermal, M: microwave, E: electrochemical, A: acidic oxidation; [2] DSSC: dye sensitized solar cell, SMOPV: small molecule organic solar cell, PSC: polymer-based organic solar cells, Si: silicon-based solar cell, QDSC: quantum dot sensitized solar cell, PerSC: perovskite-based solar cell; [3] R: reference; [4] surface groups attached after chemical treatment.

2. General Synthetic Approaches

Numerous papers describe synthetic procedures for preparing carbon and graphene quantum dots. Two main approaches can be distinguished: bottom-up and top-down synthesis. The bottom-up route builds nanostructures from small organic molecular precursors by pyrolysis, combustion or hydrothermal methods while the top-down approach is based on cutting small sheets via physical, chemical or electrochemical techniques until the required particle size is reached (Figure 2). In both cases, post treatment is done to purify or modify the surface functionalization and improve the performance of the dots. For example, the quantum yield increases after surface passivation of CDs or functionalization because the emissive traps on the nanoparticle surface disappear. Likewise, doping with heteroatoms such as nitrogen and phosphor, or metals such as Au or Mg improves the electrical conductivity and solubility of CDs and GDs [24,27].

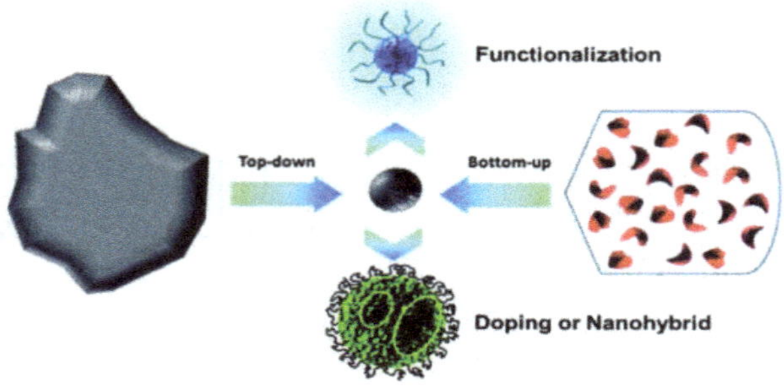

Figure 2. Schematic representation of both synthetic approaches. Reproduced with permission of [16].

In this review, we have focused exclusively on the synthetic procedures described for carbon-based dots applied in photovoltaic devices. Of course, other excellent reviews on the vast number of applications of carbon and graphene quantum dots can be found in [16,17,51], and the references cited therein.

2.1. Bottom-up Approach

2.1.1. Hydrothermal/Solvothermal Synthesis

Hydrothermal synthesis is a widespread procedure that consists of a one-step synthetic technique in which an organic precursor is heated in a Teflon line to achieve high temperature and pressure. Using various organic precursors and modifying the temperature, the optoelectronic properties of the dots are tuned. It is, thus, a low-cost, non-toxic method. In addition, hydrothermal methods produce dots with a diameter of 10 nm, which are bigger than dots produced by other techniques such as electrochemical preparations (3–5 nm).

Pioneering work by Mirtchev and coworkers introduced the use of carbon quantum dots as sensitizers in dye-sensitized solar cells (DSSC) prepared by dehydrating ɣ-butyrolactone [22]. In contrast, Yan et al. synthesized graphene dots from bromobenzoic acid using well-known Suzuki-Miayura reaction conditions (Figure 3). In order to prepare large graphene dots and avoid aggregation, they covalently attached 1,3,5-trialkyl phenyl moieties to the edge of the graphene, shielding them in the three dimensions [36]. Last but not least, Zhang et al. prepared the CDs from citric acid and ethylenediamine in aqueous solution heated for 10 h at 250 °C obtaining uniform 1–2 nm size particles [23], whereas Liu et al. synthesized the CDs combining polystyrene-co-maleic and ethylenediamine dissolved in DMF at 200 °C for 5 h [26].

The nanoparticles are nitrogen doped by this route as well. Zhang et al. used carbon tetrachloride and sodium amide as starting materials and methylbenzene as the solvent, heating at 200 °C for different periods of time to prepare well-dispersed crystalline CDs. By controlling the reaction time, the authors tuned the size and the nitrogen content of the dots in such a way that prolonged reaction times favored the incorporation of nitrogen into the carbon framework and the increase in the particle size. Regardless of the reaction time, the dots had amino functional groups on their surface [24].

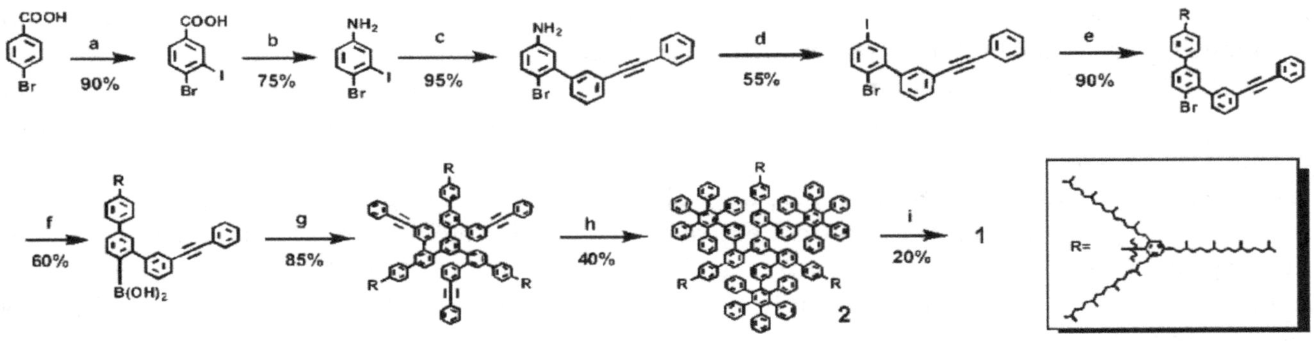

Figure 3. Suzuki reaction followed to prepare graphene dots (described as product number 1 in the reaction scheme) from bromobenzoic acid. Reproduced with permission of [36]. Steps are as follows: (**a**) $NaIO_4$, I_2, concentrated H_2SO_4, room temperature; (**b**) Heated with diphenylphosphoryl azide in triethylamine and tert-butanol at 80 °C, followed by treatment with CF_3COOH in dichloromethane at room temperature; (**c**) Suzuki condition with 3-(phenylethynyl)phenylboronic acid, $Pd(PPh_3)_4$, K_2CO_3 in water, ethanol, and toluene mixture, 60 °C; (**d**) Iodine and tert-butyl nitrite in benzene, 5 °C to room temperature; (**e**) Suzuki condition with substituted phenyl boronic acid, $Pd(PPh_3)_4$, K_2CO_3 in water, ethanol, and toluene mixture, 80 °C; (**f**) Treatment with butyllithium in tetrahydrofuran (THF) at −78 °C, then with triisopropyl borate at −78 °C, followed by treatment with acidic water at room temperature; (**g**) Suzuki condition with 1,3,5-triiodobenzene, $Pd(PPh_3)_4$, K_2CO_3 in water and toluene mixture, 80 °C; (**h**) Tetraphenylcyclopentadienone in diphenylether, 260 °C; (**i**) FeCl3 in nitromethane and dichloromethane mixture, room temperature.

2.1.2. Microwave Irradiation Synthesis

As well as the speed of the synthesis, another important advantage that microwave synthetic methods have over hydrothermal synthesis is that they can be used at lower temperatures. Dao et al. obtained high quality CDs by this synthetic approach. They mixed citric acid and urea in distilled water and the solution was then heated in a microwave oven at 700 W for 4 min. The supernatant was neutralized with sodium bicarbonate and cleaned with distilled water [27]. The resulting dots were doped with Au by chemical reduction of $HAuCl_4$ with formic acid to prepare three dimensional raspberry-like particles with a diameter of 200 nm formed by gold branches that originated high surface areas. In addition, Tsai et al. synthesised water soluble GDs by microwave irradiation using glucose as the carbon source and water as solvent heating at 700 W for 11 min. The as-prepared dots measured 3.4 nm in diameter, as observed by AFM (Atomic Force Microscopy) and TEM (Transmission Electron Microscopy) [37].

2.1.3. Soft Template Method

In this approach, reported by Kwon et al., CDs are made into an emulsion that acts as a self-assembled soft template because the size of the dots is controlled by regulating the amount of the emulsifier. Synthesis in a non aqueous medium favors organic-based surface capping and size tuning. For that, the authors mix oleylamine and octadecene with citric acid solved in water. The water droplets, stabilized by the oleylamine, are eliminated heating at 250 °C forcing the intermolecular dehydration of citric acid molecules which form polymer-like structures. Further carbonization render organic soluble carbon dots capped by oleylamine molecules that are chemically bound to the dot surface carbonyl groups. The concentration of oleylamine determines the final size of the dots [32].

2.2. *Top-down Approach*

Electrochemical Methods

Electrochemical methods make it possible to fine tune carbon nanostructures by controlling the voltage/current applied. For instance, applying a controlled bias to a bulk of carbon precursors leads to electrochemical corrosion reactions over the carbon reactants and subsequently to carbon nanostructures. It is important to notice that this particular technique does not require high temperatures, is easy to scale-up and can proceed under aqueous or non-aqueous solutions. It is one of the fastest routes for preparing graphene sheets [52]. For example, Sun et al. prepared carbon quantum dots by combining the electrochemical method with etching methods [28]. In brief, they used graphite rods as both electrodes whereas the reaction was conducted by applying an alternate bias between 100 and 150 V during 10 h in the presence of an ethanol solution of NaOH (Figure 4). Then they added $MgSO_4$ followed by stirring, deposition, centrifugation and drying of the solvent in order to obtain the uniform and monodisperse dots.

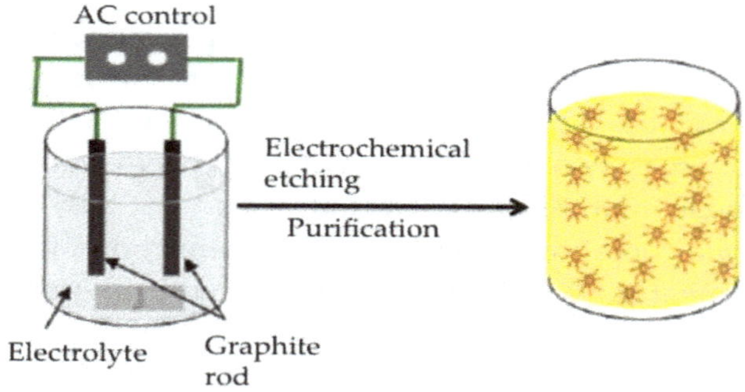

Figure 4. Schematic view of the obtention of CDs by electrochemical methods. Reproduced with permission of [28].

More recently, Zhu and co-workers obtained graphene dots. In this case, the electrolysis took place under a current intensity between 80 and 200 mA/cm^2 with a graphite rod as anode in a basic solution and Pt foil as the counter electrode. In order to finish the reaction, they added 1 mL of 80% hydrazine hydrate and stirred the solution for 8h. It was then centrifuged and dialyzed in water for one day [38,53]. Yan Li and co-workers prepared homogeneous GDs by electrochemical methods from graphene films [39]. For that, graphene films, prepared by the filtration method and treated in oxygen plasma to improve its hidrophilicity, were used as working electrodes in combination with Pt wire and Ag/AgCl that acted as counter and reference electrode, respectively, in phosphate buffer solution. After CV scan rate of 0.5 V/s within ± 3 V in 0.1 M PBS, water soluble GDs with uniform 3–5 nm size were obtained.

2.3. *Acidic Oxidation or Chemical Ablation*

In essence, this two-step procedure consists of the exfoliation of graphite powder using concentrated mineral acids and oxidizing agents under refluxing conditions. This approach, also known as Hummers method, is one of the most popular procedures described for obtaining graphite oxide. The first step is often followed by further chemical reduction to prepare the quantum dots. For example, Dutta et al. treated graphite with sodium nitrate in aqueous sulfuric acid solution with potassium permanganate stirring for four days. Once the graphite oxide was ready, ultrasonication of the sample in water produced graphene oxide, which was converted to graphene dots by reduction in hydrazine solution [40]. Pan et al. prepared GDs from graphene oxide that was transformed to graphene sheets by the Hummer's method. For the second step, they applied a hydrothermal treatment in basic solution (heating at 200 °C for 10 h at pH 8) to cut the graphene sheets into dots that were further purified by dyalisis [43] .

Carbon black has also been used as a carbon source. Chen et al. prepared GDs by oxidation of the carbon black in nitric acid under reflux conditions overnight. After cooling and centrifugation, the supernatant was heated to recover the dots [44]. An alternative source are carbon fibers, as reported by Peng et al. [49]. In this case, the fibers were sonicated and heated for 30 h at 100 °C in acidic medium. After being cooled, the mixture was diluted in water, the pH tuned to 7 and the solution dialyzed.

3. Photonic Properties

It is a remarkable fact that both structures show quantum confinement effects, which means that the energy band gap is determined by the size and shape of the structure (Figure 5). In addition, the optical properties are also influenced by the fabrication variability, which results in a wide array of sizes and surface functionalizing groups and/or defects. Therefore, the determination of the origin of the material's optical properties is one of the most controversial topics in research into carbon and graphene quantum dots.

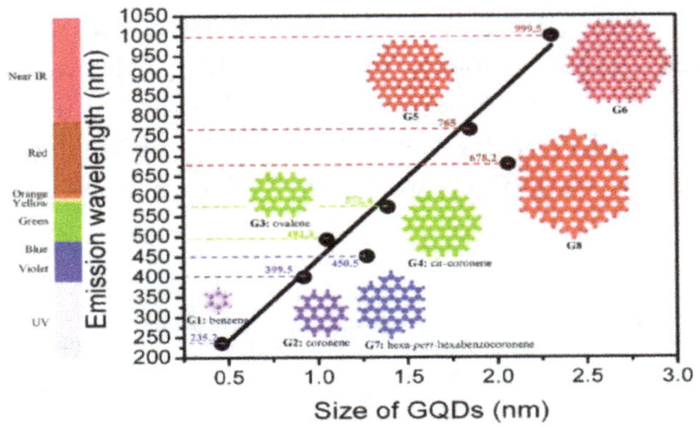

Figure 5. Estimated variation of the emission wavelength with the size for GDs. Reproduced with permission of [54].

3.1. Light Absorption

Both CDs and GDs have an absorbance band in the UV region between 260 and 320 nm assigned to the π–π^* transition of C=C bonds with sp2 hybridization and, sometimes, a weaker shoulder at 270–400 nm attributed to δ–π^* transitions of the C=O bonds, with a tail extending into the visible wavelengths. Graphene quantum dots also have extinction coefficients in the UV region from 10 to 200×10^3 $M^{-1}cm^{-1}$, which is larger than common fluorophores and comparable to other quantum dots [53,54].

3.2. Light Emission

The photoluminescence (PL) mechanism in CDs and GDs is still an open question and different processing methods cause PL of different origins. In fact, PL has been reported to be influenced by the dot size, the excitation wavelength, the degree of surface functionalization or oxidation, the pH during synthesis, the solvent polarity and the doping with heteroatoms. Both CDs and GDs show strong photoluminescent emission that is mostly exciton depedent, which means that the emission peak moves as the excitation wavelength is changed. The origin of fluorescence emission has been intensively studied and assigned to quantum confinement effects, triplet carbenes at zigzag edges or edge defects, excitonic transitions, surface states or functional groups [55–59].

4. Graphene Quantum Dots in Photovoltaics

Researchers have already found various applications for graphene dots in solar cells, mainly in silicon-based solar cells, dye-sensitised solar cells, organic solar cells (OSC) and, more recently, perovskite solar cells. Silicon diodes (either crystalline, c-Si, or amorphous, a-Si) are based on silicon p-n junctions that act both as light absorbers and charge transport carriers. Although Si diodes dominate the PV market because of their high efficiency (recently reported to be 25.6%) [2] and long lifetime, the incorporation of graphene sheets as transparent electrodes has already been explored to improve the performance of the diodes [60].

The device structure of DSSC, which are photo-electrochemical solar cells, is more complex. The electron transport layer is often based on mesoporous nanocrystalline metal oxide films, usually TiO_2 or ZnO, supported on a conducting substrate. The electron transport layer can be configured as planar, mesoporous or columnar morphologies. The mesoporous metal oxide film is sensitized to absorb visible light after the adsorption of a dye monolayer. Examples of popular dyes are Ru(II)-containing polypyridyls, porphyrins, phthalocyanines, squarines or organic dyes [61]. The device is filled with an electrolyte that regenerates the sensitizer, normally iodide/tri-iodide redox electrolyte, defined as hole transport layer (HTL) and a platinum coated counter electrode (Figure 6a). DSSCs have attracted considerable attention since the landmark paper in 1991 by Gratzel and O'Regan [62]. Because of their potential low cost, environmentally friendly components, ease of fabrication in air and such optical properties as transparency and colour, which depends on the dye selected, DSSCs have attracted attention for building-integrated photovoltaic applications. Record efficiencies of 13% have recently been achieved with the molecularly engineered porphyrin dye SM315 [63]. A solid state version of the DSSC can be achieved by replacing the liquid electrolyte with a solid hole transport material such as spiro-OMeTAD or a semiconductor polymer [64].

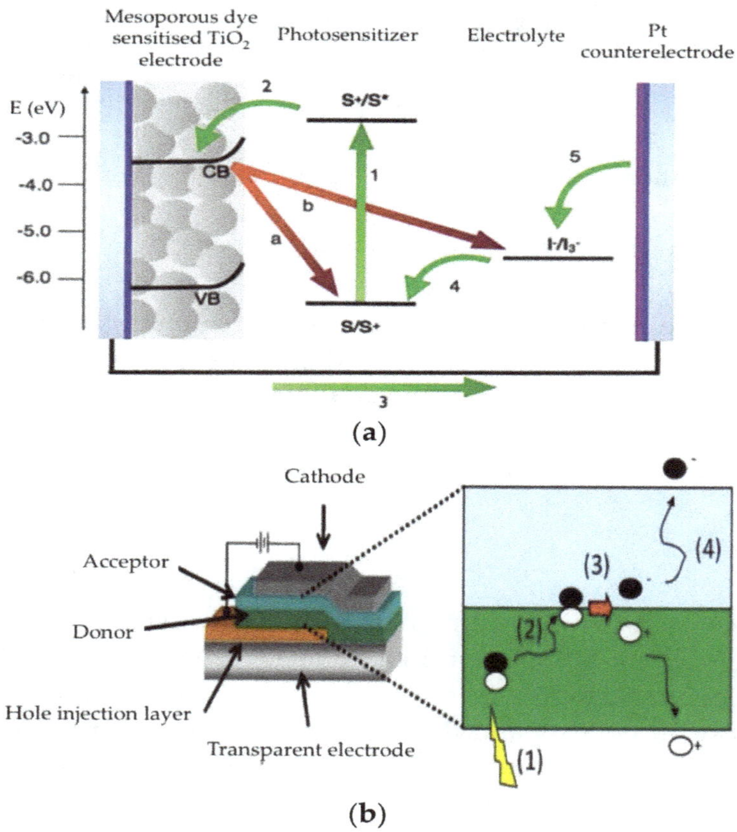

Figure 6. Schematic representation of the composition and charge transfer processes in (**a**) DSSC [(1) light absorption; (2) electron injection; (3) electron collection; (4) reduction of the oxidized dye cation by the redox couple; (5) regeneration of the electrolyte at the counterelectrode] and (**b**) OSC [(1) Light absorption and creation of an exciton; (2) exciton diffusion; (3) exciton splitting at the interface; (4) diffusion and collection of charges]. Reproduced with permission of [61,65], respectively.

Organic photovoltaics (OSCs) combine carbon-based semiconductor materials and molecules, which play the roles of light absorption and carrier transport sandwiched between selective metal electrodes. Depending on the molecular weight of the organic material, OSCs are classified as polymer (PSC) or small-molecule solar cells (SMOPV). The former are processed from solution in organic solvents to form bulk heterojunctions in the photoactive layer in conjunction with either the electron or hole acceptor material. The latter can also be processed using high vacuum techniques. It is well established that the intermixing of the donor and the acceptor optimizes the exciton separation and subsequent carrier collection (Figure 6b) [65,66]. For many years, the most efficient, and widely used, electron acceptor materials were those based on such fullerene derivatives as PCBM. In fact, record efficiencies above 11% have recently been reported using PffBT4T-derivatives as donors and C71-fullerene derivatives as acceptors [67,68]. Only recently has it been shown that other electron acceptor materials can be used to match the high efficiency obtained with fullerene derivatives [69].

In both types of solar cell, DSSC and OSC, which differ from silicon solar cells in the materials they use to transport the hole and the electron carriers, carbon nanomaterials can be easily adapted to have different roles, as described below.

4.1. Light Harvesting

Even though thin film layers of CDs and GDs have been used more as selective contacts in molecular solar cells, several groups have tried using these materials as light harvesting components. For example, Dutta and coworkers sensitized ZnO nanowires with graphene dots to prepare the structure of AZO/ZnO nanowires/GDs/TPD/Au. Graphene quantum dots participated in the charge transfer to the ZnO nanowires (nw). This is reflected in the increase of Jsc and Voc compared to the control (ZnO nw without dots) and has an efficiency of 0.2%. This low value was attributed to inefficient hole collection by TPD originated by the non-optimized thickness of the graphene layer [40]. For the deposition of GD into mesoporous layers of titania, Yan and colleagues prepared large dots functionalized with 1,3,5-trialkyl -substituted phenyl moieties (at the 2-position) at the edges of the dots to favour solubilization into common solvents and avoid aggregation. The Voc and FF of the as-prepared TiO_2/GDs/I_3^-/I^- diodes were comparable to those obtained with the widely used Ru-based sensitizer (0.48 V, 58%, respectively). However, Jsc was much lower, which was attributed to the low affinity of the dots to the oxide surface, which resulted in poor physical adsorption and subsequent poor charge injection [36]. In addition, the dot size may have prevented effective packaging on the surface. Taking into account that graphene dots have a limited spectral absorption range in the visible, co-sensitization of the device with dyes to cover all the visible range of the spectrum emerges as an effective alternative. In this regard, the work of Fang and colleagues combined GDs with the well-known N719 dye. The dots, synthesized by acidic and hydrothermal methods from graphene oxide, were surface-passivated with PEG so carboxyl and hydroxyl groups on the surface promoted the linkage to the titania surface. Tests done with different concentrations of GDs showed that higher loadings resulted in agglomeration. The best results gave an efficiency of 6.1% due to higher Jsc and Voc than the reference, which gave 5.1% [47]. In a second example provided by Mihalache et al., N3 was combined with GDs prepared by microwave-assisted synthesis. They used this method to obtain dots with higher quantum yields and a self-passivated surface with amino functional groups to improve the affinity for the titania surface. The resulting device had better Jsc than the TiO_2/N3 devices due to the expansion of the absorption range, which was confirmed by the increase in the IPCE throughout the range. However, the Voc was lower, although the overall efficiency of 2.15% was higher than the 1.92% of the reference. The efficiency improved as a result of the crossover between two mechanisms: first, a Foster Resonance Energy Transfer (FRET) dominant process in the blue part of the spectrum because of the significant overlap between the emission spectra of the GDs and the absorption spectrum of N3, and second, a charge transfer mediated by GDs towards the red part of the spectrum due to the cascaded energy level alignment of the LUMO levels of N3-GDs-TiO_2 (2.98, 3.16, 4 eV, respectively), which increased the rate of electron injection [48]. Photovoltage decay analyses confirmed the hypothesis that the GDs inhibited the back electron transfer from N3 to the electrolyte. Therefore, the dots were playing a dual role in these devices as active absorbers and reductors of the recombination reactions.

Li and coworkers tested the GDs as alternatives to the popular fullerene derivative acceptors for application in organic solar cells. They reported the preparation of monodisperse graphene dots by electrochemical methods between 3 and 5 nm in size. Surface groups such as hydroxyl, carbonyl and carboxylic acid groups facilitated dispersion in common organic solvents and subsequent mixing with polymers leading to the structure ITO/PEDOT:PSS/P3HT-GQD/Al [39]. The value of the LUMO level (4.2–4.4 eV) of the GDs led to the formation of an electron transport cascade in the system P3HT-GDs-Al. Compared to P3HT-only devices, the GDs increased the exciton separation and carrier transport leading to an efficiency of 1.28%. However, the efficiency was lower than that of devices prepared with fullerenes as electron acceptors because of lower electron affinity and the non-optimized morphology, which resulted in lower FF. Similar experiments made by Gupta et al. compared the effect of graphene dots, synthesized by acidic and hydrothermal methods, and graphene sheets, both functionalized with aniline, as electron acceptors in the structure ITO/PEDOT:PSS/P3HT:ANI-GQD/LiF/Al [46]. They combined P3HT with increasing amounts of GDs in order to optimize the devices. Results were best with 1 wt. % for which efficiencies were 1.14%. Dots gave higher Jsc values than graphene sheets because their homogenous and uniform distribution within P3HT enhanced exciton separation and transport towards electrodes, which resulted, in turn, in higher FF. Another paper by Kim et al. compared the effects of GDs with different oxidation degrees on OSC [45]. The dots, prepared using Hummers' method, were oxidized, and then hydrothermally reduced for 5 h or for 10 h before being added to the PTB7:PC71BM bulk heterojunction. During reduction, the oxygen-related functional groups were gradually removed while the size remained unaltered below 1 nm. In addition, the reduction had a negative effect on the light absorption but enhanced conductivity. After optimizing the concentration of dots in the BHJ, the researchers found that the positive effect of GDs varies with their reduction time, because Jsc increased with the oxidized dots whereas FF increased with the dots reduced for 5 h. This agreed with the observations made about the morphology and composition of the dots and shows that the functional groups, richer in oxidized GDs, play a positive role in light harvesting while sp2 carbon-richer reduced samples make a beneficial contribution to charge conductivity, decreasing the leakage current and enhancing shunt resistance and FF. The maximum efficiency, 7.6%, was thus achieved with 5 h-reduced GDs.

Finally, Tsai et al. combined GDs with n-type silicon heterojunction solar cells to expand the spectral range absorption and decrease the number of wasted photons in the UV region. To do so, they added GDs at different concentrations by solution processing on top of Ag/ITO/a-Si/ITO/Ag devices where the silicon wafer is structured as a micro pyramid [37]. The results demonstrated that the addition of GDs increased the Jsc and the FF, reaching a record efficiency of 16.55% when 0.3 wt. % concentration was used.

4.2. Counterelectrode

Platinum is the most popular material used as the counterelectrode in DSSC because its energy levels are suitable and it is easy to prepare. However, platinum is a rare precious metal and this increases the cost of the device. It is, therefore, prone to be substituted. In this regard, graphene sheets emerge as an excellent alternative nanomaterial because of their high carrier mobility, surface area and optical transparency. Examples of the use of plain graphene or composites of graphene with polymers, metals or carbon nanotubes can be found in the review by Wang et al. [70]. The defects and the functional groups of the sheets play a critical role in the electrocatalytic sites of the counterelectrode, making research on this topic necessary if understanding and efficiencies are to be increased. Chen and coworkers proposed a composite made of GDs embedded in polypyrrol (PPy) in the structure FTO/TiO_2/N719/I_3^-/I^-/GD-PPy as an effective method to lower the cost of the device. PPy is cheap and easy to produce although high charge transfer resistance has prevented it from being used in optoelectronic devices. Graphene dots containing −COOH and −OH groups on the edge interacted electrostatically with the N sites of the pyrrol, giving rise to highly porous structures. Cells were built with amounts of GDs ranging between 3% and 30%. Performance was best with 10%. The efficiency reached 5.27%, which is 20% more than when the pure PPy counterelectrode was used and is lower than when the electrode was Pt (efficiencies 4.46% and 6.02%, respectively). The amount of GD had to be finely tuned since increasing concentrations at low values increased the Jsc and the FF by reducing the internal resistance and enhancing charge transfer, whereas higher doping rates increased the charge recombination at the counterelectrode resulting in lower Jsc and Voc values [44].

4.3. Hole Collector

GDs such as HTL have been added to silicon solar cells and polymer solar cells because of their excellent charge transport properties and transparency. Since the fabrication of large area graphene sheets involves complicated deposition and transfer processes, research has also focused on solution processed GDs. Recently, Gao and coworkers reported the structure In-Ga/c-Si/GD/Au in which dots were prepared from graphene sheets with final sizes ranging between 2 and 6 nm. Epoxy, carboxyl and other oxygenous functional groups have been detected in the edges. c-Si was also passivated to improve the interaction between the two materials. Of all the options the methyl group showed the best results due to the reduction of surface carrier recombination. The diodes were prepared in air by solution processing and gave an efficiency of 6.63% which is higher than the 2.26% obtained without GDs. The dots increased Jsc and Voc because the current leakage reduced after recombination was suppressed at the anode. Although the GDs show strong absorption in the UV, the contribution to the Jsc could not be observed when the EQE was measured [42]. Moreover, the addition of GDs resulted in good stability of the c-Si/GDs cells after storage for half a year. Tsai et al. added an extra layer of PEDOT:PSS and GDs to micro-structured amorphous silicon heterojunctions leading to the configuration Al/a-Si/PEDOT:PSS-GDs/Ag. The dots, prepared by microwave methods, measured 2.9 nm, roughly 12 layers of graphene. The Jsc and FF of the diodes increased with increasing concentrations of GDs up to 0.5%, at which point the efficiency started to decline because of increased recombination reactions probably arising from the formation of GD aggregates. Therefore, a record performance of 13.22% was achieved due to the contribution of the GDs to light harvesting below 400 nm and the improvement in conductivity and the subsequent carrier collection efficiency [41].

Searching for enhanced stability and lifetime, Li and colleagues used GDs in polymer solar cells to substitute the hygroscopic PEDOT:PSS in the configuration ITO/GDs/P3HT:PCBM/LiF/Al. The dots were created by acid treatment of carbon fibers. Optimization of the HTL thickness between 1.5 and 2 nm resulted in devices that had efficiency values similar to those of the cells prepared with PEDOT:PSS, 3.5%, due to the homogeneous morphology and good conductivity of the GDs. Moreover, measurements of efficiency in air showed that decay was slower when GDs were used. The same experiments performed on small molecule solar cells based on DR3TBDT:PC71BM gave efficiencies similar to PEDOT:PSS containing devices (6.9% efficiency), thus demonstrating the capability of GDs to act as a hole collector [50].

4.4. Electron Collector

Perovskite-based solar cells have recently attracted the research community because of their broad spectral absorption and conducting properties. These molecules have been applied in planar and mesoscopic heterojunctions and have shown efficiencies over 22% [3,71]. Meanwhile, GDs have shown ultrafast hot-electron extraction faster than 15 fs through the GDs-TiO_2 interface [72], although their application in DSSC has given low efficiencies. However, to further improve performance, Zhu et al. inserted an ultrathin layer of GDs between the perovskite and the titania layer in the configuration FTO/TiO_2 dense/TiO_2 mesoporous/$CH_3NH_3PbI_3$/GDs/spiro-OMeTAD/ Au (Figure 7). The dots, prepared by electrochemical methods, measured between 5 and 10 nm and were homogeneously distributed onto the titania layer. Optimization of the thin layer thickness led to efficiencies of 10.15%, which is higher than the 8.81% reported for the reference cell without GDs. Whereas the FF and Voc showed values similar to the reference, the Jsc increased due to faster charge extraction. Involvement of the GDs in light harvesting was discarded since the strong absorption of the perovskite dominates absorption and no contribution from the GDs could be detected [38].

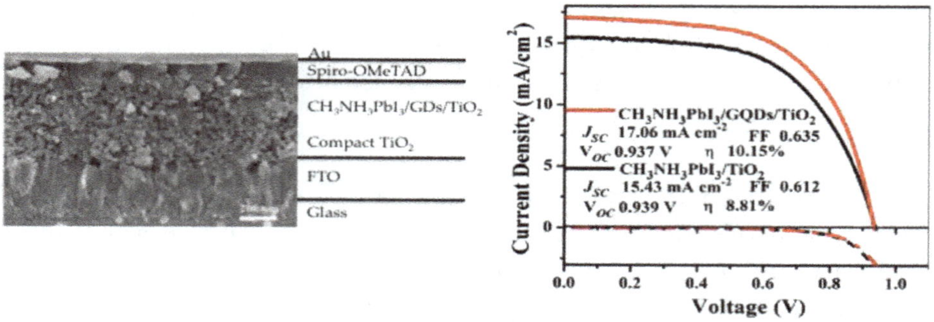

Figure 7. (**a**) Cross-sectional view of the SEM image; (**b**) JV curve of the devices comparing the effect of the insertion of the GDs. Reproduced with permission of [38].

5. Carbon Dots in Photovoltaics

The light harvesting abilities and conducting properties of carbon dots have prompted researchers to use them in a variety of roles in solar cells.

5.1. Light Harvesting

The spectral absorption features of the carbon dots in the ultraviolet region have led to their application as single absorbers in several photovoltaic cells. Briscoe and co-workers studied the construction of low-cost sustainable structured cells making use of carbon dots (CDs) obtained from biomass. They prepared the dots by hydrothermal carbonization of chitin, chitosan or glucose which led to samples with features that reflected the parent reactant. Thus, chitin and chitosan led to *N*-doped CDs (10% and 8% doping, respectively). The surface was functionalized by amides if chitin was used, amines if chitosan was used and hydroxyl if glucose was used. The differences remained during deposition onto ZnO nanorods because the best coverages were obtained with chitosan and glucose. Finally, CuSCN was added as HTL giving rise to the cell configuration FTO/ZnO nanorod/CDs/CuSCN/Au. Efficiencies were best (0.061%) with chitosan-derived CDs. It was observed that the nature of the precursor and surface functionalization heavily influences the performance of the diodes. For further optimization, the authors combined two types of CD to merge their best properties and increase optical absorption. However, the combination needed to be done with great care to prevent the series resistance from increasing and the Jsc from decreasing. Therefore, results were best with a combination of chitosan and chitin-derived carbon dots, for which efficiency was 0.077 [29].

Mirtchev et al. explored CD- DSSC with mesoporous titania. The dots were prepared by dehydratation of γ-butyrolactone and contained sulfonate, carboxylate and hydroxyl groups on the surface, thus mimicking the anchoring groups of common Ru-based sensitizers. The device was built by immersing titania in CD solution for 48 h and was completed with I_3^-/I^- as HTL to give the structure FTO/TiO_2/CDs/I_3^-/I^-/P [22]. In comparison with typical Ru-sensitizers, Jsc is the factor that limits better efficiencies because of the emissive trap sites on the surface of the dot that could act as recombination centers and because of the lower capacity of the dot to inject charges into TiO_2. The authors suggested maximizing the titania surface coverage by using smaller dots or bifunctional linker molecules to enhance the efficiency [22]. Sun et al. used a similar device configuration with titania nanotubes. The dots were prepared by electrochemical-etching methods and added to the nanotubes by impregnation for several hours. Assembly between the small dots and the titania was possible through the oxygen functional groups present on the surface of the carbon material. The device, which has a low efficiency of 0.0041%, served as proof-of-concept of the light harvesting properties of the CDs. The authors expected that optimizing the electrolyte and the electrodes would give better results [28].

Zhang et al. developed hierarchical microspheres of rutile built by uniform nanorods to prepare solar cells made of metal-free sensitizers. They synthesized nitrogen-doped carbon dots (NCDs) by one-pot solvothermal methods and anchor them to the rutile structures by means of the surface groups. The configuration of the cell was TiO_2/NCDs/I_3^-/I^-/Pt and the Jsc values were higher than those of similar devices prepared without NCDs. The final efficiency was 0.13% [25].

CDs have also been applied in nanostructured silicon solar cells. Xie et al. intended to broaden the absorption range of the silicon nanowires (Si nw) by creating core/shell heterojunctions with carbon dots. The nanoparticles were synthesized by electrochemical etching methods and added to the silicon wires to form a homogeneous and continuous shell of 23 nm corresponding to 5 layers of dots. The overall structure of the device was In-Ga/Si nw/CD/Au and reached an efficiency of 9.1% which is much higher than the references prepared with planar silicon and five layers of CDs (4.05%) or silicon nanowires without CDs (1.58%) [30]. The reasons for the enhanced performance of the device were the increase in optical absorption in the UV region and the fact that recombination was lower because of the electron blocking layer action of the CDs (Figure 8).

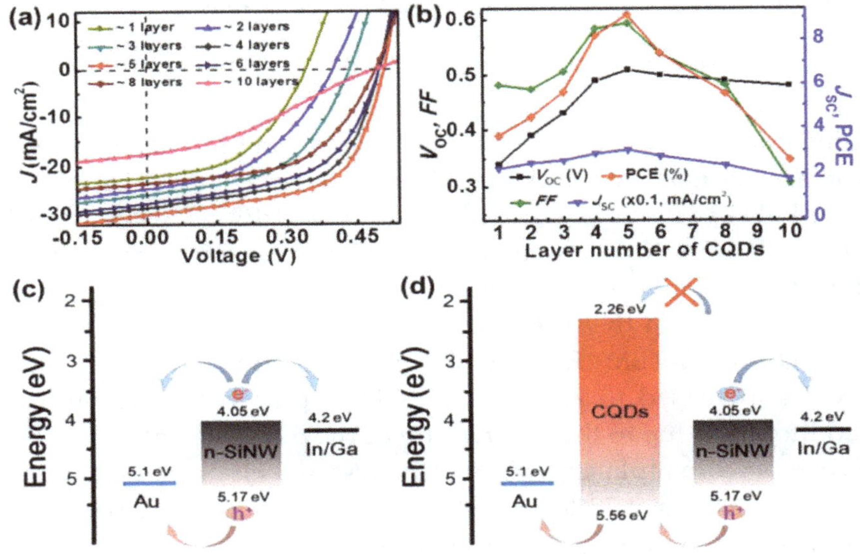

Figure 8. Variation of the JV curve (**a**) and cell parameters (**b**) with increasing layers of CDs; Energy level alignments of the cells without (**c**) and with (**d**) CDs. Reproduced with permission of [30].

An innovative approach has recently been reported by Huang et al. who prepared composites of CDs and polysiloxane to coat the substrate of the solar cells, which had the configuration CD-polysiloxane/ITO/ZnO/P3HT:PCBM/MoO_3/Ag [31]. The dots were prepared by a one-step reaction with ascorbic acid as the carbon source and KH791 as the stabilizing and passivating agent and source of the siloxane polymer. The composite contributed to light harvesting in the UV part of the spectrum and increased the efficiency by about 12% compared to the polymer:fullerene solar cell (3.18% and 2.85%, respectively). Similar observations are reported by Liu et al. who added increasing amounts of CDs to the active layer in the cell configuration ITO/TiO_2/PCDTBT:PCBM:CDs/MoO_3/Ag. The increase in absorption in the UV region, together with the improvement in charge transport resulted in enhanced FF and Jsc when 0.062 wt. % ratio was used leading to efficiencies of 7.05% [26].

5.2. Counterelectrode

Dao et al. studied different options for the counterelectrode (CE) component of quantum dot solar cells looking for lower resistance and higher reduction rates of the redox electrolyte. They compared sputtered gold, CDs and CD-containing gold particles in the ZnO nanowire/CdS/CdSe/polysulfide electrolyte/CE configuration. The Cd-Au structures were formed by a dense array of gold rods covered by small 1.2 nm CDs in a 200 nm wide raspberry-like superstructure. When applied as the CE, they showed enhanced redox activity toward the polysulfide electrolyte that increased the efficiency to 5.4% whereas CDs and the sputtered gold gave efficiencies of 0.18 and 3.6%, respectively. These results are explained by the larger surface area of the Au-CD structures and the reduced internal charge transfer resistance of the material that contributed to the increment of Jsc and the FF [27].

5.3. Hole Collection

CDs have also been tested in the charge transport layers of perovskite solar cells as alternatives to the expensive hole transporter spiro-OMeTAD in the configuration FTO/TiO_2 dense/TiO_2 mesoporous/$CH_3NH_3PbI_{3-x}Cl_x$/CDs/Au [35]. The dots were prepared by polymerization-carbonization of citric acid using p-phenylenediamine as passivating agent and deposited by solution processing onto the perovskite layer. The resulting devices performed better than the control without HTL, although Jsc, Voc and FF values were lower than those of the spiro-OMeTAD device. The poorer performance (3% vs. 8% efficiencies for the CDs and the spiro-OMeTAD-containing devices, respectively) was attributed to non-optimized device fabrication.

5.4. Electron Collection

The potential contribution of CDs to the charge transport in the solar cells led to the nanocrystals being used as electron acceptors. Kwon and coworkers tested oleylamine-capped CDs in combination with the electron donor P3HT to form the structure ITO/PEDOT:PSS/P3HT:CDs/Al [32]. Compared to the 1.99% efficiency of the P3HT/PCBM reference, the 0.23% obtained points to the insulating character of oleylamine as the origin of the lower Jsc values. Zhang et al. in addition, worked on organic solar cells and tested the ability of the CDs as electron acceptors. They prepared the configuration ITO/PEDOT:PSS/DR3TBDT:PC71BM/ETL/Al (ETL: electron transport layer) and observed that the efficiency of the devices increased to 7.67% when CDs replaced the widely used LiF in the ETL. In addition, extended lifetimes due to the air stability of the dots were also reported. When the small molecular light harvesters were replaced by P3HT:PCBM, the efficiency was also higher when CDs were used instead of LiF (3.42% vs. 3.38%, respectively) [23]. The improvement was attributed in both cases to the balance of the charge transport by decreasing the series resistance and increasing the shunt resistance resulting in the increase of charge collection.

Another strategy for enhancing the charge transport is to combine CDs with electron acceptor molecules. Narayanan et al. described a device made of quantum dots ZnS/CdS/ZnS, which act as an exciton generator, and the small molecule CuPc as an electron acceptor (Figure 9). The quantum dots absorbed light in the blue-green region of the spectrum and transferred the energy via Förster resonance to the red absorber phthalocyanine. The addition of the CDs to the heterojunction accelerated the charge transfer towards the electrode and decreased the electron recombination rate, which was reflected in the increase in IPCE. Thus, the resulting Jsc was 5.76 times higher than the reference prepared without CDs. Voc was also enhanced, and the efficiency increased to 0.35%. The carbon nanocrystals measured 16 nm and were closely connected to CDs and CuPc, as observed by HRTEM [33].

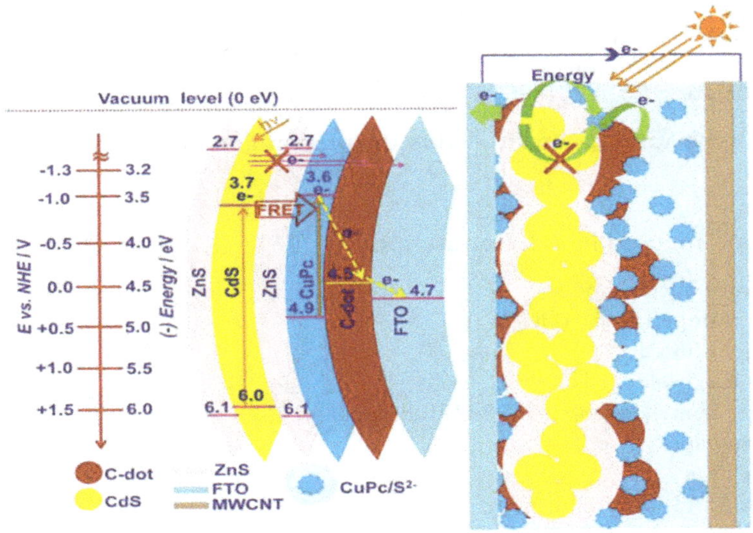

Figure 9. Energy band diagram showing possible paths for energy and charge transport and structure of the FTO/ZnS-CdS-ZnS/CDs/CuPc/S^{2-}/MWCNT devices. Reproduced with permission of [33].

Similar results were observed by Ma et al. when they added CDs to titania functionalized with the rhodamine B sensitizer in the system FTO/TiO_2/RhB/CQD/I_3^-/I^-/Pt [34]. The combination of the dots with rhodamine increased light harvesting in the UV region and suppressed electron recombination leading to 0.147% efficiency. Therefore, electrochemically generated CDs were responsible for the 7-fold increase in the Jsc.

6. Outlook and Perspectives

Carbon-based materials are an exciting challenge in the area of materials chemistry and nanotechnology. Needless to say, they are abundant and they are also inert, non-toxic and, when scaled-up, cost effective. However, at present their applications are limited due to the numerous physical and chemical phenomena that are still unexplored. This review aimed to give a general overview of the enormous potential graphene and carbon dots have in photovoltaic applications. There are, of course, more applications, but the ones discussed here will help researchers interested in exploring the boundaries of graphene and carbon nanoform research.

For instance, the absorption ability of the carbon nanostructures in the UV region complements light harvesting in those cells where absorption is confined to the visible region. The increased number of captured photons leads to the boost of the IPCE and the Jsc. On the other hand, their redox characteristics accelerate charge transfer from the absorber to the electrode. Therefore, electron recombination diminishes whereas Voc increases. These beneficial effects are influenced by the synthetic approach, which determines the size of the particles and the functional groups found on the surface and edges of the crystals. These groups have a major influence on the optical properties and the interactions with the materials of which the devices are made. In this regard, some authors have investigated on the addition of specific functionalities to enhance the interaction between the dots and other components of the device. However, the synthetic variability hinders reproducibility and affects the efficiency. The examples reported in this review highlight the need for further optimization of the structure and linkage; so consequently, the size and surface molecules need to be fine tuned if efficient devices are to be prepared. Nonetheless, these materials are expected to play an important role in energy-harvesting devices that help to decrease CO_2 emissions and lower the cost of renewable energy.

Acknowledgments: E.M.F. thanks Spanish MINECO for funding through the Ramon y Cajal fellowship RYC-2010-06787 and project MAT2012-31570. SP acknowledges financial support from ICIQ and EURECAT. EP would like to thank AGAUR for the SGR-2014-763 grant, as well as the Spanish MINECO for the SEVERO OCHOA Excellence Acreditation 2014.20188SEV-2013-0319 and the CTQ2013-47183-R project.

Author Contributions: All the authors contributed to the preparation of the manuscript.

Abbreviations

The following abbreviations are used in this manuscript:

AFM	Atomic force microscope
AZO	Aluminum-doped zinc oxide
CDs	Carbon quantum dots
DSSC	Dye-sensitised solar cells
DR3TBDT	3-ethyl rhodanine benzo[1,2-b:4,5-b′]dithiophene
ETL	Electron transport layer
GDs	Graphene dots
HTL	Hole transport layer
HRTEM	High resolution transmission electron microscopy
KH791	(N-(2-aminoethyl)-3-aminopropyl)tris-(2-ethoxy) silane
N719	Di-tetrabutylammonium cis-bis(isothiocyanato) bis (2,2′-bipyridyl -4,4′- dicarboxylato) ruthenium(II)
N3	Cis-Bis(isothiocyanato) bis(2,2′-bipyridyl-4,4′-dicarboxylato ruthenium(II)
MWCNT	Multiwall carbon nanotubes
NCDs	Nitrogen-doped carbon dots
OSC	Organic solar cells
P3HT	Poly(3-hexyl thiophene)
PEDOT:PSS	Poly(3,4-ethylenedioxythiophene) polystyrene sulfonate
PCBM	[6,6]-phenyl-C61-butyric acid methyl ester
PffBT4T-2OD	Poly [(5,6-difluoro-2,1,3-benzothiadiazol-4,7-diyl)-alt-(3,3‴–di (2-octyldodecyl)-2, 2′; 5′, 2″; 5″, 2‴-quaterthiophen-5,5‴-diyl)
Ppy	Polypyrrol
PSC	Polymer solar cell
SMOPV	Small molecule organic photovoltaics
TC71BM	[6,6]-2-Thienyl-C71-butyric acid methyl ester
TPD	NN′-diphenyl-N-N′-bis(3-methylphenyl)-1,1′-biphenyl)-4,4′-diamine
Spiro-OMeTAD	N2,N2,N2′,N2′,N7,N7,N7′,N7′-octakis(4-methoxyphenyl)-9,9′-spiro bi[9H-fluorene]-2,2′,7,7′-tetramine

References

1. Armaroli, N.; Balzani, V. Solar Electricity and Solar Fuels: Status and Perspectives in the Context of the Energy Transition. *Chem. A Eur. J.* **2016**, *22*, 32–57. [CrossRef] [PubMed]
2. Green, M.A.; Emery, K.; Hishikawa, Y.; Warta, W.; Dunlop, E.D. Solar cell efficiency tables (version 47). *Prog. Photovolt. Res. Appl.* **2016**, *24*, 3–11. [CrossRef]
3. National Center for Photovoltaics. Available online: http://www.nrel.gov/ncpv/ (accessed on 23 August 2016).
4. Albero, J.; Clifford, J.N.; Palomares, E. Quantum dot based molecular solar cells. *Coord. Chem. Rev.* **2014**, *263–264*, 53–64. [CrossRef]
5. Medintz, I.L.; Uyeda, H.T.; Goldman, E.R.; Mattoussi, H. Quantum dot bioconjugates for imaging, labelling and sensing. *Nat. Mater.* **2005**, *4*, 435–446. [CrossRef] [PubMed]
6. Nurmikko, A. What future for quantum dot-based light emitters? *Nat. Nanotechnol.* **2015**, *10*, 1001–1004. [CrossRef] [PubMed]
7. Xu, X.; Ray, R.; Gu, Y.; Ploehn, H.J.; Gearheart, L.; Raker, K.; Scrivens, W.A. Electrophoretic analysis and purification of fluorescent single-walled carbon nanotube fragments. *J. Am. Chem. Soc.* **2004**, *126*, 12736–12737. [CrossRef] [PubMed]
8. Sun, Y.P.; Zhou, B.; Lin, Y.; Wang, W.; Fernando, K.A.S.; Pathak, P.; Meziani, M.J.; Harruff, B.A.; Wang, X.; Wang, H.; et al. Quantum-sized carbon dots for bright and colorful photoluminescence. *J. Am. Chem. Soc.* **2006**, *128*, 7756–7757. [CrossRef] [PubMed]
9. Novoselov, K.S.; Geim, A.K.; Morozov, S.V.; Jiang, D.; Zhang, Y.; Dubonos, S.V.; Grigorieva, I.V.; Firsov, A.A. Electric Field Effect in Atomically Thin Carbon Films. *Science* **2011**, *306*, 666–669. [CrossRef] [PubMed]

10. Ding, C.; Zhu, A.; Tian, Y. Functional surface engineering of C-dots for fluorescent biosensing and in vivo bioimaging. *Acc. Chem. Res.* **2014**, *47*, 20–30. [CrossRef] [PubMed]
11. Li, H.; Kang, Z.; Liu, Y.; Lee, S.-T. Carbon nanodots: Synthesis, properties and applications. *J. Mater. Chem.* **2012**, *22*, 24230–24253. [CrossRef]
12. Zheng, X.T.; Ananthanarayanan, A.; Luo, K.Q.; Chen, P. Glowing graphene quantum dots and carbon dots: Properties, syntheses, and biological applications. *Small* **2015**, *11*, 1620–1636. [CrossRef] [PubMed]
13. Baker, S.N.; Baker, G.A. Luminescent carbon nanodots: Emergent nanolights. *Angew. Chem. Int. Ed.* **2010**, *49*, 6726–6744. [CrossRef] [PubMed]
14. Shen, J.; Zhu, Y.; Yang, X.; Li, C. Graphene quantum dots: Emergent nanolights for bioimaging, sensors, catalysis and photovoltaic devices. *Chem. Commun.* **2012**, *48*, 3686–3699. [CrossRef] [PubMed]
15. Ding, H.; Yu, S.-B.; Wei, J.-S.; Xiong, H.-M. Full-Color Light-Emitting Carbon Dots with a Surface-State-Controlled Luminescence Mechanism. *ACS Nano* **2015**, *10*, 484–491. [CrossRef] [PubMed]
16. Wang, Y.; Hu, A. Carbon quantum dots: Synthesis, properties and applications. *J. Mater. Chem. C* **2014**, *2*, 6921–6939. [CrossRef]
17. Li, X.M.; Rui, M.C.; Song, J.Z.; Shen, Z.H.; Zeng, H.B.; Li, X.M.; Rui, M.C.; Song, J.Z.; Shen, Z.H.; Zeng, H.B. Carbon and Graphene Quantum Dots for Optoelectronic and Energy Devices: A Review. *Adv. Funct. Mater.* **2015**, *25*, 4929–4947. [CrossRef]
18. Zhang, Z.; Zhang, J.; Chen, N.; Qu, L. Graphene quantum dots: An emerging material for energy-related applications and beyond. *Energy Environ. Sci.* **2012**, *5*, 8869–8890. [CrossRef]
19. Miao, P.; Han, K.; Tang, Y.; Wang, B.; Lin, T.; Cheng, W. Recent advances in carbon nanodots: Synthesis, properties and biomedical applications. *Nanoscale* **2015**, *7*, 1586–1595. [CrossRef] [PubMed]
20. Fan, Z.; Li, S.; Yuan, F.; Fan, L. Fluorescent graphene quantum dots for biosensing and bioimaging. *RSC Adv.* **2015**, *5*, 19773–19789. [CrossRef]
21. Van Pham, C.; Madsuha, A.F.; Nguyen, T.V.; Krueger, M. Graphene-quantum dot hybrid materials on the road to optoelectronic applications. *Synth. Met.* **2016**, *219*, 33–43. [CrossRef]
22. Mirtchev, P.; Henderson, E.J.; Soheilnia, N.; Yip, C.M.; Ozin, G.A. Solution phase synthesis of carbon quantum dots as sensitizers for nanocrystalline TiO_2 solar cells. *J. Mater. Chem.* **2012**, *22*, 1265–1269. [CrossRef]
23. Zhang, H.; Zhang, Q.; Li, M.; Kan, B.; Ni, W.; Wang, Y.; Yang, X.; Du, C.; Bc, X.W.; Chen, Y. Investigation of the enhanced performance and lifetime of organic solar cells using solution-processed carbon dots as the electron transport layers. *J. Mater. Chem. C* **2015**, *3*, 12403–12409. [CrossRef]
24. Zhang, Y.-Q.; Ma, D.-K.; Zhuang, Y.; Zhang, X.; Chen, W.; Hong, L.-L.; Yan, Q.-X.; Yu, K.; Huang, S.-M. One-pot synthesis of N-doped carbon dots with tunable luminescence properties. *J. Mater. Chem.* **2012**, *22*, 16714–16718. [CrossRef]
25. Zhang, Y.Q.; Ma, D.K.; Zhang, Y.G.; Chen, W.; Huang, S.M. N-doped carbon quantum dots for TiO_2-based photocatalysts and dye-sensitized solar cells. *Nano Energy* **2013**, *2*, 545–552. [CrossRef]
26. Liu, C.; Chang, K.; Guo, W.; Li, H.; Shen, L.; Chen, W.; Yan, D. Improving charge transport property and energy transfer with carbon quantum dots in inverted polymer solar cells. *Appl. Phys. Lett.* **2014**, *105*, 073306. [CrossRef]
27. Dao, V.D.; Kim, P.; Baek, S.; Larina, L.L.; Yong, K.; Ryoo, R.; Ko, S.H.; Choi, H.S. Facile synthesis of carbon dot-Au nanoraspberries and their application as high-performance counter electrodes in quantum dot-sensitized solar cells. *Carbon* **2016**, *96*, 139–144. [CrossRef]
28. Sun, M.; Ma, X.; Chen, X.; Sun, Y.; Cui, X.; Lin, Y. A nanocomposite of carbon quantum dots and TiO_2 nanotube arrays: Enhancing photoelectrochemical and photocatalytic properties. *RSC Adv.* **2014**, *4*, 1120–1127. [CrossRef]
29. Briscoe, J.; Marinovic, A.; Sevilla, M.; Dunn, S.; Titirici, M. Biomass-Derived Carbon Quantum Dot Sensitizers for Solid-State Nanostructured Solar Cells. *Angew. Chem. Int. Ed.* **2015**, *54*, 4463–4468. [CrossRef] [PubMed]
30. Xie, C.; Nie, B.; Zeng, L.; Liang, F.X.; Wang, M.Z.; Luo, L.; Feng, M.; Yu, Y.; Wu, C.Y.; Wu, Y.; et al. Core-shell heterojunction of silicon nanowire arrays and carbon quantum dots for photovoltaic devices and self-driven photodetectors. *ACS Nano* **2014**, *8*, 4015–4022. [CrossRef] [PubMed]
31. Huang, J.J.; Zhong, Z.F.; Rong, M.Z.; Zhou, X.; Chen, X.D.; Zhang, M.Q. An easy approach of preparing strongly luminescent carbon dots and their polymer based composites for enhancing solar cell efficiency. *Carbon* **2014**, *70*, 190–198. [CrossRef]

32. Kwon, W.; Lee, G.; Do, S.; Joo, T.; Rhee, S.W. Size-controlled soft-template synthesis of carbon nanodots toward versatile photoactive materials. *Small* **2014**, *10*, 506–513. [CrossRef] [PubMed]
33. Narayanan, R.; Deepa, M.; Srivastava, A.K. Forster resonance energy transfer and carbon dots enhance light harvesting in a solid-state quantum dot solar cell. *J. Mater. Chem. A* **2013**, *1*, 3907–3918. [CrossRef]
34. Ma, Z.; Zhang, Y.L.; Wang, L.; Ming, H.; Li, H.; Zhang, X.; Wang, F.; Liu, Y.; Kang, Z.; Lee, S.T. Bioinspired photoelectric conversion system based on carbon-quantum-dot- doped dye-semiconductor complex. *ACS Appl. Mater. Interfaces* **2013**, *5*, 5080–5084. [CrossRef] [PubMed]
35. Paulo, S.; Stoica, G.; Cambarau, W.; Martinez-Ferrero, E.; Palomares, E. Carbon quantum dots as new hole transport material for perovskite solar cells. *Synth. Met.* **2016**. [CrossRef]
36. Yan, X.; Cui, X.; Li, B.; Li, L.S. Large, solution-processable graphene quantum dots as light absorbers for photovoltaics. *Nano Lett.* **2010**, *10*, 1869–1873. [CrossRef] [PubMed]
37. Tsai, M.-L.; Tu, W.-C.; Tang, L.; Wei, T.-C.; Wei, W.-R.; Lau, S.P.; Chen, L.-J.; He, J.-H. Efficiency Enhancement of Silicon Heterojunction Solar Cells via Photon Management Using Graphene Quantum Dot as Downconverters. *Nano Lett.* **2016**, *16*, 309–313. [CrossRef] [PubMed]
38. Zhu, Z.; Ma, J.; Wang, Z.; Mu, C.; Fan, Z.; Du, L.; Bai, Y.; Fan, L.; Yan, H.; Phillips, D.L.; et al. Efficiency enhancement of perovskite solar cells through fast electron extraction: The role of graphene quantum dots. *J. Am. Chem. Soc.* **2014**, *136*, 3760–3763. [CrossRef] [PubMed]
39. Li, Y.; Hu, Y.; Zhao, Y.; Shi, G.; Deng, L.; Hou, Y.; Qu, L. An electrochemical avenue to green-luminescent graphene quantum dots as potential electron-acceptors for photovoltaics. *Adv. Mater.* **2011**, *23*, 776–780. [CrossRef] [PubMed]
40. Dutta, M.; Sarkar, S.; Ghosh, T.; Basak, D. ZnO/graphene quantum dot solid-state solar cell. *J. Phys. Chem. C* **2012**, *116*, 20127–20131. [CrossRef]
41. Tsai, M.-L.; Wei, W.-R.; Tang, L.; Chang, H.-C.; Tai, S.-H.; Yang, P.-K.; Lau, S.P.; Chen, L.-J.; He, J.-H. 13% Efficiency Si Hybrid Solar Cells via Concurrent Improvement in Optical and Electrical Properties by Employing Graphene Quantum Dots. *ACS Nano* **2016**, *10*, 815–821. [CrossRef] [PubMed]
42. Gao, P.; Ding, K.; Wang, Y.; Ruan, K.; Diao, S.; Zhang, Q.; Sun, B.; Jie, J. Crystalline Si/graphene quantum dots heterojunction solar cells. *J. Phys. Chem. C* **2014**, *118*, 5164–5171. [CrossRef]
43. Pan, D.; Zhang, J.; Li, Z.; Wu, M. Hydrothermal route for cutting graphene sheets into blue-luminescent graphene quantum dots. *Adv. Mater.* **2010**, *22*, 734–738. [CrossRef] [PubMed]
44. Chen, L.; Guo, C.X.; Zhang, Q.; Lei, Y.; Xie, J.; Ee, S.; Guai, G.; Song, Q.; Li, C.M. Graphene quantum-dot-doped polypyrrole counter electrode for high-performance dye-sensitized solar cells. *ACS Appl. Mater. Interfaces* **2013**, *5*, 2047–2052. [CrossRef] [PubMed]
45. Kim, J.K.; Park, M.J.; Kim, S.J.; Wang, D.H.; Cho, S.P.; Bae, S.; Park, J.H.; Hong, B.H. Balancing light absorptivity and carrier conductivity of graphene quantum dots for high-efficiency bulk heterojunction solar cells. *ACS Nano* **2013**, *7*, 7207–7212. [CrossRef] [PubMed]
46. Gupta, V.; Chaudhary, N.; Srivastava, R.; Sharma, G.D.; Bhardwaj, R.; Chand, S. Luminscent graphene quantum dots for organic photovoltaic devices. *J. Am. Chem. Soc.* **2011**, *133*, 9960–9963. [CrossRef] [PubMed]
47. Fang, X.; Li, M.; Guo, K.; Li, J.; Pan, M.; Bai, L.; Luoshan, M.; Zhao, X. Graphene quantum dots optimization of dye-sensitized solar cells. *Electrochim. Acta* **2014**, *137*, 634–638. [CrossRef]
48. Mihalache, I.; Radoi, A.; Mihaila, M.; Munteanu, C.; Marin, A.; Danila, M.; Kusko, M.; Kusko, C. Charge and energy transfer interplay in hybrid sensitized solar cells mediated by graphene quantum dots. *Electrochim. Acta* **2015**, *153*, 306–315. [CrossRef]
49. Peng, J.; Gao, W.; Gupta, B.K.; Liu, Z.; Romero-Aburto, R.; Ge, L.; Song, L.; Alemany, L.B.; Zhan, X.; Gao, G.; et al. Graphene quantum dots derived from carbon fibers. *Nano Lett.* **2012**, *12*, 844–849. [CrossRef] [PubMed]
50. Li, M.; Ni, W.; Kan, B.; Wan, X.; Zhang, L.; Zhang, Q.; Long, G.; Zuo, Y.; Chen, Y. Graphene quantum dots as the hole transport layer material for high-performance organic solar cells. *Phys. Chem. Chem. Phys.* **2013**, *15*, 18973–18978. [CrossRef] [PubMed]
51. Li, L.; Wu, G.; Yang, G.; Peng, J.; Zhao, J.; Zhu, J.-J. Focusing on luminescent graphene quantum dots: Current status and future perspectives. *Nanoscale* **2013**, *5*, 4015–4039. [CrossRef] [PubMed]
52. Low, C.T.J.; Walsh, F.C.; Chakrabarti, M.H.; Hashim, M.A.; Hussain, M.A. Electrochemical approaches to the production of graphene flakes and their potential applications. *Carbon* **2013**, *54*, 1–21. [CrossRef]

53. Zhang, M.; Bai, L.; Shang, W.; Xie, W.; Ma, H.; Fu, Y.; Fang, D.; Sun, H.; Fan, L.; Han, M.; et al. Facile synthesis of water-soluble, highly fluorescent graphene quantum dots as a robust biological label for stem cells. *J. Mater. Chem.* **2012**, *22*, 7461–7467. [CrossRef]
54. Sk, M.A.; Ananthanarayanan, A.; Huang, L.; Lim, K.H.; Chen, P. Revealing the tunable photoluminescence properties of graphene quantum dots. *J. Mater. Chem. C* **2014**, *2*, 6954–6960. [CrossRef]
55. Song, Y.; Zhu, S.; Yang, B. Bioimaging based on fluorescent carbon dots. *RSC Adv.* **2014**, *4*, 27184–27200. [CrossRef]
56. Gan, Z.; Xiong, S.; Wu, X.; Xu, T.; Zhu, X.; Gan, X.; Guo, J.; Shen, J.; Sun, L.; Chu, P.K. Mechanism of photoluminescence from chemically derived graphene oxide: Role of chemical reduction. *Adv. Opt. Mater.* **2013**, *1*, 926–932. [CrossRef]
57. Hong, G.; Diao, S.; Antaris, A.L.; Dai, H. Carbon Nanomaterials for Biological Imaging and Nanomedicinal Therapy. *Chem. Rev.* **2015**, *115*, 10816–10906. [CrossRef] [PubMed]
58. Gan, Z.; Xu, H.; Hao, Y. Mechanism for excitation-dependent photoluminescence from graphene quantum dots and other graphene oxide derivates: Consensus, debates and challenges. *Nanoscale* **2016**, 7794–7807. [CrossRef] [PubMed]
59. Ritter, K.A.; Lyding, J.W. The influence of edge structure on the electronic properties of graphene quantum dots and nanoribbons. *Nat. Mater.* **2009**, *8*, 235–242. [CrossRef] [PubMed]
60. Xie, C.; Zhang, X.; Wu, Y.; Zhang, X.; Zhang, X.; Wang, Y.; Zhang, W.; Gao, P.; Han, Y.; Jie, J. Surface passivation and band engineering: A way toward high efficiency graphene-planar Si solar cells. *J. Mater. Chem. A* **2013**, *1*, 8567–8574. [CrossRef]
61. Clifford, J.N.; Martínez-Ferrero, E.; Viterisi, A.; Palomares, E. Sensitizer molecular structure-device efficiency relationship in dye sensitized solar cells. *Chem. Soc. Rev.* **2011**, *40*, 1635–1646. [CrossRef] [PubMed]
62. O'Regan, B.; Gratzel, M. A low-cost, high-efficiency solar cell based on dye-sensitized colloidal TiO_2 films. *Nature* **1991**, *353*, 737–740. [CrossRef]
63. Mathew, S.; Yella, A.; Gao, P.; Humphry-Baker, R.; Curchod, B.F.E.; Ashari-Astani, N.; Tavernelli, I.; Rothlisberger, U.; Khaja, N.; Grätzel, M. Dye-sensitized solar cells with 13% efficiency achieved through the molecular engineering of porphyrin sensitizers. *Nat. Chem.* **2014**, *6*, 242–247. [CrossRef] [PubMed]
64. Weickert, J.; Dunbar, R.B.; Hesse, H.C.; Wiedemann, W.; Schmidt-Mende, L. Nanostructured organic and hybrid solar cells. *Adv. Mater.* **2011**, *23*, 1810–1828. [CrossRef] [PubMed]
65. Brabec, C.J.; Heeney, M.; McCulloch, I.; Nelson, J. Influence of blend microstructure on bulk heterojunction organic photovoltaic performance. *Chem. Soc. Rev.* **2011**, *40*, 1185–1199. [CrossRef] [PubMed]
66. Lu, L.; Zheng, T.; Wu, Q.; Schneider, A.M.; Zhao, D.; Yu, L. Recent Advances in Bulk Heterojunction Polymer Solar Cells. *Chem. Rev.* **2015**, *115*, 12666–12731. [CrossRef] [PubMed]
67. Liu, Y.; Zhao, J.; Li, Z.; Mu, C.; Ma, W.; Hu, H.; Jiang, K.; Lin, H.; Ade, H.; Yan, H. Aggregation and morphology control enables multiple cases of high-efficiency polymer solar cells. *Nat. Commun.* **2014**, *5*, 5293. [CrossRef] [PubMed]
68. Zhao, J.; Li, Y.; Yang, G.; Jiang, K.; Lin, H.; Ade, H.; Ma, W.; Yan, H. Efficient organic solar cells processed from hydrocarbon solvents. *Nat. Energy* **2016**, *1*, 15027. [CrossRef]
69. Zhao, W.; Qian, D.; Zhang, S.; Li, S.; Inganäs, O.; Gao, F.; Hou, J. Fullerene-Free Polymer Solar Cells with over 11% Efficiency and Excellent Thermal Stability. *Adv. Mater.* **2016**, *28*, 4737–4739. [CrossRef] [PubMed]
70. Wang, H.; Hu, Y.H. Graphene as a counter electrode material for dye-sensitized solar cells. *Energy Environ. Sci.* **2012**, *5*, 8182–8188. [CrossRef]
71. Yang, W.S.; Noh, J.H.; Jeon, N.J.; Kim, Y.C.; Ryu, S.; Seo, J.; Seok, S., II. High-performance photovoltaic perovskite layers fabricated through intramolecular exchange. *Science* **2015**, *348*, 1234–1237. [CrossRef] [PubMed]
72. Williams, K.J.; Nelson, C.A.; Yan, X.; Li, L.-S.; Zhu, X. Hot Electron Injection from Graphene Quantum Dots to TiO_2. *ACS Nano* **2013**, *7*, 1388–1394. [CrossRef] [PubMed]

15

Improvement of Exciton Collection and Light-Harvesting Range in Ternary Blend Polymer Solar Cells based on Two Non-Fullerene Acceptors

Yanbin Wang [1,*], Changlong Zhuang [1], Yawen Fang [1], Hyung Do Kim [2], Huang Yu [1], Biaobing Wang [1,*] and Hideo Ohkita [2,*]

1 Jiangsu Key Laboratory of Environmentally Friendly Polymeric Materials, School of Materials Science and Engineering, Jiangsu Collaborative Innovation Center of Photovolatic Science and Engineering, Changzhou University, Changzhou 213164, Jiangsu, China; 19085204172@smail.cczu.edu.cn (C.Z.); 18000472@smail.cczu.edu.cn (Y.F.); 18000198@smail.cczu.edu.cn (H.Y.)

2 Department of Polymer Chemistry, Graduate School of Engineering, Kyoto University, Katsura, Nishikyo, Kyoto 615-8510, Japan; hyungdokim@photo.polym.kyoto-u.ac.jp

* Correspondence: wangyanbin@cczu.edu.cn (Y.W.); biaobing@cczu.edu.cn (B.W.); ohkita@photo.polym.kyoto-u.ac.jp (H.O.)

Abstract: A non-fullerene molecule named Y6 was incorporated into a binary blend of PBDB-T and IT-M to further enhance photon harvesting in the near-infrared (near-IR) region. Compared with PBDB-T/IT-M binary blend devices, PBDB-T/IT-M/Y6 ternary blend devices exhibited an improved short-circuit current density (J_{SC}) from 15.34 to 19.09 mA cm^{-2}. As a result, the power conversion efficiency (PCE) increased from 10.65% to 12.50%. With an increasing weight ratio of Y6, the external quantum efficiency (EQE) was enhanced at around 825 nm, which is ascribed to the absorption of Y6. At the same time, EQE was also enhanced at around 600–700 nm, which is ascribed to the absorption of IT-M, although the optical absorption intensity of IT-M decreased with increasing weight ratio of Y6. This is because of the efficient energy transfer from IT-M to Y6, which can collect the IT-M exciton lost in the PBDB-T/IT-M binary blend. Interestingly, the EQE spectra of PBDB-T/IT-M/Y6 ternary blend devices were not only increased but also red-shifted in the near-IR region with increasing weight ratio of Y6. This finding suggests that the absorption spectrum of Y6 is dependent on the weight ratio of Y6, which is probably due to different aggregation states depending on the weight ratio. This aggregate property of Y6 was also studied in terms of surface energy.

Keywords: exciton harvesting; ternary blend solar cells; non-fullerene; energy transfer; surface energy

---

1. Introduction

Polymer solar cells have been studied widely because of their excellent advantages, such as flexibility, being light weight, and involving simple large-scale fabrication [1–5]. Typically, the photoactive layer of polymer solar cells is composed of one donor and one acceptor. Currently, conjugated polymers are widely employed as a donor material, and fullerene derivatives or non-fullerene derivatives are employed as an acceptor material in most cases. Fullerene acceptors have been dominated in the past two decades, and offered the highest power conversion efficiencies (PCEs) [6–10]. In 2015, Zhan et al. designed a non-fullerene acceptor named IT-IC, and fabricated the device based on a low-bandgap polymer PCE10 and ITIC. As a result, they obtained a PCE of 6.80%, which was higher than that of the devices based on PCE10 and PCBM [11]. Since then, much more attention has been paid to non-fullerene acceptors [12–15], and a record PCE of 18% has been reported very recently [16]. Among the non-fullerene acceptors, ((2,2′-((2*Z*,2′*Z*)-((12,13-bis(2-ethylhexyl)-3,9-diundecyl-12,13-dihydro-[1,2,5]thiadiazolo [3,4-*e*]thieno[2,″3′′:4′,5′]thieno[2′,3′:4,5]pyrrolo[3,2-*g*]thieno [2′,3′:4,5]thieno[3,2-*b*] indole-2,10-diyl)bis(methanylylidene))bis(5,6-difluoro-3-oxo-2,3-dihydro-1*H*-indene-2,1-diyliden e))dimalononitrile),Y6 has been most widely studied recently, and hence a PCE of more than 16% has been reported by several groups [17–21]. The high PCE of devices based on Y6 is partly ascribed to its wide light-harvesting range of up to 1000 nm. However, there is a limitation in binary blend polymer solar cells even though non-fullerene acceptors exhibit a good light-harvesting property. This is because the optical absorption bandwidth of organic photoactive materials is typically limited to be about 200 nm at most. Therefore, it is still difficult even for polymer/non-fullerene binary solar cells to effectively harvest many more photons over a wide wavelength range from the visible to the near-IR region at the same time.

Ternary blend polymer solar cells could be a simple and effective way to expand the optical absorption range [22–34]. Consequently, improved photovoltaic performance has been reported for ternary blend polymer solar cells [24–34]. The photoactive layer of ternary blend polymer solar cells is composed of one donor and two acceptors or two donors and one acceptor. In most cases, it is desirable to keep the ratio of donor to acceptor in the ternary blend the same as that in the optimized binary blend cell because the hole and electron transport would be balanced. In other words, with increasing weight ratio of the third component, the photon harvesting of the original component should decrease [35–39]. For polymer/fullerene solar cells, such a trade-off relation has been overcome by efficient energy transfer, which improves the exciton harvesting in ternary blend solar cells [24,25,28,40]. For example, when a low-bandgap polymer, poly[(4,4-bis(2-ethylhexyl)-dithieno[3,2-*b*:2′,3′-*d*]silole)-2,6-diyl-*alt*-(2,1,3-benzothiadiazole)-4,7-diyl] (PSBTBT), was incorporated into binary blend devices based on poly(3-hexylthiophene) (P3HT) and a fullerene derivative (PCBM), the photocurrent was improved not only at the PSBTBT absorption band in the near-IR region but also at the P3HT absorption band in the visible region, although the optical absorption of P3HT rather decreased. This is because P3HT excitons lost in P3HT/PCBM binary blends can be collected by an efficient energy transfer from P3HT to PSBTBT, followed by an efficient charge transfer to PCBM [28].

In this study, a low-bandgap non-fullerene molecule named Y6 was incorporated into a binary blend of poly[[4,8-bis[5-(2-ethylhexyl)-2-thienyl]benzo[1,2-*b*:4,5-*b*′]dithiophene-2,6-diyl]-2,5-thiophenediyl [5,7-bis(2-ethylhex-yl)-4,8-dioxo-4*H*,8*H*-benzo[1,2-*c*:4,5-*c*′]dithiophene-1,3-diyl]] (PBDB-T) and 3,9-bis (2-methylene-(3-(1,1-dicyanomethylene)-6/7-methyl)-indanone))-5,5,11,11-tetrakis(4-*n*-hexylphenyl)-dithieno[2,3-*d*:2′,3′-*d*′]-*s*-indaceno[1,2-*b*:5,6-*b*′]dithiophene (IT-M) to further improve the photon harvesting efficiency in the near-IR range. Figure 1 shows the chemical structures and the energy-level diagram of the three materials employed in this study. The complementary absorption also gives a large spectral overlap between the IT-M fluorescence and the Y6 absorption, which could enhance the exciton harvesting of IT-M by an efficient energy transfer from IT-M to Y6. Based on this strategy, a high short-circuit current density (J_{SC}) and a high fill factor (FF) were obtained at the same time, resulting in an improved PCE of 12.5%, which is even higher than those of both individual binary solar cells based on PBDB-T/IT-M and PBDB-T/Y6. We also found that PBDB-T/IT-M/Y6 ternary blends exhibit increased and red-shifted absorption in the near-IR region with increasing weight ratio of Y6. In order to address the origin of this spectral change, the absorption spectra of Y6 in different polymer matrices were studied in terms of surface energy.

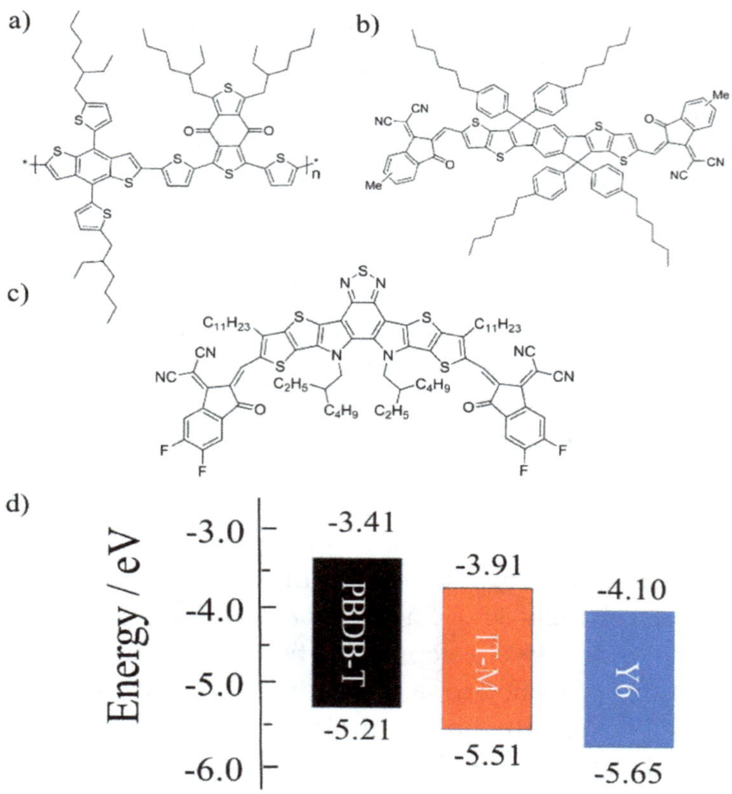

Figure 1. Chemical structures of materials employed in this study: (**a**) PBDB-T, (**b**) IT-M, and (**c**) Y6. (**d**) Energy level diagram of photovoltaic materials used in this study [11,18].

2. Materials and Methods

2.1. Materials

A conjugated polymer PBDB-T (Number average molecular weight M_n = 29 000 g mol^{-1}, polydispersity index PDI = 1.5), non-fullerene molecules IT-M (Purity: 99%) and Y6 (Purity: 99%), and poly[9,9-bis(3′-(*N*,*N*-dimethyl)-*N*-ethylammonium-propyl-2,7-fluorene)-*alt*-2,7-(9,9-dioctyl fluorine)] dibromide (PFN-Br) (M_n = 50,000 g mol^{-1}, PDI = 2.5) were purchased from Solarmer Materials, Incorporated (Beijing, China). These materials were used without further purification.

2.2. Device Fabrication

Non-fullerene-based ternary blend polymer solar cells were fabricated as follows. Indium–tin-oxide (ITO)-coated glass substrates (10 Ω per square) were rinsed by ultrasonication in toluene, acetone, and ethanol for 15 min in sequence. The cleaned substrates were dried under nitrogen gas and then treated with a UV–O_3 cleaner for 30 min. A hole-transporting buffer layer (40 nm) of PEDOT:PSS (AI4083) was spin coated onto the cleaned substrates at a spin rate of 3000 rpm for 60 s and then dried on a hot plate at 140 °C for 10 min in air. Prior to the spin coating, the solution of PEDOT:PSS was filtered with a PTFE syringe filter (pore size: 0.45 μm). The blend active layer was prepared on the ITO/PEDOT:PSS-coated substrate by spin coating at a spin rate of 2200 rpm for 60 s, and subsequently annealed on a hot plate at 140 °C for 10 min in the nitrogen atmosphere. A blend solution of PBDB-T/IT-M/Y6 was prepared by dissolving donor polymers (PBDB-T) and acceptors (IT-M and Y6) (1 : 1 by weight) with a composition of 10 : x : 10 − x mg in 1 mL of chlorobenzene with 1% volume ratio of 1,8-diiodooctane (DIO). The blend solution was stirred at 60 °C overnight. Note that the weight fraction of Y6 was optimized in the range 5–50 wt %. The thickness of blend films was ~100 nm. A PFN-Br buffer layer (~5 nm) was prepared on the active layer by spin coating at a spin rate of 3000 rpm for 60 s from a solution of PFN-Br (0.5 mg) in 1 mL anhydrous methanol. Finally, 100 nm of aluminum top electrode was thermally evaporated on top of the PFN-Br layer under vacuum at 2.5×10^{-4} Pa. Consequently, the device layered structure obtained was as follows: ITO/PEDOT:PSS/PBDB-T : IT-M : Y6/PFN-Br/Al. The effective area of the device was 0.07 cm^2.

2.3. Measurements

J–V characteristics were measured with a direct-current (DC) voltage and current source/monitor (Keithley, 2611B, Cleveland, USA) in the dark and under illumination with an AM 1.5G simulated solar light with 100 mW cm^{-2}. The light intensity was corrected with a calibrated silicon photodiode reference cell (Bunkou Keiki, BS-520, Tokyo, Japan). External quantum efficiency (EQE) spectra were measured with a spectral response measurement system (Bunkou Keiki, ECT-250D). The power of the incident monochromatic light was kept under 0.05 mW cm^{-2}, which was measured with a calibrated silicon reference cell (Bunkou Keiki, BS-520BK, Tokyo, Japan).

The ionization potential of PBDB-T, IT-M, and Y6 films was measured with a photoelectron yield spectrometer (Riken Keiki, AC-3, Tokyo, Japan). All the neat films were fabricated by spin coating from each chlorobenzene solution on the ITO substrate. The threshold energy for the photoelectron emission was estimated on the basis of the cubic root of the photoelectron yield plotted against the incident photon energy, as reported previously [24].

Absorption and photoluminescence (PL) spectra were measured at room temperature with a spectrophotometer (Hitachi, U-4100, Tokyo, Japan) and a spectrofluorometer (Horiba Jobin Yvon, NanoLog, Kyoto, Japna) equipped with a photomultiplier tube (Hamamatsu Photonics, R928P, Hamamatsu, Japan) and a liquid-nitrogen-cooled InGaAs near-IR array detector (Horiba Jobin Yvon, Symphony II, Kyoto, Japan), respectively.

The surface energy γ_X of the material X was evaluated from a contact angle θ_X, as reported previously [27,32,41]. The contact angle θ_X was measured for an ultrapure water droplet on the material film at room temperature. The interfacial energy γ_{AB} between materials A and B was evaluated from γ_A and γ_B by the Neumann's Equation.

3. Results

3.1. Optoelectronic Properties

As shown in Figure 2a, the donor polymer PBDB-T exhibits absorption bands in the visible region from 450 to 710 nm, the non-fullerene acceptor IT-M exhibits an absorption band in the visible to near-IR region from 500 to 800 nm, and the non-fullerene acceptor Y6 exhibits absorption bands mainly in the near-IR region with an absorption tail extending up to 1000 nm. In other words, these three photovoltaic materials show a complementary absorption from the visible to the near-IR range. On the other hand, as shown in Figure 2b, PBDB-T exhibits a PL peak at around 690 nm, IT-M exhibits a PL peak at around 770 nm and a shoulder peak at around 830 nm, and Y6 exhibits a PL peak at around 940 nm. Obviously, the PL peaks of PBDB-T and IT-M were located in the absorption range of IT-M and Y6, respectively. In other words, there is a good spectral overlap between the PL of PBDB-T and the absorption of IT-M, and between the PL of IT-M and the absorption of Y6, respectively, suggesting that the energy transfer from PBDB-T to IT-M and from IT-M to Y6 could occur. On the other hand, as shown in Figure 1d, cascade energy structures would be formed in PBDB-T/IT-M/Y6 blend films, which are beneficial for the charge transfer among the three materials. In other words, there is a competition of energy transfer with charge transfer at the interfaces of PBDB-T/IT-M and of IT-M/Y6.

Figure 3 shows the PL spectra of neat and blend films with different compositions upon photoexcitation of IT-M mainly at 710 nm. For the IT-M/Y6 binary blend film, the PL from IT-M was strongly quenched, and instead the PL from Y6 was clearly observed, indicating an efficient energy transfer from IT-M to Y6. For the PBDB-T/IT-M binary blend film, on the other hand, the PL from IT-M was quenched to ~10% relative to that of the IT-M neat film, suggesting that about 10% of IT-M excitons are radiatively deactivated to the ground state before arriving at the PBDB-T/IT-M interface. For the PBDB-T/IT-M/Y6 ternary blend film, no PL was observed by the addition of only 10 wt % of Y6 molecules into the PBDB-T/IT-M binary blends. In other words, the PL from IT-M was completely quenched and no PL from Y6 was observed. This is most probably because the 10% of IT-M excitons that would be lost in the absence of Y6 are collected to the Y6 domains by an energy transfer followed by a charge transfer to IT-M or PBDB-T.

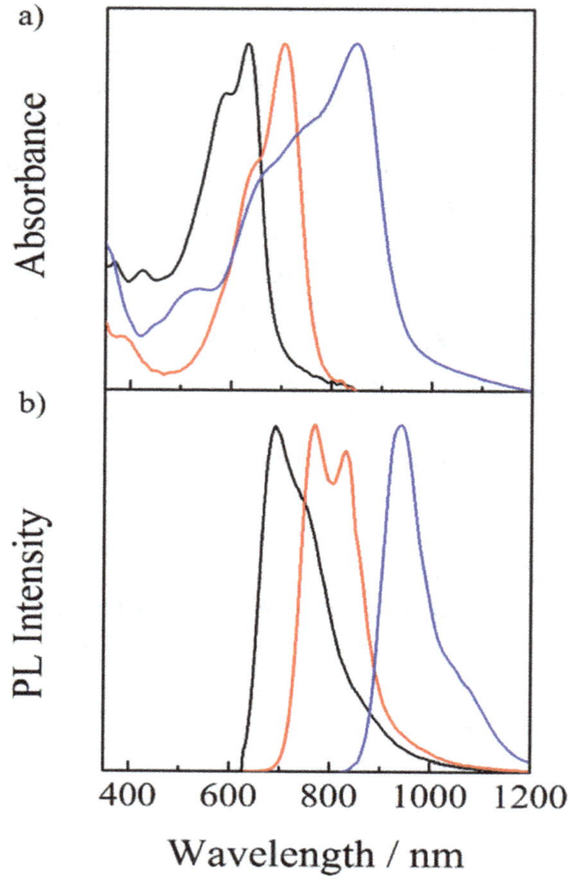

Figure 2. (**a**) Normalized UV-visible absorption and (**b**) PL spectra of PBDB-T (black lines), IT-M (red lines), and Y6 (blue lines) neat films excited at 550, 710, and 750 nm, respectively.

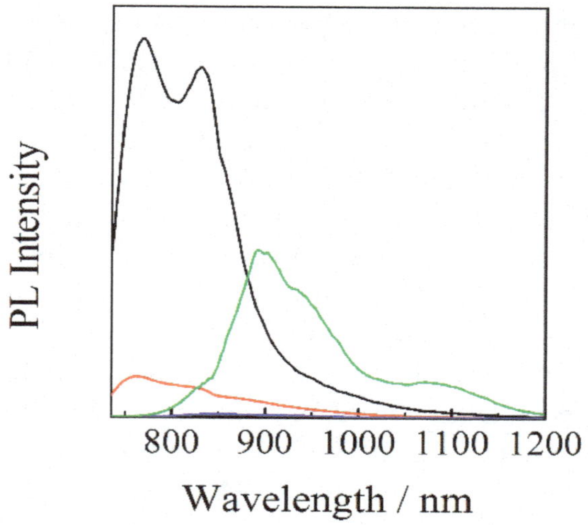

Figure 3. PL spectra of IT-M neat (black line), PBDB-T/IT-M (1 : 1) binary blend (red line), PBDB-T/IT-M/Y6 (1 : 0.8 : 0.2) ternary blend (blue line), and IT-M/Y6 (1 : 1) binary blend (green line) films. All the films were excited at 710 nm, and the PL intensity of all the films was corrected for variation in the absorption at an excitation wavelength of 710 nm. The PL intensity of Y6 in the IT-M/Y6 binary blend was corrected by subtracting the PL intensity due to the direct excitation of Y6 at 710 nm.

3.2. *J–V Characteristics*

In order to discuss the sensitization effect of Y6, we fabricated binary and ternary blend polymer solar cells with a structure of ITO/PEDOT:PSS/active layers/PFN-Br/Al under the same conditions. The device parameters are summarized in Table 1. The overall donor to acceptor ratio was maintained at 1 : 1 in this study. As shown in Figure 4, the PBDB-T/IT-M binary control device gave a short-circuit current density (J_{SC}) of 15.34 mA cm^{-2}, an open-circuit voltage (V_{OC}) of 0.946 V, a fill factor (FF) of 0.734, and a power conversion efficiency (PCE) of 10.65%, which are comparable to those reported previously [26,42]. With the incorporation of 10 wt % Y6 into the binary blend, the J_{SC} was obviously enhanced up to 19.09 mA cm^{-2}, which is much higher than that of the PBDB-T/IT-M binary control device and is approaching that of the PBDB-T/Y6 binary device. The FF was slightly decreased compared with that of the PBDB-T/IT-M binary control device, but it is much higher than that of the PBDB-T/Y6 binary device. V_{OC} was decreased compared with that of the PBDB-T/IT-M binary control device, but it is much higher than that of the PBDB-T/Y6 binary device. This is because the lowest unoccupied molecular orbital energy level of Y6 is much deeper than that of IT-M. As a result, the PCE was improved from 10.65% for the PBDB-T/IT-M binary control devices to 12.50% for the PBDB-T/IT-M/Y6 ternary blend devices, which is also much higher than that of the PBDB-T/Y6 binary blend devices. Further addition of Y6 rather decreased the photocurrent generation and hence degraded the overall photovoltaic performance.

Table 1. Photovoltaic parameters of PBDB-T/IT-M/Y6 ternary blend polymer solar cells with different compositions.

PBDB-T/IT-M/Y6	J_{SC}/mA cm^{-2}	V_{OC}/V	FF	PCE [a]/%
1 : 1 : 0	15.34 (15.02 ± 0.32)	0.946 (0.944 ± 0.002)	0.734 (0.718 ± 0.016)	10.65 (10.28 ± 0.27)
1 : 0.9 : 0.1	17.68 (17.24 ± 0.44)	0.917 (0.916 ± 0.001)	0.728 (0.701 ± 0.027)	11.72 (11.34 ± 0.38)
1 : 0.8 : 0.2	19.09 (18.72 ± 0.37)	0.902 (0.901 ± 0.001)	0.726 (0.713 ± 0.023)	12.50 (12.25 ± 0.22)
1 : 0.7 : 0.3	18.80 (18.37 ± 0.43)	0.872 (0.870 ± 0.001)	0.711 (0.693 ± 0.018)	11.65 (11.62 ± 0.33)
1 : 0 : 1	19.27 (19.05 ± 0.22)	0.689 (0.687 ± 0.002)	0.647 (0.625 ± 0.022)	8.60 (8.34 ± 0.26)

[a] The average values were obtained from 10 devices.

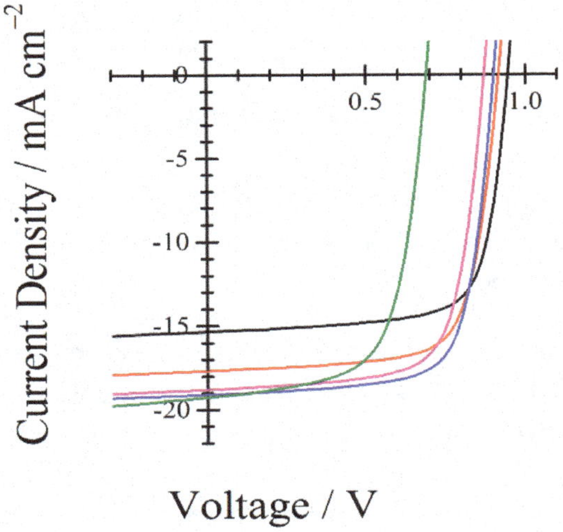

Figure 4. *J–V* characteristics of PBDB-T/IT-M/Y6 ternary blend polymer solar cells with different compositions: 1 : 1 : 0 (black line), 1 : 0.9 : 0.1 (red line), 1 : 0.8 : 0.2 (blue line), 1 : 0.7 : 0.3 (purple line), and 1 : 0 : 1 (green line).

3.3. *External Quantum Efficiency (EQE) Spectra*

In order to address the origin of enhancement in J_{SC}, we measured the absorption and EQE spectra of PBDB-T/IT-M/Y6 ternary blend solar cells with different compositions. As shown in Figure 5a, the absorption was increased at around 790 nm by the addition of Y6 into the binary blend of

PBDB-T/IT-M. Correspondingly, as shown in Figure 5b, EQE was enhanced near the Y6 absorption region. Instead, the absorption was decreased at around 700 nm, which is ascribed to the decrease of IT-M in the ternary blend. Interestingly, the EQE ascribed to IT-M was rather increased even though the absorption was decreased. More specifically, the EQE was increased at 700 nm from 72% to 82% by the incorporation of 10 wt % of Y6 into PBDB-T/IT-M binary blends, although the absorption efficiency was decreased at 700 nm from 92% to 87%. This is because there is efficient energy transfer from IT-M to Y6, as will be discussed detail later. Interestingly, as shown in Figure 5a,b, the absorption range was red-shifted with increasing weight ratio of Y6 in the ternary blend. Correspondingly, the EQE range was also red-shifted. This spectral change is probably due to the aggregation of Y6 in the ternary blend, which is dependent upon the weight fraction of Y6, as will be discussed later.

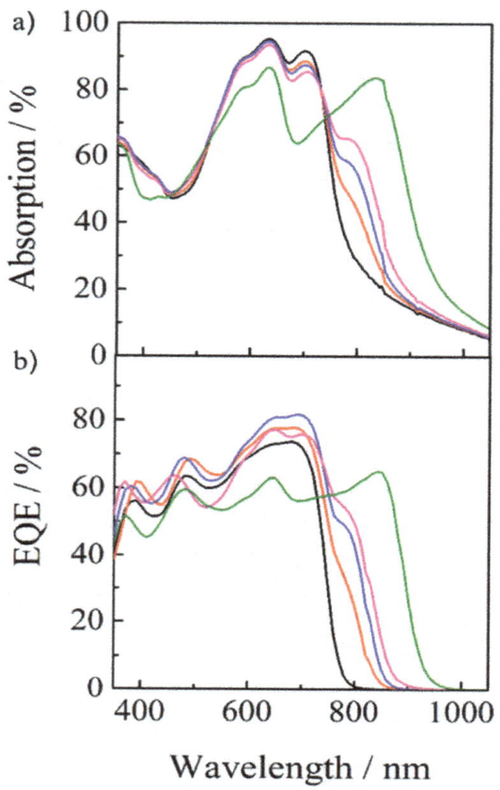

Figure 5. (**a**) UV-visible absorption and (**b**) EQE spectra of PBDB-T/IT-M/Y6 ternary blend films with different blend ratios: 1 : 1 : 0 (black line), 1 : 0.9 : 0.1 (red line), 1 : 0.8 : 0.2 (blue line), 1 : 0.7 : 0.3 (purple line), and 1 : 0 : 1 (green line).

In order to discuss the origin of such a spectral shift, we measured the absorption spectra of Y6 doped in different polymer films. As shown in Figure 6, the absorption spectra were different even though the Y6 weight fraction was the same (5 wt %), suggesting different aggregation states of Y6. In the chlorobenzene solution, Y6 isolated molecules exhibit an absorption band at around 730 nm. On the other hand, Y6 neat films exhibit absorption at around 843 nm, which is red-shifted by more than 100 nm compared to the absorption of Y6 in a chlorobenzene solution. In regiorandom poly(3-hexylthiophene) (RRa-P3HT) films, Y6 exhibits an absorption band at around 765 nm, which is similar to that of Y6 in a chlorobenzene solution, suggesting that Y6 are likely to be relatively homogeneously distributed in RRa-P3HT films. In polystyrene (PS) films, on the other hand, Y6 exhibits an absorption band at around 828 nm, which is rather similar to that in Y6 neat films, suggesting that Y6 molecules form aggregates similar to those in Y6 neat films. These findings suggest that Y6 molecules are more aggregated in these polymer films in the increasing order of PS, PBDB-T, and RRa-P3HT. These different aggregates will be discussed in terms of surface energy.

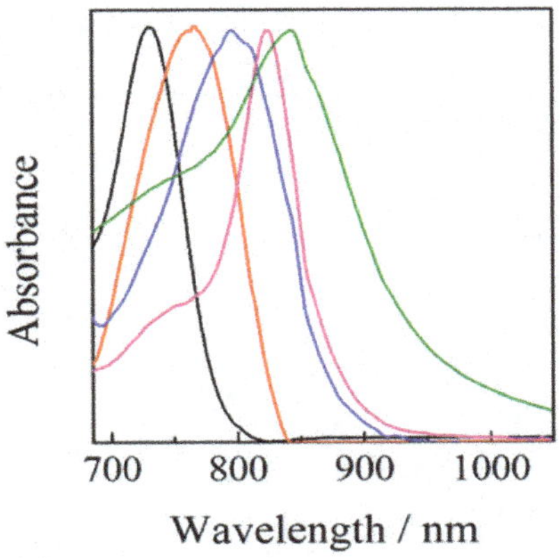

Figure 6. Normalized UV-visible absorption spectra of Y6 in a chlorobenzene solution (black line), Y6 neat film (green line), Y6 doped in RRa-P3HT (red line), PBDB-T (blue line), and polystyrene (PS) (purple line) films with a weight fraction of 5 wt %. The absorption spectra of Y6 in different polymer films were corrected by subtracting the absorption of polymer neat films.

4. Discussion

As mentioned above, PBDB-T/IT-M/Y6 ternary blend polymer solar cells exhibited the improved PCE compared to binary blend control cells. This is mainly due to the improved photocurrent by the addition of a near-IR non-fullerene molecule (Y6). Here, we discuss two mechanisms for the improved photocurrent: one is the improved exciton-harvesting of IT-M due to energy transfer and the other is the improved photon-harvesting by the addition of Y6.

Firstly, we focus on the energy transfer from IT-M to Y6. As shown in Figure 5, EQE was increased at around 700 nm while the absorption was rather decreased at around 700 nm, which is mainly ascribed to IT-M absorption. Thus, the internal quantum efficiency should be improved at the IT-M absorption. As shown in Figure 2, there is a good spectral overlap between the absorption of Y6 and the PL of IT-M, suggesting an efficient energy transfer from IT-M to Y6. Indeed, as shown in Figure 3, PL was observed from Y6 in the IT-M/Y6 binary blend even though IT-M was selectively excited, indicating an efficient energy transfer from IT-M to Y6. On the other hand, the PL of IT-M was quenched to 10% for the PBDB-T/IT-M binary blend, while it was completely quenched for the PBDB-T/IT-M/Y6 ternary blend, suggesting that the 10% IT-M excitons that would be lost in the PBDB-T/IT-M binary blend are efficiently collected to Y6 by energy transfer, as observed for the PL spectra of the IT-M/Y6 ternary blend. Subsequently, an efficient charge transfer occurs from Y6 to IT-M or PBDB-T. The improved exciton quenching efficiency reasonably explains the change of EQE and absorption.

Next, we focus on the improved photon harvesting by the addition of Y6. As shown in Figure 5, an additional EQE signal was observed at around 790 nm, which is consistent with an additional absorption due to the incorporation of Y6 into the PBDB-T/IT-M binary blend. This finding suggests that Y6 molecules contribute to the photocurrent generation in the ternary blend solar cell. Furthermore, the additional absorption was increased in intensity and also red-shifted in the absorption tail with increasing weight ratio of Y6. This is probable due to different aggregate states of Y6 depending on the weight fraction of Y6 in PBDB-T/IT-M/Y6 ternary blends.

In order to study this spectral property, the absorption spectra of Y6 were measured for different polymer matrices. As mentioned before, with the addition of 5 wt % of Y6 into RRa-P3HT, PBDB-T, and PS polymer matrices, the absorption peaks of Y6 were estimated to be 765 nm for Y6 in RRa-P3HT, 796 nm for Y6 in PBDB-T, and 828 nm for Y6 in PS films. This result suggests that the dispersed states in these three materials are different. With increasing weight ratio of Y6, as shown in Figure 7, the absorption peaks of Y6 in these three binary blends were red-shifted and finally saturated to a neat film state. In more detail, the peak shift was saturated at around 70 wt % for Y6 in RRa-P3HT, at around 50 wt % for Y6 in PBDB-T, and at around 20 wt % for Y6 in PS films. These results, again, indicate that Y6 molecules are likely to form aggregates more easily in the order of PS, PBDB-T, and RRa-P3HT films.

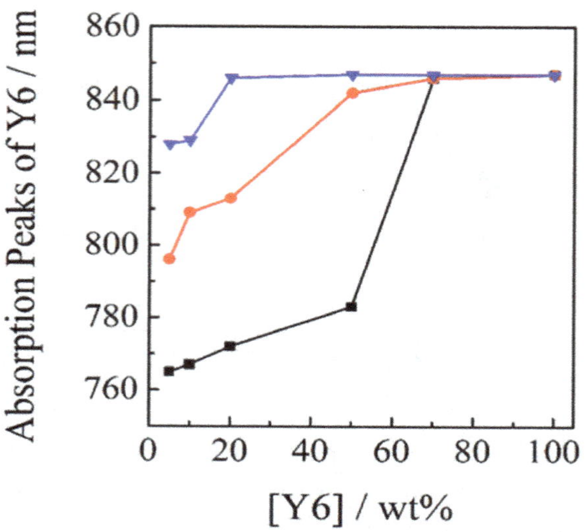

Figure 7. Peak wavelength of the Y6 absorption in different polymer matrices plotted against the weight fraction of Y6: RRa-P3HT (black line), PBDB-T (red line), and PS (blue line).

We therefore discuss the spectral property of Y6 molecules in terms of surface energy. Table 2 summarizes the surface energy of the materials used in this study. The compatibility of two different materials can be predicted on the basis of their surface energy: a similar surface energy indicates good compatibility, while a largely different surface energy indicates poor compatibility [43,44]. As shown in Table 2, the difference in surface energy $\Delta\gamma$ is as small as 2.6 mJ m^{-2} for Y6 and RRa-P3HT, which is the smallest, and the $\Delta\gamma$ is as large as 8.3 mJ m^{-2} for Y6 and PS, which is the largest. The trend in $\Delta\gamma$ is consistent with the absorption peak shift observed for Y6 in these polymer films. We thus conclude that dye aggregation can be controlled by the careful selection of material combinations.

Table 2. Surface energy (γ) of materials used in this study.

Materials	Y6	RRa-P3HT	PBDB-T	PS
γ/mJ m^{-2}	17.3	19.9	22.9	25.6

Finally, we consider the future direction for highly efficient non-fullerene polymer solar cells based on Y6. As shown in Figure 2, Y6 molecules exhibit a good optical absorption in the near-IR region up to 1000 nm, which is beneficial as a near-IR sensitizer for wide- and middle-bandgap photoactive materials. On the other hand, Y6 molecules also exhibit a large spectral change depending on the matrix material. In particular, the absorption of Y6 was most red-shifted in PS films with a large surface energy, resulting in the extension of the absorption tail up to 1000 nm. As shown in Figure 5a, the absorption peak of Y6 is 790 nm in PBDB-T/IT-M/Y6 ternary blend films with a Y6 fraction of 10 wt %. If we select a conjugated polymer with a higher surface energy like PS, Y6 molecules would exhibit a more red-shifted absorption band, resulting in a higher J_{SC}. We therefore believe that this finding is generally applicable to other photoactive materials with a large spectral change depending on the aggregation state.

5. Conclusions

In summary, PBDB-T/IT-M/Y6 ternary blend polymer solar cells exhibited J_{SC} increased from 15.34 to 19.09 mA cm^{-2}, and hence PCE improved from 10.65% to 12.5% compared to PBDB-T/IT-M binary blend polymer solar cells. The increased J_{SC} is partly due to the improved light harvesting by the additional absorption of Y6 in the near-IR region as well as the improved exciton harvesting by efficient energy transfer from IT-M to Y6. Interestingly, the absorption and EQE spectra due to Y6 were red-shifted with increasing weight ratio of Y6. We found that this special shift in absorption is dependent on the Y6 aggregation state, which is sensitive to the surface energy of component materials. The absorption of Y6 was more red-shifted in polymer films with a surface energy more different from Y6, resulting in more efficient light harvesting up to 1000 nm. We thus believe that this finding is generally applicable to other photoactive materials with a large spectral change depending on the aggregation state.

Author Contributions: Y.W., B.W., H.D.K., and H.O. designed the research objective. Y.W. performed statistical analysis and drafted the initial article. C.Z., and Y.F. contributed to the section content and edited subsequent drafts, while all authors reviewed and provided feedstock on the submitted article. All authors have read and agreed to the published version of the manuscript.

Acknowledgments: We appreciate the Jiangsu Students' platform for innovation and entrepreneurship training program.

References

1. Nayak, P.K.; Mahesh, S.; Snaith, H.J.; Cahen, D. Photovoltaic solar cell technologies: Analysing the state of the art. *Nat. Rev. Mater.* **2019**, *4*, 269–285. [CrossRef]
2. Hou, W.; Xiao, Y.; Han, G.; Lin, J.Y. The applications of polymers in solar cells: A review. *Polymers* **2019**, *11*, 143. [CrossRef]
3. Li, Y.; Xu, G.; Cui, C.; Li, Y. Flexible and semitransparent organic solar cells. *Adv. Energy Mater.* **2018**, *8*, 1701791. [CrossRef]
4. Mateker, W.R.; McGehee, M.D. Progress in understanding degradation mechanisms and improving stability in organic photovoltaics. *Adv. Mater.* **2017**, *29*, 1603940. [CrossRef]
5. Dennler, G.; Scharber, M.C.; Brabec, C.J. Polymer-fullerene bulk-heterojunction solar cells. *Adv. Mater.* **2009**, *21*, 1323–1338. [CrossRef]
6. He, Z.; Xiao, B.; Liu, F.; Wu, H.; Yang, Y.; Xiao, S.; Wang, C.; Russell, T.P.; Cao, Y. Status and prospects for ternary organic photovoltaics. *Nat. Photonics* **2015**, *9*, 174–179. [CrossRef]
7. Liu, Y.; Zhao, J.; Li, Z.; Mu, C.; Ma, W.; Hu, H.; Jiang, K.; Lin, H.; Ade, H.; Yan, H. Aggregation and morphology control enables multiple cases of high-efficiency polymer solar cells. *Nat. Commun.* **2014**, *5*, 6293. [CrossRef]
8. Liang, Y.; Xu, Z.; Xia, J.; Tsai, S.T.; Wu, Y.; Li, G.; Ray, C.; Yu, L. For the bright future—bulk heterojunction polymer solar cells with power conversion efficiency of 7.4%. *Adv. Mater.* **2010**, *22*, E135–E138. [CrossRef]
9. Park, S.H.; Roy, A.; Beaupre, S.; Cho, S.; Coates, N.; Moon, J.S.; Moses, D.; Leclerc, M.; Lee, K.; Heeger, A.J. Bulk heterojunction solar cells with internal quantum efficiency approaching 100%. *Nat. Photonics* **2009**, *3*, 297–303. [CrossRef]
10. Li, G.; Shrotriya, V.; Huang, J.; Yao, Y.; Moriarty, T.; Emery, K.; Yang, Y. High-efficiency solution processable polymer photovoltaic cells by self-organization of polymer blends. *Nat. Mater.* **2005**, *4*, 864–868. [CrossRef]
11. Lin, Y.; Wang, J.; Zhang, Z.G.; Bai, H.; Li, Y.; Zhu, D.; Zhan, X. An electron acceptor challenging fullerenes for efficient polymer solar cells. *Adv. Mater.* **2015**, *27*, 1170–1174. [CrossRef]
12. Genene, Z.; Mammo, W.; Wang, E.; Andersson, M.R. Recent advances in n-type polymers for all-polymer solar cells. *Adv. Mater.* **2019**, *31*, 1807275. [CrossRef]
13. Wadsworth, A.; Moser, M.; Marks, A.; Little, M.S.; Gasparini, N.; Brabec, C.J.; Baran, D.; McCulloch, I. Critical review of the molecular design progress in non-fullerene electron acceptors towards commercially viable organic solar cells. *Chem. Soc. Rev.* **2019**, *48*, 1596–1625. [CrossRef]
14. Hou, J.; Inganäs, O.; Friend, R.H.; Gao, F. Organic solar cells based on non-fullerene acceptors. *Nat. Mater.* **2018**, *17*, 119–128. [CrossRef]
15. Chen, W.; Zhang, Q. Recent progress in non-fullerene small molecule acceptors in organic solar cells (OSCs). *J. Mater. Chem. C* **2017**, *5*, 1275–1302. [CrossRef]
16. Liu, Q.; Jiang, Y.; Jin, Y.; Qin, J.; Xu, J.; Li, W.; Xiong, J.; Liu, J.; Xiao, Z.; Sun, K.; et al. 18% efficiency organic solar cells. *Sci. Bull.* **2020**. [CrossRef]

17. Fan, B.; Zhang, D.; Li, M.; Zhong, W.; Zeng, Z.; Ying, L.; Huang, F.; Cao, Y. Achieving over 16% efficiency for single-junction organic solar cells. *Sci. China Chem.* **2019**, *62*, 746–752. [CrossRef]
18. Yuan, J.; Zhang, Y.; Zhou, L.; Zhang, G.; Yip, H.L.; Lau, T.K.; Lu, X.; Zhu, C.; Peng, H.; Johnson, P.A.; et al. Single-junction organic solar cell with over 15% efficiency using fused-ring acceptor with electron-deficient core. *Joule* **2019**, *3*, 1140–1151. [CrossRef]
19. Xu, X.; Feng, K.; Bi, Z.; Ma, W.; Zhang, G.; Peng, Q. Single-junction polymer solar cells with 16.35% efficiency enabled by a platinum (II) complexation strategy. *Adv. Mater.* **2019**, *31*, 1901872. [CrossRef]
20. Liu, B.; Wang, Y.; Chen, P.; Zhang, X.; Sun, H.; Tang, Y.; Liao, Q.; Huang, J.; Wang, H.; Meng, H.; et al. Boosting efficiency and stability of organic solar cells using ultralow-cost BiOCl nanoplates as hole transporting layers. *ACS Appl. Mater. Interfaces* **2019**, *11*, 33505–33514. [CrossRef]
21. Chang, Y.; Lau, T.K.; Pan, M.A.; Lu, X.; Yan, H.; Zhan, C. The synergy of host–guest nonfullerene acceptors enables 16%-efficiency polymer solar cells with increased open-circuit voltage and fill-factor. *Mater. Horiz.* **2019**, *6*, 2094–2102. [CrossRef]
22. Gasparini, N.; Salleo, A.; McCulloch, I.; Baran, D. The role of the third component in ternary organic solar cells. *Nat. Rev. Mater.* **2019**, *4*, 229–242. [CrossRef]
23. Rasi, D.D.C.; Janssen, R.A.J. Advances in solution-processed multijunction organic solar cells. *Adv. Mater.* **2019**, *31*, 1806499. [CrossRef] [PubMed]
24. Wang, Y.; Wang, T.; Chen, J.; Kim, H.D.; Gao, P.; Wang, B.; Iriguchi, R.; Ohkita, H. Quadrupolar D–A–D diketopyrrolopyrrole-based small molecule for ternary blend polymer solar cells. *Dyes Pig.* **2018**, *158*, 213–218. [CrossRef]
25. Wang, Y.; Chen, J.; Kim, H.D.; Wang, B.; Iriguchi, R.; Ohkita, H. Ternary blend solar cells based on a conjugated polymer with diketopyrrolopyrrole and carbazole units. *Front. Energy Res.* **2018**, *6*, 113. [CrossRef]
26. Wang, Y.; Kim, H.D.; Wang, B.; Ohkita, H. Visible sensitization for non-fullerene polymer solar cells using a wide bandgap polymer. *J. Photopolym. Sci. Technol.* **2018**, *31*, 177–181. [CrossRef]
27. Xu, H.; Ohkita, H.; Tamai, Y.; Benten, H.; Ito, S. Interface engineering for ternary blend polymer solar cells with a heterostructured near-IR dye. *Adv. Mater.* **2015**, *27*, 5868–5874. [CrossRef]
28. Wang, Y.; Ohkita, H.; Benten, H.; Ito, S. Highly efficient exciton harvesting and charge transport in ternary blend solar cells based on wide- and low-bandgap polymers. *Phys. Chem. Chem. Phys.* **2015**, *17*, 27217–27224. [CrossRef]
29. Wang, Y.; Zheng, B.; Tamai, Y.; Ohkita, H.; Benten, H.; Ito, S. Dye sensitization in the visible region for low-bandgap polymer solar cells. *J. Electrochem. Soc.* **2014**, *161*, D3093–D3096. [CrossRef]
30. Xu, H.; Ohkita, H.; Hirata, T.; Benten, H.; Ito, S. Near-IR dye sensitization of polymer blend solar cells. *Polymer* **2014**, *55*, 2856–2860. [CrossRef]
31. Xu, H.; Wada, T.; Ohkita, H.; Benten, H.; Ito, S. Dye sensitization of polymer/fullerene solar cells incorporating bulky phthalocyanines. *Electrochim. Acta* **2013**, *100*, 214–219. [CrossRef]
32. Honda, S.; Ohkita, H.; Benten, H.; Ito, S. Selective dye loading at the heterojunction in polymer/fullerene solar cells. *Adv. Energy Mater.* **2011**, *1*, 588–598. [CrossRef]
33. Honda, S.; Ohkita, H.; Benten, H.; Ito, S. Multi-colored dye sensitization of polymer/fullerene bulk heterojunction solar cells. *Chem. Commun.* **2010**, *46*, 6596–6598. [CrossRef] [PubMed]
34. Honda, S.; Nogami, T.; Ohkita, H.; Benten, H.; Ito, S. Improvement of the light-harvesting efficiency in polymer/fullerene bulk heterojunction solar cells by interfacial dye modification. *ACS Appl. Mater. Interfaces* **2009**, *4*, 804–810. [CrossRef]
35. Ameri, T.; Min, J.; Li, N.; Machui, F.; Baran, D.; Forster, M.; Schottler, K.J.; Dolfen, D.; Scherf, U.; Brabec, C.J. Performance enhancement of the P3HT/PCBM solar cells through NIR sensitization using a small-bandgap polymer. *Adv. Energy Mater.* **2012**, *2*, 1198–1202. [CrossRef]
36. Li, J.; Liang, Z.; Peng, Y.; Lv, J.; Ma, X.; Wang, Y.; Xia, Y. 36% Enhanced efficiency of ternary organic solar cells by doping a NT-based polymer as an electron-cascade donor. *Polymers* **2018**, *10*, 703. [CrossRef]
37. Zhang, M.; Zhang, F.; An, Q.; Sun, Q.; Wang, J.; Li, L.; Wang, W.; Zhang, J. High efficient ternary polymer solar cells based on absorption complementary materials as electron donor. *Sol. Energy Mater. Sol. Cells* **2015**, *141*, 154–161. [CrossRef]
38. Shen, W.; Chen, W.; Zhu, D.; Zhang, J.; Xu, X.; Jiang, H.; Wang, T.; Wang, E.; Yang, R. High-performance ternary polymer solar cells from a structurally similar polymer alloy. *J. Mater. Chem. A* **2017**, *5*, 12400–12406. [CrossRef]

39. An, Q.; Zhang, F.; Zhang, J.; Tang, W.; Wang, Z.; Li, L.; Xu, Z.; Teng, F.; Wang, Y. Enhanced performance of polymer solar cells through sensitization by a narrow band gap polymer. *Sol. Energy Mater. Sol. Cells* **2013**, *118*, 30–35. [CrossRef]
40. Wang, Y.; Ohkita, H.; Benten, H.; Ito, S. Efficient exciton harvesting through long-range energy transfer. *ChemPhysChem* **2015**, *16*, 1263–1267. [CrossRef]
41. Sumita, M.; Sakata, K.; Asai, S.; Miyasaka, K.; Nakagawa, H. Dispersion of fillers and the electrical conductivity of polymer blends filled with carbon black. *Polym. Bull.* **1991**, *25*, 265–271. [CrossRef]
42. Zhao, W.; Li, S.; Zhang, S.; Liu, X.; Hou, J. Ternary polymer solar cells based on two acceptors and one donor for achieving 12.2% efficiency. *Adv. Mater.* **2017**, *29*, 1604059. [CrossRef] [PubMed]
43. Huang, J.H.; Hsiao, Y.S.; Richard, E.; Chen, C.C.; Chen, P.; Li, G.; Chu, C.W.; Yang, Y. The investigation of donor-acceptor compatibility in bulk-heterojunction polymer systems. *Appl. Phys. Lett.* **2013**, *103*, 043304. [CrossRef]
44. Du, X.; Lin, H.; Chen, X.; Tao, S.; Zheng, C.; Zhang, X. Ternary organic solar cells with a phase-modulated surface distribution via the addition of a small molecular luminescent dye to obtain a high efficiency over 10.5%. *Nanoscale* **2018**, *10*, 16455–16467. [CrossRef] [PubMed]

16

Ultraviolet Plasmonic Aluminium Nanoparticles for Highly Efficient Light Incoupling on Silicon Solar Cells

Yinan Zhang [1,*], Boyuan Cai [2] and Baohua Jia [1,*]

1 Centre for Micro-Photonics, Faculty of Science, Engineering and Technology, Swinburne University of Technology, Hawthorn, Victoria 3122, Australia
2 Institute of Photonics Technology, Jinan University, Guangzhou 510632, China; caiboyuan@jnu.edu.cn
* Correspondance: yinanzhang@swin.edu.au (Y.Z.); bjia@swin.edu.au.com (B.J.)

Academic Editors: Guanying Chen, Zhijun Ning and Hans Agren

Abstract: Plasmonic metal nanoparticles supporting localized surface plasmon resonances have attracted a great deal of interest in boosting the light absorption in solar cells. Among the various plasmonic materials, the aluminium nanoparticles recently have become a rising star due to their unique ultraviolet plasmonic resonances, low cost, earth-abundance and high compatibility with the complementary metal-oxide semiconductor (CMOS) manufacturing process. Here, we report some key factors that determine the light incoupling of aluminium nanoparticles located on the front side of silicon solar cells. We first numerically study the scattering and absorption properties of the aluminium nanoparticles and the influence of the nanoparticle shape, size, surface coverage and the spacing layer on the light incoupling using the finite difference time domain method. Then, we experimentally integrate 100-nm aluminium nanoparticles on the front side of silicon solar cells with varying silicon nitride thicknesses. This study provides the fundamental insights for designing aluminium nanoparticle-based light trapping on solar cells.

Keywords: ultraviolet plasmonics; aluminium nanoparticles; light incoupling; light trapping; solar cells

---

1. Introduction

Light management is of critical importance for constructing high efficiency solar cells. Conventionally, micro-scale textured surfaces, such as pyramid structure, have been widely employed to improve the light absorption in Si solar cells. Textured surfaces not only reduce the surface reflection by increasing the chance of the reflected light bounced back to the Si layer but also enhance the light trapping through the redirection of the incident light. Although they have been well demonstrated and applied on the commercial solar cells, they induce a few drawbacks as well. Firstly, textured surfaces increase the surface area of the solar cells, leading to an increase of the surface recombination and thus the reduction of the solar cell performance. Secondly, they increase the total volume of the depletion junction region. Thirdly, they are not suitable for the light management in ultra-thin solar cells, which have been identified as one of the most effective strategies for the cost reduction of the solar cells.

Recently, nano-scale light management strategies, particularly the plasmonic nanoparticles (NPs) and nanostructures have emerged and attracted a great deal of interest [1–3]. Metal NPs, supporting localized surface plasmons, exhibit both far-field scattering and near-field light concentration due to the collective oscillation of the free electrons in the metal [4,5]. Both effects can contribute to the light absorption enhancement of solar cells. By tuning the geometry parameters of the NPs, the strength of light scattering/concentration and the resonance wavelength can be adjusted accordingly. Nanoparticles with a size smaller than 20 nm show strong near-field light concentration in their immediate vicinity, which can potentially enhance the light absorption in this region. However, so far, all of the thicknesses of the active layers of the solar cells are a few times larger than the near-field light concentration regions of the NPs. Therefore, the near-field light concentration is hardly able to increase the absorption of the solar cells. On the other hand, the scattering of the particles has been the main mechanism used for light absorption enhancement. Nanoparticles larger than 100 nm show strong scattering strength. Once they are placed on the surface of the solar cells, light will be preferentially scattered into the high-index substrates. This increases the photon absorption in solar cells in at least two ways. Firstly, the preferential scattering reduces the light reflection due to the optical impedance matching, leading to an enhanced light incoupling into the solar cell. Secondly, light is redistributed inside the solar cells as a result of the scattering, leading to an increased light path length, which is particularly useful for the weakly absorbed near-bandgap energy.

At the early stage of the plasmonic solar cell research, the noble metal Ag and Au NPs were widely studied and demonstrated to be effective in trapping the near bandgap energy due to higher radiative efficiency [6–25]. However, a transmission reduction is always found at the short wavelength of the usable solar spectrum (300–1200 nm) due to the Fano effect, *i.e.*, the destructive interference between the incident light and the scattered modes at the wavelengths below the surface plasmon resonances (SPRs) [26]. Furthermore, the noble metal Ag and Au are expensive, limiting their real-life applications. Al NPs, with their SPRs spanning from visible to ultraviolet region, can potentially blueshift this reduction away from the usable solar spectrum, thereby leading to a broadband light incoupling [27–30]. Together with the low cost, earth-abundance, and their high compatibility with the complementary metal-oxide semiconductor (CMOS) manufacturing process, Al NPs hold great promises for the next-generation plasmonic solar cells. It should be mentioned here that embedding the NPs in a thin dielectric layer can also avoid this detrimental Fano effect, providing another way to mitigate this light incoupling reduction at the short wavelengths [31].

So far, Al NPs have been applied to many types of solar cells, such as Si wafer solar cells [32], thin-film GaAs solar cells [33], thin-film Si solar cells [34,35], GaInP/GaInAs/Ge triple junction solar cells [36], dye-sensitized solar cells [37], and organic solar cells [38], making them a rising star in this field. However, there is no systemic study on the critical parameters of the Al NP array, which is of importance in designing and optimizing their light incoupling. Here, in this paper, we report some key factors that determine the light incoupling by the Al NPs located on the front side of Si solar cells. Figure 1 schematically shows the structure of the solar cells, consisting of the Al NPs, the SiN_x spacing layer, and the Si layer. We first numerically study the scattering and absorption properties of the Al NPs and the influence of the NP shape, size, surface coverage, and the spacing layer on the light incoupling using the finite difference time domain (FDTD) method. Followed this, we experimentally integrate 100-nm Al NPs on the front side of Si solar cells with varying SiN_x spacing layer thicknesses.

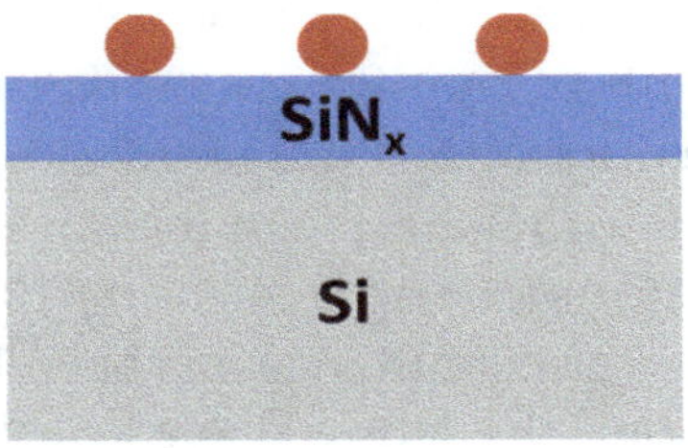

Figure 1. Schematic diagram of the investigated solar cell structure, consisting of the Al nanoparticles (NPs), the SiN_x spacing layer, and the Si layer with the red spheres representing the metal NPs.

2. Results and Discussion

2.1. Broadband Light Incoupling by Al NPs

To investigate the light incoupling properties, it is important to understand the scattering and absorption properties of the Al NPs in comparison to other widely used metal NPs, including Ag and Au. Figure 2a,b illustrate the normalized scattering and absorption cross sections of the three types of metal NPs (100-nm diameter) on top of a Si layer, with the arrows indicating their dipole resonance wavelengths. Clearly, the SPRs of the Ag and Au NPs both lie in the visible ranges at around 400 nm and 550 nm, respectively, whereas the Al NP shows an ultraviolet resonance at around 300 nm. The absorption cross sections of the NPs (Figure 2b) demonstrate an associated absorption loss, particularly at the resonance region. It should be noted that the slightly large absorption at around 800 nm for Al NPs is introduced by the interband transition.

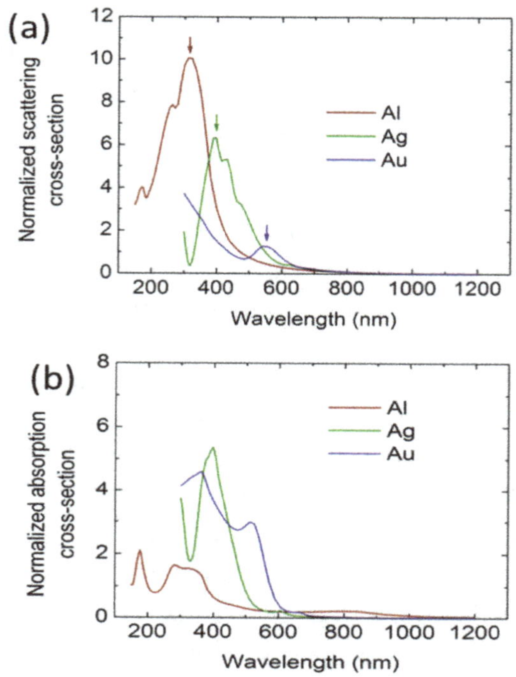

Figure 2. Calculated normalized scattering (**a**) and absorption (**b**) cross sections of the Al NPs with a 100-nm diameter on top of a Si layer, in comparison with Ag and Au NPs.

To demonstrate the superiority of the Al NPs in light incoupling, we simulated the light transmittance into the solar cells integrated with an ordered periodic NP array with a 100-nm diameter and 10% surface coverage (280-nm pitch) for all the three materials: Al, Ag, and Au. In an ordered array, far-field diffraction occurs, leading to a redistribution of the scattered light and change of the light incoupling spectra, whereas the total scattering and incoupling of the NPs in a random array is generally the sum of that for each NP, provided that the surface coverage is low enough [23]. Figure 3a shows the normalized transmittance of the solar cell integrated with NPs. As can be seen, the light transmittance into Si has been increased at wavelengths above the SPRs for Ag and Au NPs, with the largest enhancement up to 43% at 510 nm and 26% at 600 nm, respectively. However, at short wavelengths, the light transmittance has been largely reduced. For Al NPs, the light transmittance has been increased among the entire wavelengths from 300 to 1200 nm without any reduction, demonstrating a broadband light incoupling enhancement. Figure 3b shows that Al NPs also perform better in terms of loss control at the wavelengths below 600 nm, except at around 800 nm with a minor absorption.

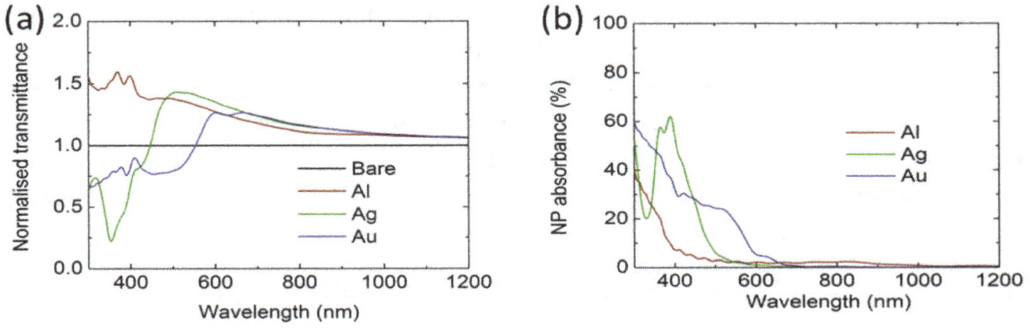

Figure 3. (**a**) Calculated normalized light transmittance of Si solar cells integrated with an array of Al NPs (100-nm diameter and 10% surface coverage), compared with the Si solar cells integrated with Ag and Au NPs. (**b**) Calculated absorption losses in the NP arrays.

2.2. *Shape Study of the Al NPs*

We have demonstrated that the Al NPs show broadband light incoupling, compared with Ag and Au NPs. In this section, we study the influence of the shape of the Al NPs on the light incoupling in the Si wafer solar cells. We simulated the light transmittance of a Si wafer with an ordered hemispherical (100-nm diameter) and cubic (100-nm width) Al NP array with 10% surface coverage, with the results shown in Figure 4. It is remarkable and interesting that the transmittance of the Si wafer with the hemispherical and cubic Al NPs reduces at the longer wavelengths (Figure 4a), compared with that of the bare Si wafer, which is attributed to the redshifts of the Fano resonance for the dipole scattering modes. To understand this, we calculated the scattering cross sections of the hemispherical and cubic Al NP on top of a Si layer, as shown in Figure 4b. Clearly, the dipole resonance shifts to around 1050 nm and 960 nm for the hemispherical and cubic NPs due to the large contact area between the NPs and the high index Si substrate. Therefore, the light incoupling reduction shifts to the long wavelengths. This is distinct from the broadband light incoupling of the 100-nm spherical Al NPs with only a point contact with the Si layers.

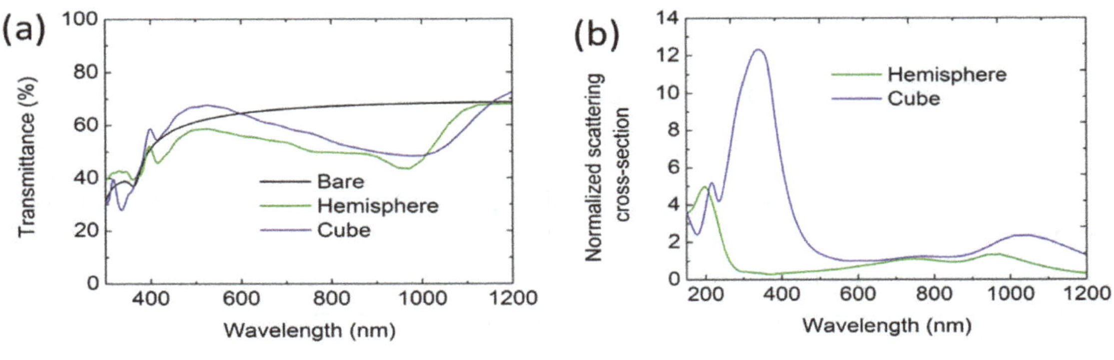

Figure 4. (**a**) Calculated transmittance of the Si solar cells with an array of hemispherical and cubic Al NPs (100-nm diameter/width and 10% surface coverage), referenced to that of a bare Si layer. (**b**) Calculated normalized scattering cross sections of the corresponding Al NPs on top of the Si layer.

2.3. *Size and Surface Coverage Effect*

Since spherical Al NPs have been demonstrated to be more effective in light incoupling than the other shapes, this section studies the two key parameters in optimizing the spherical Al NP array, *i.e.*, the diameter and the surface coverage. Before investigating their influence on the light incoupling, we calculated the scattering and absorption of the Al NPs with different sizes. Figure 5a,b give the normalized scattering and absorption cross sections of the spherical Al NPs with varying diameters. As can be seen, the dipole resonance scattering (indicated with arrows) redshifts and broadens when the particle size increases. At the same time, higher order resonances are excited at the short wavelengths and also redshift. For the absorption, it is found that the interband absorption peaks at around 200 nm do not shift, but the strength reduces when the particle size increases. However, the absorption at around 800 nm increases when the particle size increases, which is distinct from the phenomena observed in the widely used Ag and Au NPs. This is possibly due to the fact that the dipole resonances shift to this region, leading to a resonance-related near-field light confinement, which overlays onto the interband absorption of the Al NPs.

Figure 6a,b present the light transmittance of the Si solar cells integrated with an array of Al NPs under the configurations of varying diameter at 10% surface coverage and increasing surface coverage from 10% to 70% for a 100-nm particle size. Compared with 100-nm Al NPs, 200-nm and 300-nm Al NPs induce a transmission reduction region at short wavelengths due to the redshifts of the dipole scattering-related Fano resonance (Figure 5a). However, the transmission becomes larger at the longer wavelengths owing to the increased scattering cross section when the particle size increases. With the surface coverage increasing, the transmission increases at first and then decreases, reaching a maximum value (>90%) at 600 nm for 40% surface coverage. When the surface coverage increases beyond 40%, the particle interplay becomes severe, leading to redshifts of the dipole scattering modes; therefore, the Fano resonance-related transmission reduction occurs. The optimization of the size and the surface coverage would be a result of the spectra shifts matching the peak intensity of the standard AM1.5G solar photon fluxes. The optimized enhancement could be around 28.7% for the Al NPs under the configuration of a 150-nm diameter and 30% surface coverage. By further tuning the particle height and width, the enhancement can be as high as 30%. This is far higher than that induced by the optimized Ag (27.2%) and Au (14.5%) NPs.

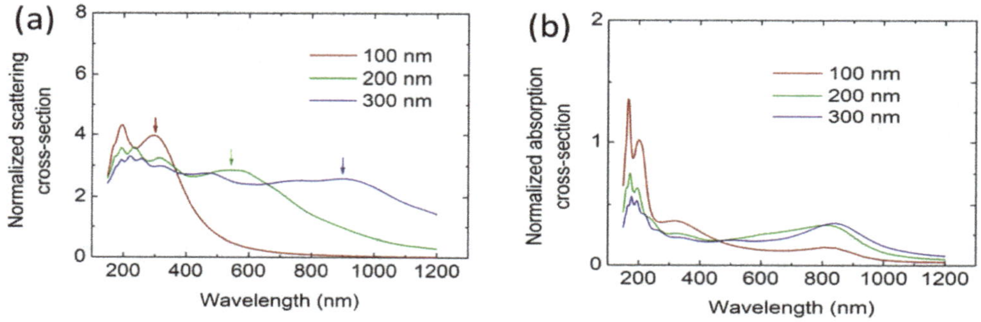

Figure 5. Calculated normalized scattering (**a**) and absorption (**b**) cross sections of the Al NPs in the air with 100-nm, 200-nm and 300-nm diameters, respectively.

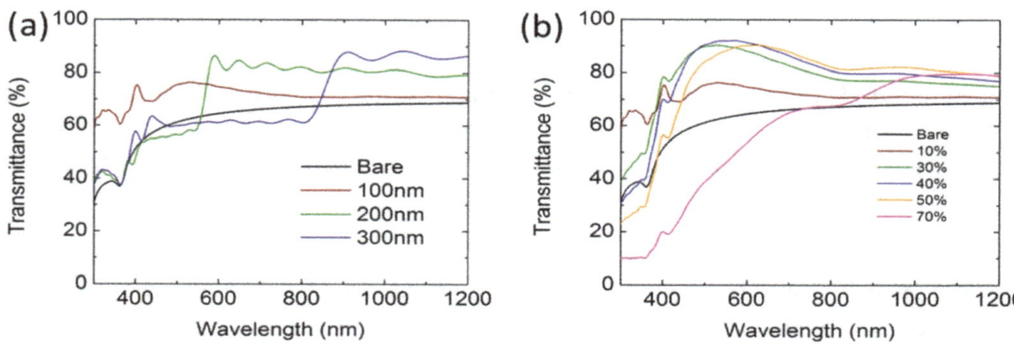

Figure 6. Calculated light transmittance into the Si wafer under the configurations of (**a**) varying diameter at 10% surface coverage and (**b**) increasing surface coverage from 10% to 70% for a 100-nm particle size.

2.4. Study on the Thickness of SiN_x Spacing Layer

The spacing layer between the Al NPs and the Si layer plays an important role in determining the incoupling efficiency induced by the Al NPs since it affects the distance between the excited dipole moment in the NPs and the underlying substrate. This section gives some study on the influence of the spacing layer thickness on the Al NPs incoupling, using the SiN_x spacing layer as an example. The light transmittance into the Si layer by the Al NP array, using an example configuration of a 100-nm diameter and 10% surface coverage, was calculated with varying thicknesses of the SiN_x layer from 20 nm to 120 nm. Figure 7a shows the transmittance spectra of the Si wafer integrated with the Al NPs on top of 20-nm SiN_x. As shown, the 20-nm SiN_x boosts the transmittance across the entire wavelength band from 300 nm to 1200 nm. It acts as an antireflection layer, making use of the destructive interference between the reflected light from the air/SiN_x and SiN_x/Si interfaces. By adding the Al NPs, the transmittance is further increased, with an enhancement up to 25% at the wavelength around 400 nm compared with that of the solar cells with 20-nm SiN_x solely. Figure 7b gives the integrated transmittance at different thicknesses of SiN_x layer by weighting the transmittance to the AM1.5G solar photon fluxes. Clearly, as the SiN_x thickness increases from 20 nm to 120 nm, the integrated transmittance firstly increases and then reduces, reaching a maximum transmittance at 80-nm SiN_x for both the cells with and without NPs. However, the enhancement (the ratio of the integrated transmittance of the solar cells with NPs to that without NPs) reduces from around 12% to 2%, as shown in the inset of Figure 7b, demonstrating that thicker spacing layer reduces the light incoupling efficiency.

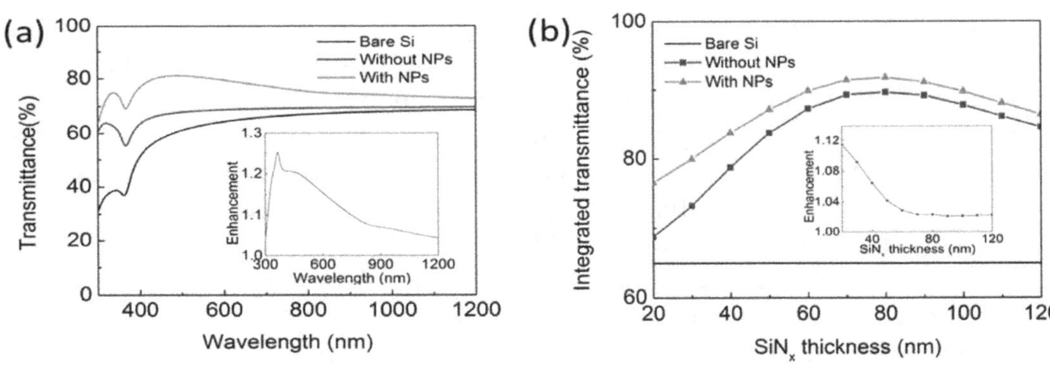

Figure 7. (**a**) Calculated light transmittance spectra of the Si wafer integrated with the Al NP array (100-nm diameter and 10% surface coverage) on top of the 20-nm SiN_x, referenced to that without Al NPs and the bare Si. (**b**) Calculated integrated light transmittance of the Si wafer with the Al NPs as a function of the SiN_x spacing layer thickness. The inset figures show the relative enhancement.

2.5. Experimental Demonstration

As a demonstration, we experimentally fabricated the Al and Ag NPs with a 100-nm average size and integrated them on the front side of the bare Si solar cells by the spray-coating method. Before the integration, we measured their morphologies and optical properties. Figure 8a,b present the scanning electron images (SEMs) of the prepared Al and Ag NPs, with their extinction spectra shown in Figure 8c,d, respectively. As shown, the particles are randomly distributed on the surface of the Si and aggregate in some regions, which slightly affects the particle resonances and the light incoupling. The SPRs for Al and Ag NPs lie in below 300 nm and around 400 nm, respectively, which agree well with the above-mentioned simulation.

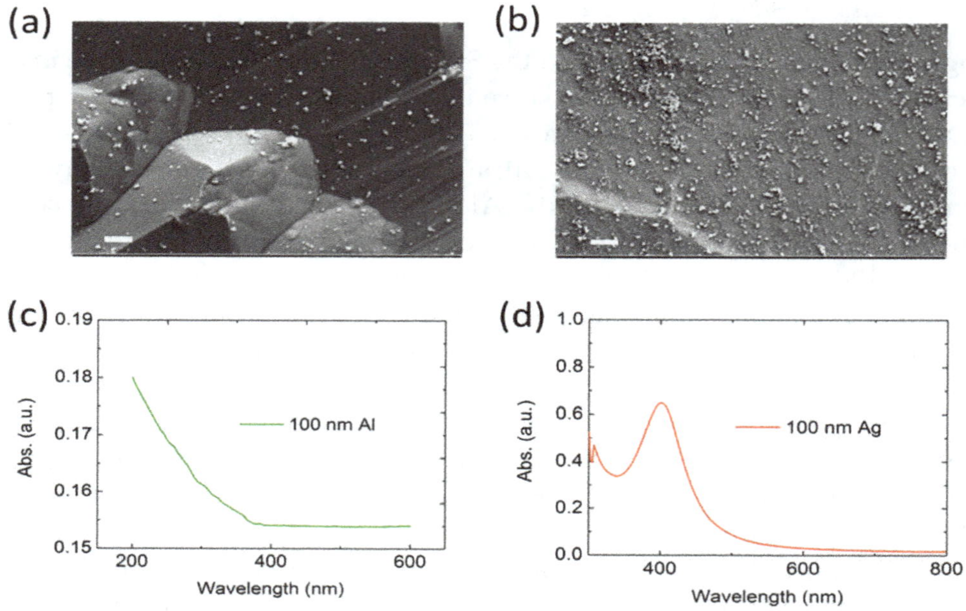

Figure 8. Scanning electron images (SEM) of the fabricated Al (**a**) and Ag (**b**) NPs (Scale bar: 1 μm). The UV-VIS-NIR spectra of the Al (**c**) and Ag (**d**) NPs in deionized water solution.

Figure 9a,b show the reflectance and the normalized external quantum efficiency (EQE) of the solar cells integrated with Al and Ag NPs without the SiN_x layer, respectively. It is noted that the reflectance for both solar cells reduces among the entire wavelength region. However, the EQE enhancement for the Al NPs is broadband over the entire spectrum with an enhancement up to 40% at the short wavelengths, whereas that for the Ag NPs is found to reduce at the short wavelengths below 400 nm. This agrees well with the simulated light transmittance shown in Figure 3a. Then, the Al NPs were integrated on the front surface of the solar cells with various SiN_x thicknesses at 20 nm, 40 nm, 60 nm, and 80 nm, respectively. The EQE results and the corresponding photocurrent density are shown in Figure 10a,b, respectively. As can be seen, the EQE is increased among the entire wavelengths from 300 nm to 1200 nm, even with a 20-nm-thick SiN_x layer, agreeing well with the light transmittance shown in Figure 7a. Accordingly, the photocurrent was increased from 26.2 mA/cm^2 to 28.3 mA/cm^2, representing an enhancement of 8%. In Figure 10b, the photocurrents of the solar cells integrated with and without Al NPs both increase when the SiN_x thickness increases from 20 nm to 80 nm. However, the enhancement induced by the Al NPs shown in the inset figure shows a saturated trend due to a reduced light incoupling. The largest photocurrent achieved is 36 mA/cm^2 with Al NPs on top of the 80-nm-thick SiN_x layer.

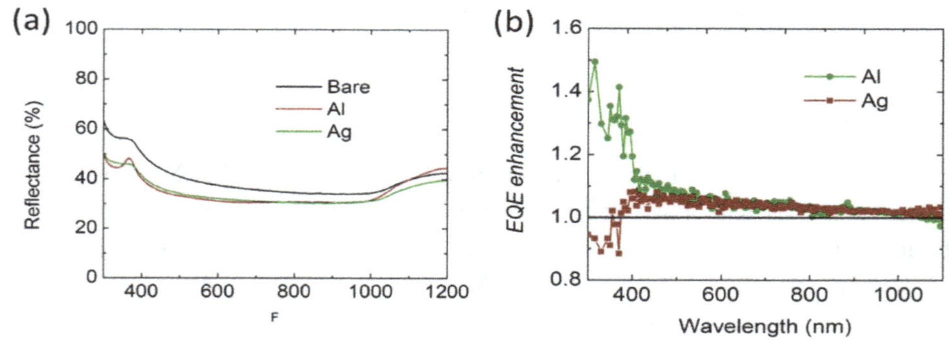

Figure 9. (**a**) Measured reflectance of the Si solar cells integrated with Al and Ag NPs. (**b**) Measured external quantum efficiency (EQE) enhancement of the solar cells integrated with Al and Ag NPs, relative to that without NPs.

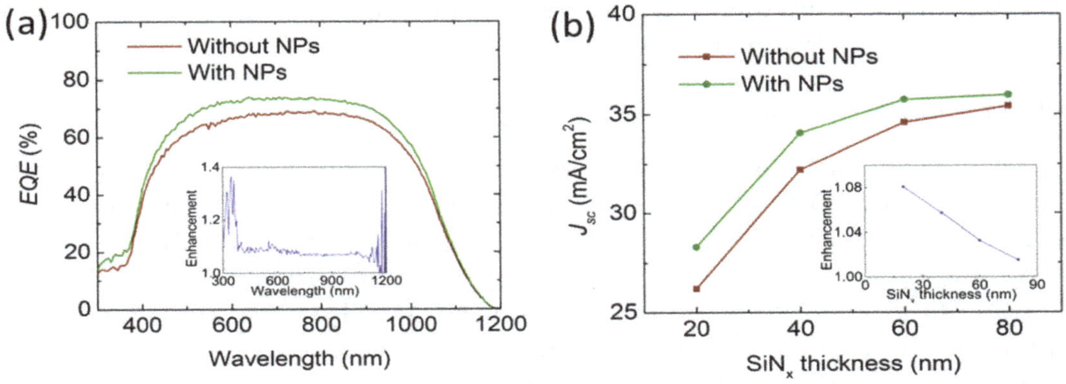

Figure 10. (**a**) Measured EQE of the solar cells with the Al NPs on top of a 20-nm SiN_x spacing layer, compared with that without Al NPs. (**b**) The corresponding photocurrent density as a function of the SiN_x layer thickness. The inset figures are the relative enhancement.

3. Materials and Methods

3.1. Numerical Simulations

Lumerical FDTD solutions (Lumerical Solutions, Inc., Vancouver, Canada) was employed to perform the optical simulation. For the scattering/absorption cross-section calculation, a 3-D total-field scattered-field (TFSF) source from 300 nm to 1200 nm was used to surround a NP. Two analysis groups, comprising 6 transmission monitors for each group in 3-D, were placed inside and outside the light source box, respectively. The inside and outside groups obtain the absorption power and the scattering power, respectively. The boundary conditions were set as the perfect matcher layers (PMLs) to prevent the scattering light bouncing back to the NPs from the boundaries, leading to an interference effect. For the light incoupling calculation, a spherical NP was placed on the surface of a planar Si slab with and without a SiN_x spacing layer and illuminated under a 300–1200-nm plane wave source weighted against the AM1.5G solar spectrum. PML boundary conditions were used in the incident direction to prevent an interference effect, and periodic boundary conditions (PBCs) were used in the lateral direction to simulate an ordered array of NPs.

3.2. NP Fabrication and Characterization

Due to the strong activeness of the Al NPs, which can be easily oxidized, it is challenging to synthesize Al NPs using the conventional chemical reduction method. To overcome this challenge, we developed a novel fabrication process based on the conventional thermal evaporation and annealing process. An ultra-thin Al film, with a few nanometers in thickness, was evaporated on the NaCl powder substrate, followed by a high temperature annealing process. Then, the powder was dissolved in the deionized water, and the Al NPs were separated from the water solution by centrifugation. The particle size can be adjusted by tuning the thickness of the Al film and the annealing condition. In this paper, 10-nm Al film was evaporated and annealed in a nitrogen atmosphere at 400 °C to form the Al NPs with the desired size (around 100 nm in diameter). As a remark, the Al itself is easy to be oxidized, leading to a thin layer of alumina on the surface of the Al NPs, preventing it from being further oxidized. This thin layer of alumina is only a few nanometers, as shown in [29], and would not affect the optical properties of the Al NPs much. For the purpose of comparison, we also synthesized 100-nm Ag NPs by the $NaBH_4$ reduction of silver nitrate solution [39]. A 5-mL deionized water solution with 5 mM $AgNO_3$ and 5 mM sodium citrate were prepared. Next, the suspension was subjected to the sonication. A 0.6-mL portion of 50 mM freshly prepared $NaBH_4$ was injected quickly at room temperature, followed by 30 s of sonication. Then, the solution was centrifuged at 5000 rpm for 10 min, and the supernatant was removed and the precipitate, containing the Ag NPs was redispersed in deionized water. The UV-VIS-IR spectrum of the NPs in the solution was measured by the spectrophotometer (Lambda 1050, PerkinElmer, Waltham, MA, USA).

3.3. Solar Cell Fabrication and Characterization

The planar Si solar cells were fabricated by the following processes. 200-μm thick Si wafers with resistivities around 1–3 Ω·cm were used. Firstly, the saw-damaged surface of the wafer with around 20 μm were etched off by dipping the wafers into the HNO_3/HF solution. After the standard RCA cleaning (as developed at Radio Corporation of America) process, the n^+ emitters were formed by a phosphorus diffusion process using the $POCl_3$ source at the temperature around 850 °C. The phosphosilicate glass introduced by the diffusion process was removed by the HF solution. Then, Al paste was screen printed on the rear surface and fired at a temperature around 800 °C for a few seconds to form the back surface field (BSF) and the back contact. An Al front contact grid was formed by the photolithography. After that, the SiN_x ARC was deposited by the plasmon-enhanced chemical vapour deposition (PECVD) system at 350 °C. The EQE of the solar cells was measured across 300–1200 nm using PV Measurements QEX10 (PV Measurements, Boulder, CO, USA). The photocurrent density was calculated by integrating the EQE to the AM1.5G solar photon fluxes.

4. Conclusions

In conclusion, we have demonstrated highly efficient light incoupling by Al NPs and have revealed the key factors that determine the incoupling efficiency, including the shape, size, surface coverage, and the spacing layer through numerical simulation. Furthermore, we experimentally compared the performance of the Si solar cells integrated with and without the Al NPs on top of the SiN_x spacing layer at varying thicknesses.

Acknowledgments: Baohua Jia thanks the Australia Research Council for its support (DP150102972).

Author Contributions: Y.Z. and B.J. conceived and designed the simulations and experiments; Y.Z. performed the simulations and experiments; Y.Z., B.C. and B.J. analyzed the data; Y.Z. wrote the paper, B.C. and B.J. revised the paper.

References

1. Atwater, H.A.; Polman, A. Plasmonics for improved photovoltaic devices. *Nat. Mater.* **2010**, *9*, 205–213. [CrossRef] [PubMed]
2. Green, M.A.; Pillai, S. Harnessing plasmonics for solar cells. *Nat. Photonics* **2012**, *6*, 130–132. [CrossRef]
3. Pillai, S.; Green, M.A. Plasmonics for photovoltaic applications. *Sol. Energy Mater. Sol. Cells* **2010**, *94*, 1481–1486. [CrossRef]
4. Kelly, K.L.; Coronado, E.; Zhao, L.L.; Schatz, G.C. The Optical Properties of Metal Nanoparticles: The Influence of Size, Shape, and Dielectric Environment. *J. Phys. Chem. B* **2003**, *107*, 668–677. [CrossRef]
5. Maier, S.A.; Atwater, H.A. Plasmonics: Localization and guiding of electromagnetic energy in metal/dielectric structures. *J. Appl. Phys.* **2005**, *98*, 11101–11110. [CrossRef]
6. Catchpole, K.R.; Polman, A. Plasmonic solar cells. *Opt. Express* **2008**, *16*, 21793–21800. [CrossRef] [PubMed]
7. Catchpole, K.R.; Polman, A. Design principles for particle plasmon enhanced solar cells. *Appl. Phys. Lett.* **2008**, *93*, 191113–191113. [CrossRef]
8. Akimov, Y.A.; Koh, W.S.; Ostrikov, K. Enhancement of optical absorption in thin-film solar cells through the excitation of higher-order nanoparticle plasmon modes. *Opt. Express* **2009**, *17*, 10195–10205. [CrossRef] [PubMed]
9. Beck, F.J.; Polman, A.; Catchpole, K.R. Tunable light trapping for solar cells using localized surface plasmons. *J. Appl. Phys.* **2009**, *105*. [CrossRef]
10. Mokkapati, S.; Beck, F.J.; Polman, A.; Catchpole, K.R. Designing periodic arrays of metal nanoparticles for light-trapping applications in solar cells. *Appl. Phys. Lett.* **2009**, *95*. [CrossRef]

11. Beck, F.J.; Mokkapati, S.; Catchpole, K.R. Plasmonic light-trapping for Si solar cells using self-assembled, Ag nanoparticles. *Prog. Photovolt. Res. Appl.* **2010**, *18*, 500–504. [CrossRef]
12. Hagglund, C.; Zach, M.; Petersson, G.; Kasemo, B. Electromagnetic coupling of light into a silicon solar cell by nanodisk plasmons. *Appl. Phys. Lett.* **2008**, *92*. [CrossRef]
13. Moulin, E.; Sukmanowski, J.; Luo, P.; Carius, R.; Royer, F.X.; Stiebig, H. Improved light absorption in thin-film silicon solar cells by integration of silver nanoparticles. *J. Non-Cryst. Solids* **2008**, *354*, 2488–2491. [CrossRef]
14. Moulin, E.; Sukmanowski, J.; Schulte, M.; Gordijn, A.; Royer, F.X.; Stiebig, H. Thin-film silicon solar cells with integrated silver nanoparticles. *Thin Solid Films* **2008**, *516*, 6813–6817. [CrossRef]
15. Henson, J.; DiMaria, J.; Paiella, R. Influence of nanoparticle height on plasmonic resonance wavelength and electromagnetic field enhancement in two-dimensional arrays. *J. Appl. Phys.* **2009**, *106*, 93111–93116. [CrossRef]
16. Eminian, C.; Haug, F.J.; Cubero, O.; Niquille, X.; Ballif, C. Photocurrent enhancement in thin film amorphous silicon solar cells with silver nanoparticles. *Prog. Photovolt. Res. Appl.* **2011**, *19*, 260–265. [CrossRef]
17. Qu, D.; Liu, F.; Yu, J.; Xie, W.; Xu, Q.; Li, X.; Huang, Y. Plasmonic core-shell gold nanoparticle enhanced optical absorption in photovoltaic devices. *Appl. Phys. Lett.* **2011**, *98*, 113119–113113. [CrossRef]
18. Spinelli, P.; Hebbink, M.; de Waele, R.; Black, L.; Lenzmann, F.; Polman, A. Optical Impedance Matching Using Coupled Plasmonic Nanoparticle Arrays. *Nano Lett.* **2011**, *11*, 1760–1765. [CrossRef] [PubMed]
19. Ho, W.; Su, S.; Lee, Y.; Syu, H.; Lin, C. Performance-Enhanced Textured Silicon Solar Cells Based on Plasmonic Light Scattering Using Silver and Indium Nanoparticles. *Materials* **2015**, *8*, 6668–6676. [CrossRef]
20. Lare, M.; Lenzmann, F.; Verschuuren, M.; Polman, A. Mode coupling by plasmonic surface scatters in thin-film silicon solar cells. *Appl. Phys. Lett.* **2012**, *101*, 221110. [CrossRef]
21. Dao, V.; Choi, H. Highly-Efficient Plasmon-Enhanced Dye-Sensitized Solar Cells Created by Means of Dry Plasma Reduction. *Nanomaterials* **2016**, *6*. [CrossRef]
22. Yu, J.; Shao, W.; Zhou, Y.; Wang, H.; Liu, X.; Xu, X. Nano Ag-enhanced energy conversion efficiency in standard commercial pc-Si solar cells and numerical simulations with finite difference time domain method. *Appl. Phys. Lett.* **2013**, *103*. [CrossRef]
23. Temple, T.; Bagnall, D. Broadband scattering of the solar spectrum by spherical metal nanoparticfles. *Prog. Photovolt. Res. Appl.* **2012**, *21*, 600–611. [CrossRef]
24. Temple, T.; Bagnall, D. Optical properties of gold and aluminium nanoparticles for silicon solar cell applications. *J. Appl. Phys.* **2011**, *109*. [CrossRef]
25. Cai, B.; Li, X.; Zhang, Y.; Jia, B. Significant light absorption enhancement in silicon thin film tandem solar cells with metallic nanoparticles. *Nanotechnology* **2016**, *27*. [CrossRef] [PubMed]
26. Lukyanchuk, B.; Zheludev, N.I.; Maier, S.A.; Halas, N.J.; Nordlander, P.; Giessen, H.; Chong, C.T. The Fano resonance in plasmonic nanostructures and metamaterials. *Nat. Mater.* **2010**, *9*, 707–715. [CrossRef] [PubMed]
27. Ekinci, Y.; Solak, H.H.; Löffler, J.F. Plasmon resonances of aluminum nanoparticles and nanorods. *J. Appl. Phys.* **2008**, *104*. [CrossRef]
28. Knight, M.W.; King, N.S.; Liu, L.; Everitt, H.O.; Nordlander, P.; Halas, N.J. Aluminum for Plasmonics. *ACS Nano* **2014**, *8*, 834–840. [CrossRef] [PubMed]
29. Langhammer, C.; Schwind, M.; Kasemo, B.; Zorić, I. Localized Surface Plasmon Resonances in Aluminum Nanodisks. *Nano Lett.* **2008**, *8*, 1461–1471. [CrossRef] [PubMed]
30. Maidecchi, G.; Gonella, G.; Zaccaria, R.P.; Moroni, R.; Anghinolfi, L.; Giglia, A.; Nannarone, S.; Mattera, L.; Dai, H.-L.; Canepa, M.; *et al.* Deep Ultraviolet Plasmon Resonance in Aluminum Nanoparticle Arrays. *ACS Nano* **2013**, *7*, 5834–5841. [CrossRef] [PubMed]
31. Powell, A.; Wincott, M.; Watt, A.; Assender, H.; Smith, J. Controlling the optical scattering of plasmonic nanoparticles using a thin dielectric layer. *J. Appl. Phys.* **2013**, *113*, 184311. [CrossRef]
32. Villesen, T.; Uhrenfeldt, C.; Johansen, B.; Larsen, A. Self-assembled Al nanoparticles on Si and fused silica, and their application for Si solar cells. *Nanotechnology* **2013**, *24*. [CrossRef] [PubMed]
33. Hylton, N.P.; Li, X.F.; Giannini, V.; Lee, K.-H.; Ekins-Daukes, N.J.; Loo, J.; Vercruysse, D.; van Dorpe, P.; Sodabanlu, H.; Sugiyama, M.; *et al.* Loss mitigation in plasmonic solar cells: Aluminium nanoparticles for broadband photocurrent enhancements in GaAs photodiodes. *Sci. Rep.* **2013**, *3*. [CrossRef] [PubMed]
34. Akimov, Y.; Koh, W. Design of Plasmonic Nanoparticles for Efficient Subwavelength Light Trapping in Thin-Film Solar Cells. *Plasmonics* **2011**, *6*, 155–161. [CrossRef]

35. Akimov, Y.; Koh, W. Resonant and nonresonant plasmonic nanoparticle enhancement for thin-film silicon solar cells. *Nanotechnology* **2010**, *21*. [CrossRef] [PubMed]
36. Yang, L.; Pillai, S.; Green, M.A. Can plasmonic Al nanoparticles improve absorption in triple junction solar cells? *Sci. Rep.* **2015**, *5*, 11852. [CrossRef] [PubMed]
37. Xu, Q.; Liu, F.; Liu, Y.; Meng, W.; Cui, K.; Feng, X.; Zhang, W.; Huang, Y. Aluminum plasmonic nanoparticles enhanced dye sensitized solar cells. *Opt. Express* **2014**, *22*, A301–A310. [CrossRef] [PubMed]
38. Kochergin, V.; Neely, L.; Jao, C.-Y.; Robinson, H.D. Aluminum plasmonic nanostructures for improved absorption in organic photovoltaic devices. *Appl. Phys. Lett.* **2011**, *98*. [CrossRef]
39. Orbaek, A.W.; McHale, M.M.; Barron, A.R. Synthesis and Characterization of Silver Nanoparticles for an Undergraduate Laboratory. *J. Chem. Educ.* **2015**, *92*, 339–344. [CrossRef]

17

Morphology-Controlled High-Efficiency Small Molecule Organic Solar Cells without Additive Solvent Treatment

Il Ku Kim [1,*], Jun Hyung Jo [2] and Jung-Ho Yun [3]

1 National Photonics Semiconductor Lab, National Photonics Semiconductor Inc., Suwon-Si 17113, Korea
2 School of Information and Communication Technology, Griffith University, Southport QLD 4222, Australia; j.jo@griffith.edu.au
3 Nanomaterials Centre, School of Chemical Engineering, University of Queensland, Brisbane QLD 4072, Australia; j.yun1@uq.edu.au
* Correspondence: cto.np@npsemi.com

Academic Editor: Guanying Chen

Abstract: This paper focuses on nano-morphology-controlled small-molecule organic solar cells without solvent treatment for high power-conversion efficiencies (PCEs). The maximum high PCE reaches up to 7.22% with a bulk-heterojunction (BHJ) thickness of 320 nm. This high efficiency was obtained by eliminating solvent additives such as 1,8-diiodooctane (DIO) to find an alternative way to control the domain sizes in the BHJ layer. Furthermore, the generalized transfer matrix method (GTMM) analysis has been applied to confirm the effects of applying a different thickness of BHJs for organic solar cells from 100 to 320 nm, respectively. Finally, the study showed an alternative way to achieve high PCE organic solar cells without additive solvent treatments to control the morphology of the bulk-heterojunction.

Keywords: small molecule; organic solar cell; bulk-heterojunction; optical simulation

---

1. Introduction

During the last decades, bulk-heterojunction (BHJ) small-molecule organic solar cells (OSCs) have received more research attention [1–9]. BHJ OSCs have interpenetrating networks in a blend of conjugated organic donors and fullerene-derivative acceptors for their alternative potential in obtaining low-cost clean energy solutions [1–6]. Among the many research issues, surface morphology control is an essential part of the spin-coating process to form thin films to get high-efficiency OSCs [2]. In the BHJ photoactive layer, the ultrafast photo-induced charge transfer at the interface of the phase-separated acceptor and donor is also needed for the high performance of OSCs before charge recombination has happened [10–12]. However, despite polymeric BHJ OSCs reported with high power conversion efficiencies (PCEs), others have pointed out some drawbacks. For instance, the organic polymer has batch-to-batch variations, molecular weight differences, polydispersity and impurity which could lead to a major obstacle for high-performance organic solar cells [2,13–15]. To address the above issues, small-molecule BHJs (SM BHJs) were introduced [1–5,8].

In contrast, it has been reported that solution-processed small-molecule BHJ solar cells have a well-defined molecular structure with intermediate dimensions [1,2]. Examples of solution-processed

SM BHJ OSCs with PCEs ranging from 6.7% to 8.01% have been reported with a 1,8-diiodooctane (DIO) additive solution [1–4]. These results were mainly obtained from the morphological control of the BHJ layer in terms of the additive solvent treatment. However, there are major concerns surrounding morphology control by using an additive solution such as the optimization of solution ratio control, and a limited number of applicable matching solutions (*i.e.*, chlorobenzene, 1,2-dichlorobenzene) [1–4]. In addition, the usage of a solvent additive affects unexpected solar cell performance. To avoid this, a study suggests the treatment of the active layer by drying speed or washout with methanol [16]. However, this is also not the fundamental method to solve the additional solvent problem issue. Therefore, it does not always guarantee a high efficiency of OSCs. Furthermore, strong aggregation is another problem during the spin-coating process used to produce a high quality of thin film for SM BHJs [2].

In this paper, we report solution-processed high-efficiency SM BHJ OSCs without a DIO additive solvent treatment. For that, a well-developed small-molecular donor (p-DTS(FBTTh$_2$)$_2$) will be applied [1–5]. This small-molecule donor has good characteristics of excellent solubility in organic solvents, strong optical absorption (600–800 nm) and a good hole mobility ($\approx$0.1 cm^2/Vs) [2]. Furthermore, a generalized transfer matrix method (GTMM) for the optical modeling analysis result will be introduced to confirm the behavior of light absorption in the BHJ photoactive layer of different thicknesses [17–19].

2. Experimental Section

To address the particular issues mentioned above, OSCs fabricated through the incorporation of a donor-acceptor blend of p-DTS(FBTTh$_2$)$_2$ and [6,6]-phenyl-C$_{71}$-butyric-acid-methyl-ester (PC$_{70}$BM) to form a BHJ photoactive layer. Chemical structures for OSC showed in Figure 1a. p-DTS(FBTTh$_2$)$_2$ and PC$_{70}$BM were dissolved in chlorobenzene (CB) and stirred over 24 h with a total concentration of 50 mg/mL. Indium tin oxide (ITO)-coated glass substrates were cleaned sequentially by ultrasonic treatment in Alconox detergent, deionized water, acetone and isopropyl alcohol. A polymeric conducting thin layer of PEDOT:PSS (40 nm) was spun-cast on top of the ITO-coated glass substrate. Before spin-coating of BHJ thin film, co-dissolved donor-acceptor blend was heated at 100 °C for 30 min.

Then the p-DTS(FBTTh$_2$)$_2$:PC$_{70}$BM BHJ layer was spin-cast from the heated and blended solution with different spin speed to form a different thickness of BHJs from 300 to 2000 rpm. During the solution preparation and film formation processes, 1,8-diiodooctane (DIO) additive solvent treatment was avoided. Finally, Ca (5 nm) and Al (100 nm) electrodes were deposited on top of BHJ photoactive layer (Figure 1b). Post-annealing process is also not applied for all devices. The energy band diagram of SM BHJ OSC showed in Figure 1c. Electrical characterization of all OSCs had done in the air after encapsulation. A commercial *J-V* and external quantum efficiency (EQE) characterization system was supplied by PV Measurement Inc. (Boulder, CO, USA) to obtain the data. A GTMM analysis was accomplished to calculate and analyse the multi-layered interface OSCs. For transfer matrix analysis, optical constants of all layers have been obtained by the ellipsometry method. Finally, the nano-morphology of BHJ film was investigated by atomic force microscopy (AFM).

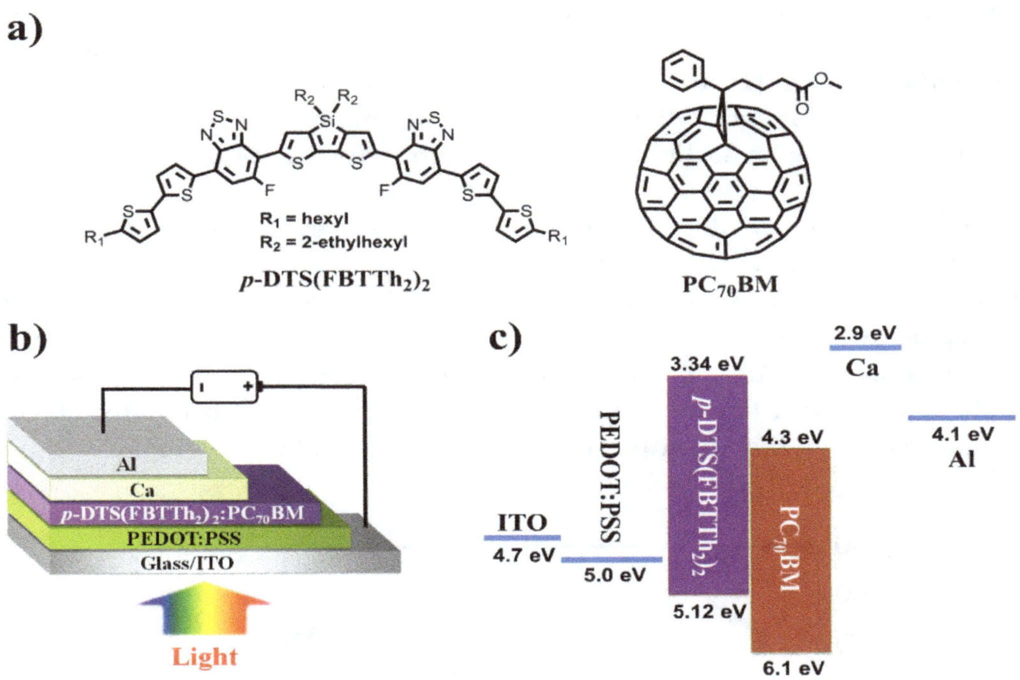

Figure 1. (**a**) Chemical structures of *p*-DTS(FBTTh$_2$)$_2$ and PC$_{70}$BM; (**b**) device architecture for small-molecule bulk-heterojunction organic solar cells (SM BHJ OSCs); (**c**) band diagram of SM BHJ OSCs.

3. Results and Discussion

SM BHJ OSCs have been tested under simulated 100 mW/cm^2 Air Mass (AM) 1.5G illumination. The optimized ratio of the small molecule donor (*p*-DTS(FBTTh$_2$)$_2$) to PC$_{70}$BM was chosen to be the same as the reported ratio (60:40 *w/w*) [1]. Device current density/voltage (*J*-*V*) characteristics of SM BHJ OSCs, in terms of BHJ without DIO treatment, with different photoactive thicknesses are shown in Figure 2.

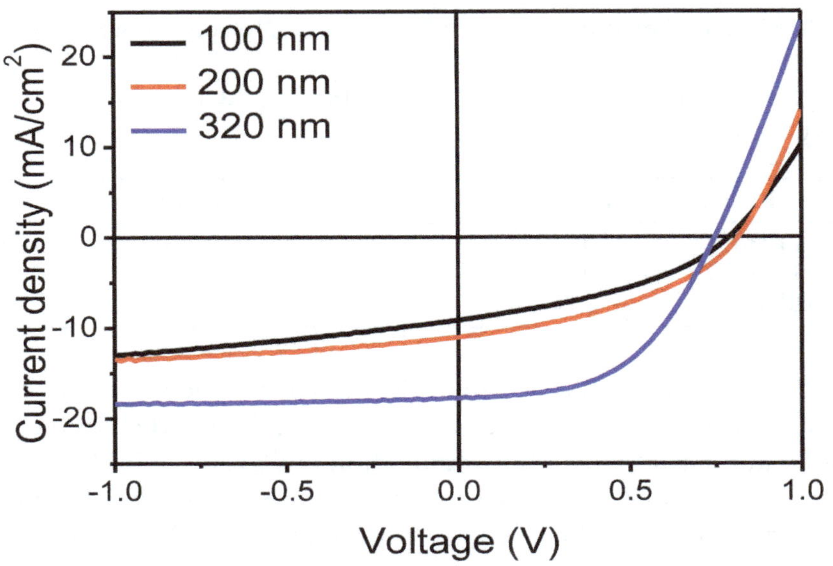

Figure 2. Measured density/voltage (*J*-*V*) curves of SM BHJ OSCs.

A V_{oc} as high as 0.74 V was observed in all devices. Combined with its high J_{sc} and fill factor (*FF*), a high power conversion efficiency (PCE) of 6.83% on average was measured with 320-nm-thick devices and the highest PCE was measured at 7.22% (V_{oc} = 0.74 V, J_{sc} = 18.23 mA/cm^2, FF = 0.54). The measured PCE of our device (2.78%) is much greater than the reported PCE value (1.8%) with the same BHJ thickness of 100 nm [1]. It is strong evidence to confirm that an additive solvent such as DIO is not an essential part for controlling the device performance, while the majority of the research groups have focused on using an additive solvent to control the p-DTS(FBTTh$_2$)$_2$ and small molecule donor materials.

The EQE spectra were obtained to confirm the accuracy of the SM BHJ OSC's photo-generated J-V result. The results are shown in Figure 3. Significantly, an EQE over 70% was achieved with a 320 nm SM BHJ photoactive layer from a wavelength range of 350–700 nm. A 1.5 G spectrum was applied to calculate the J_{sc} value by integrating the EQE data. The calculated J_{sc} values are in good agreement with the directly measured values. Furthermore, these EQE spectra confirm Beer-Lambert law by increasing the photoactive layer thicknesses. It is good to note that EQE spectra results obtained much better results to compare with the literature [1,3,4,8]. A summary of the SM BHJ OSCs' performance is shown in Table 1.

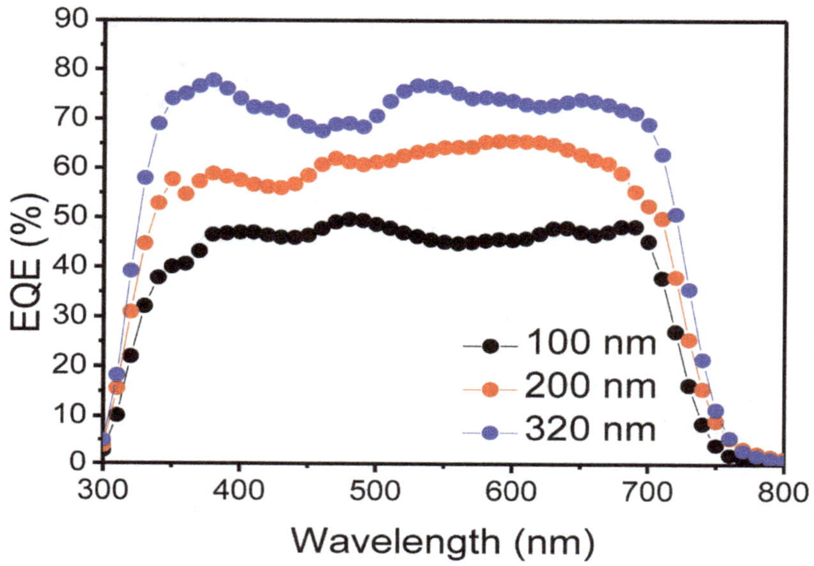

Figure 3. Measured external quantum efficiency (EQE) spectra of SM BHJ OSCs.

Table 1. A summary of SM BHJ OSC performances with changes of BHJ thickness. (PCE: Power conversion efficiency.)

BHJ Thickness	V_{oc} (V)	J_{sc} (mA/cm^2)	Fill Factor	PCE (%)
100 nm	0.78	9.2	0.39	2.78
200 nm	0.80	11.1	0.41	3.64
320 nm	0.74	17.8	0.52	6.83

Figure 4 shows the calculated charge generation rate (in s^{-1}·cm^{-3}) of SM BHJ OSCs obtained by the generalized transfer matrix method (GTMM). These calculated charge generation rate results provide a further explanation as to why a thicker junction of OSC has a higher J_{sc} value than a thinner junction of OSCs. From Figure 4a, the peak charge generation rate is located in the middle of the SM BHJ. In comparison with Figure 4b,c, the charge generation rate has the highest value in 100-nm-thick BHJ. However, under the thin junction circumstance, there is interference between the forward direction waves by absorption from the glass side and the backward direction waves by reflections from the electrodes in the photoactive layer. Therefore, the interference effect brings a low J_{sc} value in thin junction OSCs. By increasing the photoactive layer thickness containing BHJ, the OSC follows

Beer-Lambert law. Therefore, the photo-generation rate is exponentially decreased by increasing the photoactive layer thickness.

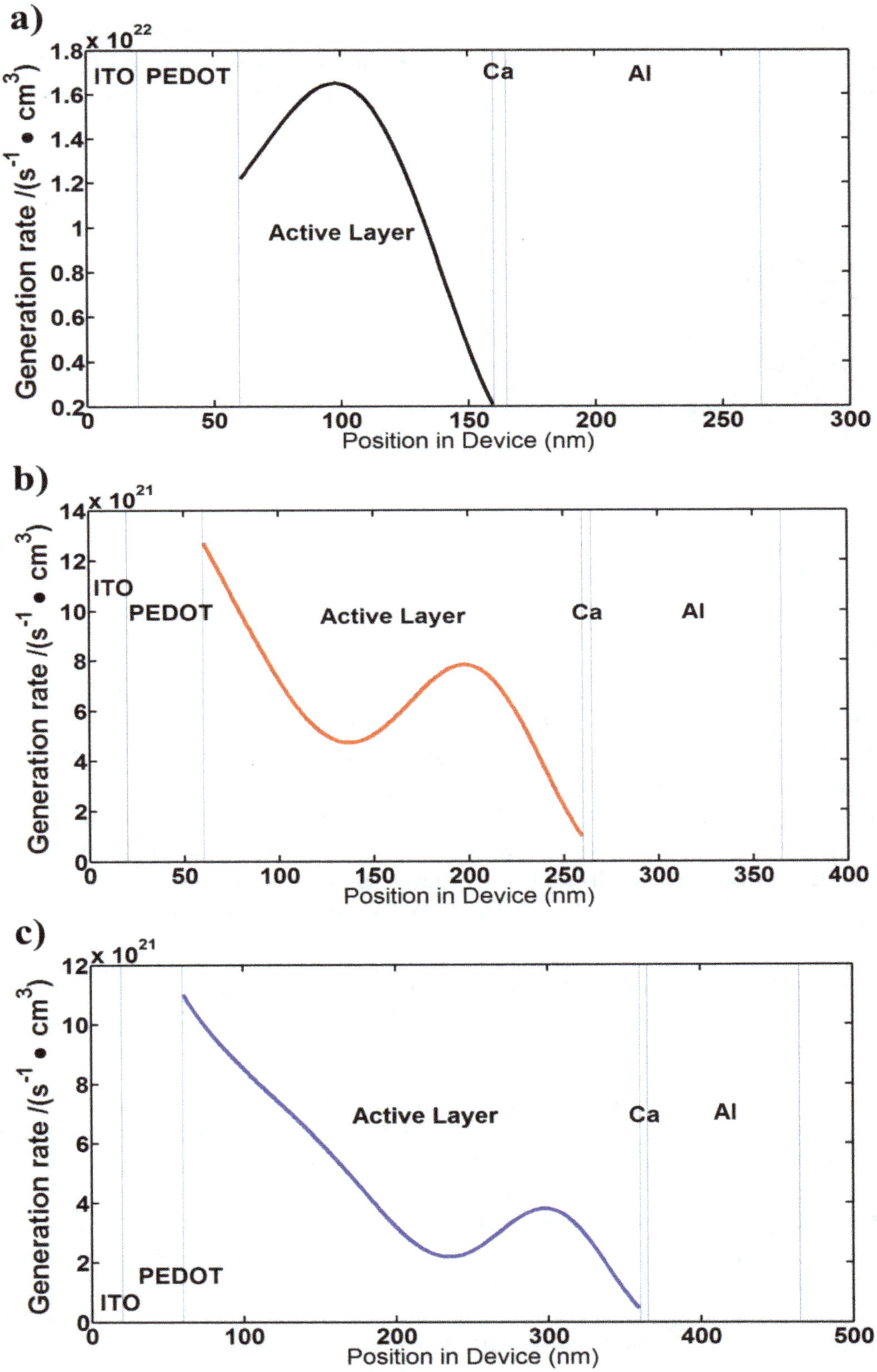

Figure 4. Simulation results of charge generation rates of SM BHJ OSCs: (**a**) BHJ active layer = 100 nm; (**b**) BHJ active layer = 200 nm; (**c**) BHJ active layer = 320 nm.

An AFM two-dimensional (2D) topography image of the *p*-DTS($FBTTh_2$)$_2$:$PC_{70}BM$ BHJ film (thickness = 320 nm) is shown in Figure 5a. As reported, conjugated small molecules tend to aggregate strongly and form a crystalline structure [1–4,8,20]. Therefore, we assumed that the smaller domains in the device allow for a higher donor-acceptor interface area [1–4,8]. Finally, it has more efficient generations of charge carriers. From the literature, it has been suggested that below 15 nm, morphology control of the high efficiency of the small molecular solar cell is guaranteed [1–4,8]. In Figure 5b, it confirms that avoiding solvent treatment is effectively working in the control of the BHJ domains within 3 nm. Therefore, this would be strong and robust evidence of the morphology control of BHJ films without DIO solvent treatment.

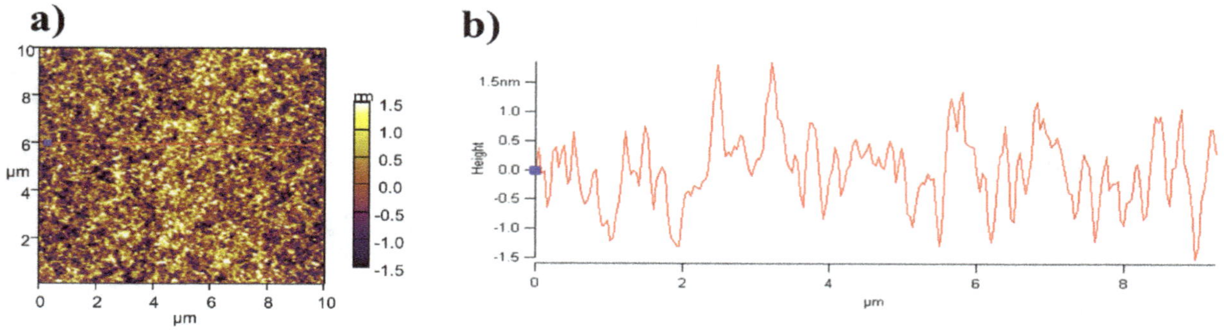

Figure 5. Measured atomic force microscopy (AFM) image of BHJ with a thickness of 320 nm: (**a**) top view of AFM; (**b**) measured cross-section of BHJ.

4. Conclusions

In conclusion, we studied the high performance of small-molecule organic solar cells without additive solvent treatment on the control of the photoactive layer in nano-morphology. Significantly, we have obtained small-molecule organic solar cells with a comparable PCE of 7.22% by eliminating the DIO additional solvent treatment. Also, GTMM analysis was accomplished to confirm the effect of the thickness variations of SM BHJ OSCs. Lastly, AFM measurement also confirms that smaller domains have achieved efficient charge generation. These results provide significant progress in showing that solution-processed small-molecule organic solar cells without solvent treatment can be an alternative method to having a high-PCE device with polymeric and/or small-molecule counterparts.

Author Contributions: I.K.K. and J.H.J. conceived and designed the experiments; I.K.K. performed the experiments; I.K.K. and J.H.J. analyzed the data; J.H.Y. contributed AFM analysis; I.K.K. wrote the paper.

References

1. Van der Poll, T.S.; Love, J.A.; Nguyen, T.Q.; Bazan, G.C. Non-Basic High-Performance Molecules for Solution-Processed Organic Solar Cells. *Adv. Mater.* **2012**, *24*, 3646–3649. [CrossRef] [PubMed]
2. Sun, Y.; Welch, G.C.; Leong, W.L.; Takacs, C.J.; Bazan, G.C.; Heeger, A.J. Solution-processed small-molecule solar cells with 6.7% efficiency. *Nat. Mater.* **2012**, *11*, 44–48. [CrossRef] [PubMed]
3. Kyaw, A.K.K.; Wang, D.H.; Gupta, V.; Leong, W.L.; Ke, L.; Bazan, G.C.; Heeger, A.J. Intensity dependence of current-voltage characteristics and recombination in high-efficiency solution-processed small-molecule solar cells. *ACS Nano* **2013**, *7*, 4569–4577. [CrossRef] [PubMed]
4. Gupta, V.; Kyaw, A.K.K.; Wang, D.H.; Chand, S.; Bazan, G.C.; Heeger, A.J. Barium: An efficient cathode layer for bulk-heterojunction solar cells. *Sci. Rep.* **2013**, *3*. [CrossRef] [PubMed]
5. Zhang, Q.; Kan, B.; Liu, F.; Long, G.; Wan, X.; Chen, X.; Zuo, Y.; Ni, W.; Zhang, H.; Li, M. Small-molecule solar cells with efficiency over 9%. *Nat. Photon.* **2015**, *9*, 35–41. [CrossRef]
6. Zhao, J.; Li, Y.; Yang, G.; Jiang, K.; Lin, H.; Ade, H.; Ma, W.; Yan, H. Efficient organic solar cells processed from hydrocarbon solvents. *Nat. Energy* **2016**, *1*. [CrossRef]

7. Li, W.; Hendriks, K.H.; Roelofs, W.; Kim, Y.; Wienk, M.M.; Janssen, R.A. Efficient small bandgap polymer solar cells with high fill factors for 300 nm thick films. *Adv. Mater.* **2013**, *25*, 3182–3186. [CrossRef] [PubMed]
8. Kyaw, A.K.K.; Wang, D.H.; Gupta, V.; Zhang, J.; Chand, S.; Bazan, G.C.; Heeger, A.J. Efficient Solution-Processed Small-Molecule Solar Cells with Inverted Structure. *Adv. Mater.* **2013**, *25*, 2397–2402. [CrossRef] [PubMed]
9. Sharenko, A.; Proctor, C.M.; van der Poll, T.S.; Henson, Z.B.; Nguyen, T.Q.; Bazan, G.C. A High-Performing Solution-Processed Small Molecule: Perylene Diimide Bulk Heterojunction Solar Cell. *Adv. Mater.* **2013**, *25*, 4403–4406. [CrossRef] [PubMed]
10. Koster, L.; Smits, E.; Mihailetchi, V.; Blom, P. Device model for the operation of polymer/fullerene bulk heterojunction solar cells. *Phys. Rev. B* **2005**, *72*. [CrossRef]
11. Street, R.; Cowan, S.; Heeger, A. Experimental test for geminate recombination applied to organic solar cells. *Phys. Rev. B* **2010**, *82*. [CrossRef]
12. Cowan, S.R.; Street, R.; Cho, S.; Heeger, A. Transient photoconductivity in polymer bulk heterojunction solar cells: competition between sweep-out and recombination. *Phys. Rev. B* **2011**, *83*. [CrossRef]
13. Walker, B.; Kim, C.; Nguyen, T.Q. Small molecule solution-processed bulk heterojunction solar cells. *Chem. Mater.* **2010**, *23*, 470–482. [CrossRef]
14. Coffin, R.C.; Peet, J.; Rogers, J.; Bazan, G.C. Streamlined microwave-assisted preparation of narrow-bandgap conjugated polymers for high-performance bulk heterojunction solar cells. *Nat. Chem.* **2009**, *1*, 657–661. [CrossRef] [PubMed]
15. Tong, M.; Cho, S.; Rogers, J.T.; Schmidt, K.; Hsu, B.B.Y.; Moses, D.; Coffin, R.C.; Kramer, E.J.; Bazan, G.C.; Heeger, A.J. Higher molecular weight leads to improved photoresponsivity, charge transport and interfacial ordering in a narrow bandgap semiconducting polymer. *Adv. Func. Mater.* **2010**, *20*, 3959–3965. [CrossRef]
16. Ye, L.; Jing, Y.; Guo, X.; Sun, H.; Zhang, S.; Zhang, M.; Huo, L.; Hou, J. Remove the residual additives toward enhanced efficiency with higher reproducibility in polymer solar cells. *J. Phys. Chem. C* **2013**, *117*, 14920–14928. [CrossRef]
17. Burkhard, G.F.; Hoke, E.T.; McGehee, M.D. Accounting for interference, scattering, and electrode absorption to make accurate internal quantum efficiency measurements in organic and other thin solar cells. *Adv. Mater.* **2010**, *22*, 3293–3297. [CrossRef] [PubMed]
18. Zhao, X.; Li, Z.; Zhu, T.; Mi, B.; Gao, Z.; Huang, W. Structure optimization of organic planar heterojunction solar cells. *J. Phys. D* **2013**, *46*. [CrossRef]
19. Jo, J.H.; Chun, Y.-S.; Kim, I. Optical Modeling-Assisted Characterization of Polymer: Fullerene Photodiodes. *IEEE Photonics J.* **2014**, *6*, 1–7.
20. Liu, Y.; Zhao, J.; Li, Z.; Mu, C.; Ma, W.; Hu, H.; Jiang, K.; Lin, H.; Ade, H.; Yan, H. Aggregation and morphology control enables multiple cases of high-efficiency polymer solar cells. *Nat. Commun.* **2014**, *5*. [CrossRef] [PubMed]

18

Ag Nanoparticle–Functionalized Open-Ended Freestanding TiO_2 Nanotube Arrays with a Scattering Layer for Improved Energy Conversion Efficiency in Dye-Sensitized Solar Cells

Won-Yeop Rho [1,2,†], Myeung-Hwan Chun [2,†], Ho-Sub Kim [2], Hyung-Mo Kim [1], Jung Sang Suh [2] and Bong-Hyun Jun [1,*]

1 Department of Bioscience and Biotechnology, Konkuk University, Seoul 143-701, Korea; rho72@snu.ac.kr (W.-Y.R.); hmkim0109@konkuk.ac.kr (H.-M.K.)

2 Department of Chemistry, Seoul National University, Seoul 151-747, Korea; hwanmc@hanmail.net (M.-H.C.); hosub@snu.ac.kr (H.-S.K.); jssuh@snu.ac.kr (J.S.S.)

* Correspondence: bjun@konkuk.ac.kr

† These authors contribute equally to this work.

Academic Editors: Guanying Chen, Zhijun Ning and Hans Agren

Abstract: Dye-sensitized solar cells (DSSCs) were fabricated using open-ended freestanding TiO_2 nanotube arrays functionalized with Ag nanoparticles (NPs) in the channel to create a plasmonic effect, and then coated with large TiO_2 NPs to create a scattering effect in order to improve energy conversion efficiency. Compared to closed-ended freestanding TiO_2 nanotube array–based DSSCs without Ag or large TiO_2 NPs, the energy conversion efficiency of closed-ended DSSCs improved by 9.21% (actual efficiency, from 5.86% to 6.40%) with Ag NPs, 6.48% (actual efficiency, from 5.86% to 6.24%) with TiO_2 NPs, and 14.50% (actual efficiency, from 5.86% to 6.71%) with both Ag NPs and TiO_2 NPs. By introducing Ag NPs and/or large TiO_2 NPs to open-ended freestanding TiO_2 nanotube array–based DSSCs, the energy conversion efficiency was improved by 9.15% (actual efficiency, from 6.12% to 6.68%) with Ag NPs and 8.17% (actual efficiency, from 6.12% to 6.62%) with TiO_2 NPs, and by 15.20% (actual efficiency, from 6.12% to 7.05%) with both Ag NPs and TiO_2 NPs. Moreover, compared to closed-ended freestanding TiO_2 nanotube arrays, the energy conversion efficiency of open-ended freestanding TiO_2 nanotube arrays increased from 6.71% to 7.05%. We demonstrate that each component—Ag NPs, TiO_2 NPs, and open-ended freestanding TiO_2 nanotube arrays—enhanced the energy conversion efficiency, and the use of a combination of all components in DSSCs resulted in the highest energy conversion efficiency.

Keywords: open-ended freestanding TiO_2 nanotube arrays; dye-sensitized solar cells; plasmonic; scattering; anodization

---

1. Introduction

Since the original work by O'Regan and Grätzel in 1991 [1], dye-sensitized solar cells (DSSCs) have been investigated extensively because of their high energy conversion efficiency and low cost [2–9]. Generally, mesoporous TiO_2 nanoparticle (NP) films and ruthenium sensitizers are used for DSSCs [2–4,10–16]. However, the efficiency of mesoporous TiO_2 NP film–based DSSCs is limited by grain boundaries, defects, and numerous trapping sites. Moreover, mesoporous TiO_2 NP films can cause charge recombination and mobility [17,18].

TiO_2 nanotubes, which enhance electron transport and charge separation by creating direct pathways and accelerating charge transfer between interfaces, have great potential to overcome the problems of mesoporous TiO_2 NP films [19–22]. TiO_2 nanotubes can be prepared by a hydrothermal method [23] or an electrochemical method [24], known as anodization. TiO_2 nanotube arrays prepared by anodization have a well-ordered and vertically oriented tubular structure that facilitates a high degree of electron transport and less charge recombination than mesoporous TiO_2 NP films [25–27]. There is much room for improvement in the energy conversion efficiency of current DSSCs based on TiO_2 nanotube arrays compared to the relatively extensively researched mesoporous TiO_2 NP film–based DSSCs [28].

To date, several approaches for improving the energy conversion efficiency of TiO_2 nanotube array–based DSSCs have been reported. Metal NPs, which can harvest light via surface plasmon resonance (SPR), have been used to enhance the energy conversion efficiency of DSSCs by introducing Au or Ag NPs into TiO_2 nanotube arrays [29–32]. Barrier layers remove TiO_2 nanotube arrays, so open-ended TiO_2 nanotube arrays, which can also be classified as arrays of columnar nanopores, have been used for DSSCs to provide increased energy conversion efficiency [33]. Moreover, the energy conversion efficiency of TiO_2 nanotube array–based DSSCs can be further increased by introducing a scattering layer to the active layer [34].

So far, TiO_2 nanotubes that make use of a scattering layer [34] or plasmonic materials [14] have been reported, but a scattering layer with plasmonic materials has not been used in TiO_2 nanotube–based DSSCs. In this study, we report the development of freestanding TiO_2 nanotube arrays filled with Ag NPs and large TiO_2 NPs, which improve the energy conversion efficiency of DSSCs. Furthermore, we compare the effects of Ag NPs and large TiO_2 NPs in open- and closed-ended freestanding TiO_2 nanotube arrays in DSSCs. The energy conversion efficiencies of the following eight types of DSSCs were compared: closed-ended freestanding TiO_2 nanotube arrays with/without Ag NPs and/or a TiO_2 scattering layer and open-ended freestanding TiO_2 nanotube arrays with/without Ag NPs and/or a TiO_2 scattering layer.

2. Results and Discussion

2.1. Structure of DSSCs with Freestanding TiO_2 Nanotube Arrays with Channels Containing Ag NPs

Figure 1 illustrates the fabrication of DSSCs with Ag NPs and large TiO_2 NPs to enable plasmonic and scattering effects in open-ended freestanding TiO_2 nanotube array–based DSSCs. Ti plates were anodized and then annealed at 500 °C for 1 h to prepare anatase TiO_2 nanotube arrays. After carrying out secondary anodization, the TiO_2 nanotube arrays were easily detached from the Ti plates. TiO_2 nanotube arrays, once separated from the Ti plates, are termed "closed-ended freestanding TiO_2 nanotube arrays". Freestanding TiO_2 nanotube arrays have a barrier layer at the bottom that disturbs electron transport and electrolyte diffusion. This barrier layer was removed using the ion-milling method with several minutes of Ar^+ bombardment to yield "open-ended freestanding TiO_2 nanotube arrays". The closed- and open-ended freestanding TiO_2 nanotube arrays were transferred to fluorine-doped tin oxide (FTO) glass using TiO_2 paste and annealed to enhance the adhesion between the closed- and open-ended freestanding TiO_2 nanotube arrays and the fluorine-doped tin oxide (FTO) glass. To improve the energy conversion efficiency by the plasmonic effect, Ag NPs were embedded in the channel of freestanding TiO_2 nanotube arrays using 254 nm ultraviolet (UV) irradiation with aqueous silver nitrate. To further enhance the energy conversion efficiency, large TiO_2 NPs (400 nm) as a scattering layer were coated onto the active layer by the doctor blade method. This substrate was sandwiched with the counter electrode and filled with electrolyte. The active area of the DSSCs was ~0.25 cm^2.

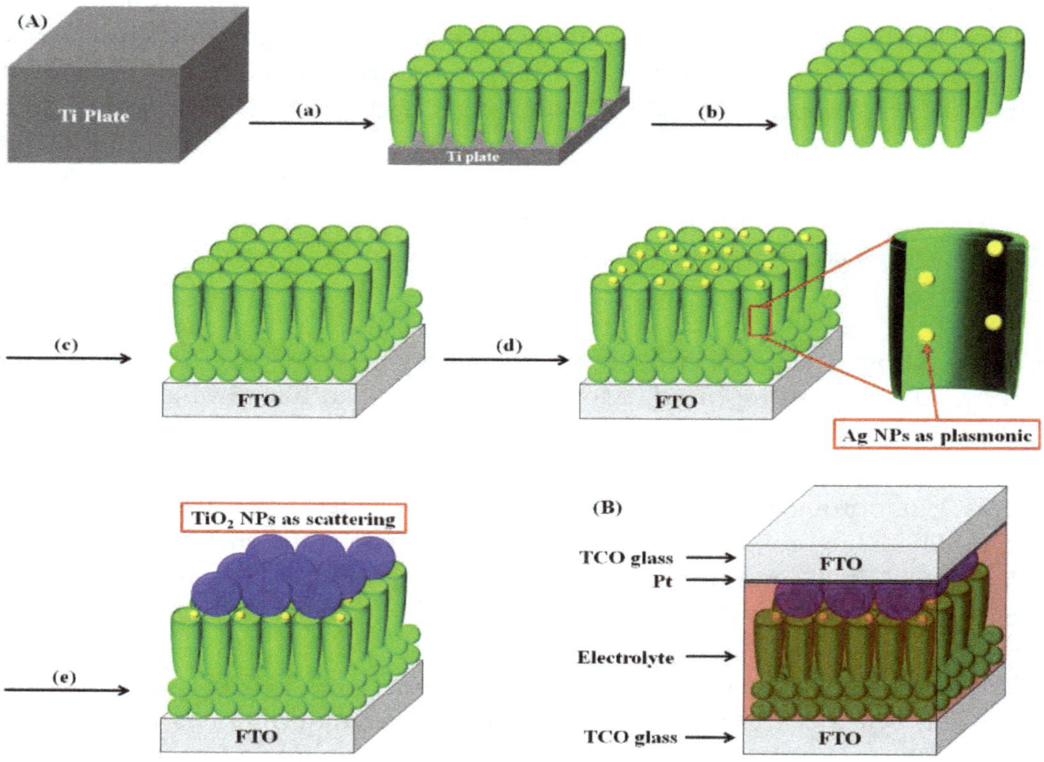

Figure 1. Overall scheme of dye-sensitized solar cells (DSSCs) with open-ended freestanding TiO_2 nanotube arrays with Ag nanoparticles (NPs) and large TiO_2 NPs. (**A**) (a) Ti anodization for TiO_2 nanotube arrays; (b) freestanding TiO_2 nanotube arrays and etching by ion milling; (c) transference of open-ended freestanding TiO_2 nanotube arrays onto fluorine-doped tin oxide (FTO) glass; (d) formation of Ag NPs by ultraviolet (UV) irradiation; and (e) introduction of large TiO_2 NPs. (**B**) Structure of a DSSC with freestanding TiO_2 nanotube arrays and large TiO_2 NPs.

2.2. Characterization of Freestanding TiO_2 Nanotube Arrays with Channels Containing Ag NPs

Field emission scanning electron microscope (FE-SEM) images of freestanding TiO_2 nanotube (TNT) arrays are shown in Figure 2. The top side of the freestanding TiO_2 nanotube arrays had 100-nm-diameter pores, as shown in Figure 2a. The bottom layer of closed-ended freestanding TiO_2 nanotube arrays before ion milling lacked pores, as shown in Figure 2b. However, after ion milling, 20-nm-diameter pores were evident on the bottom layer of open-ended freestanding TiO_2 nanotube arrays, as shown in Figure 2c. Open-ended TNT arrays can be prepared by chemical etching [35] or physical etching [33,36]. In the chemical etching method, the bottom layers of TNT arrays were easily removed by the etchant. However, the surface morphology and length of TNT arrays were also dissolved in etchant and TNT arrays are fragile when they are attached to a substrate because of their amorphous crystallinity. In the physical etching method, the bottom layer of TNT arrays was removed by the plasma or ion milling process, which is not simple. However, the surface morphology and length of TNT arrays are not damaged in the process and they are very stable when they are attached to a substrate because TNT arrays have the ability to change crystallinity from the amorphous to the anatase phase. After UV irradiation using a silver source, ~30 nm Ag NPs were seen in the channels of freestanding TiO_2 nanotubes in high-angle annular dark-field (HAADF) images, as shown in Figure 2d. The length of the TiO_2 nanotubes was ~22 μm and the length of the scattering layer, which consisted of 400 nm TiO_2 NPs, was ~10 μm.

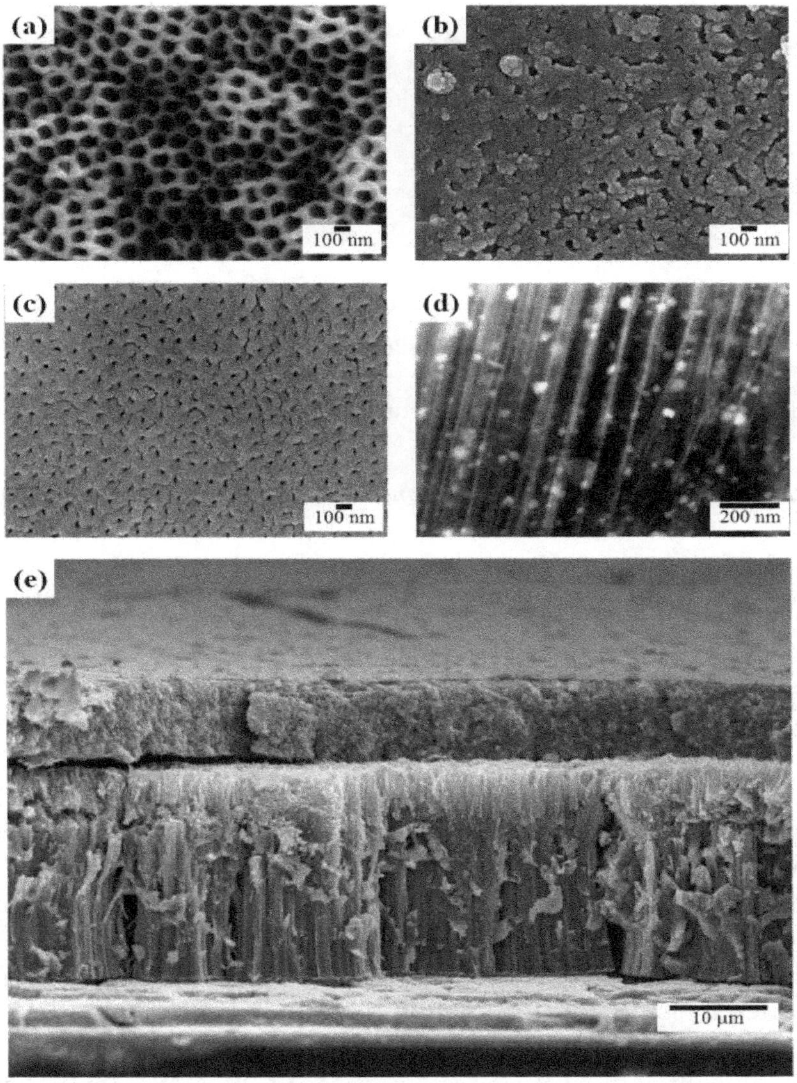

Figure 2. Field emission scanning electron microscope (FE-SEM) images of the (**a**) top, (**b**) bottom, and (**c**) bottom of post–ion milling freestanding TiO_2 nanotube arrays; (**d**) a high-angle annular dark-field (HAADF) image of Ag NPs in the channel of TiO_2 nanotube arrays; and (**e**) a side view of the active layer with freestanding TiO_2 nanotube arrays and a scattering layer.

The ultraviolet-visible (UV-vis) spectrum of Ag NPs in the channels of freestanding TiO_2 nanotubes is shown in Figure 3. A broad absorption peak centered at 402 nm was observed. The value is different from what it would be in the general solution phase, which is a 420 nm UV absorbance from 30 nm Ag NPs. This discrepancy may stem from different synthesis and measurement conditions used in this study [37–39]; the Ag NPs were synthesized by UV irradiation (at 254 nm) without adding a stabilizer and the Ag NPs were measured under dry conditions. The absorption band of Ag NPs is matched with the dye. cis-diisothiocyanato-bis(2,2′-bipyridyl-4,4′-dicarboxylato) ruthenium(II) bis(tetrabutylammonium) (N719) has two visible absorption bands, 390 nm and 531 nm, [40] that were affected by the plasmon band. Moreover, the shell of Ag NPs was prepared with $TiCl_4$ to prevent the trapping of electrons by Ag NPs and to enable better electron transport in the channel of the TiO_2 nanotube arrays.

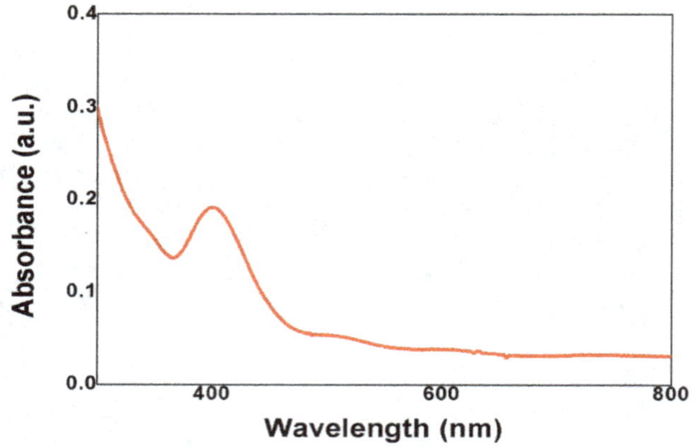

Figure 3. Ultraviolet-visible (UV-vis) spectrum of Ag NP-functionalized TiO_2 nanotubes.

2.3. DSSCs with Closed-Ended Freestanding TiO_2 Nanotube Arrays with Channels Containing Ag NPs and Large TiO_2 NPs

The photocurrent-voltage curves of DSSCs fabricated using closed-ended freestanding TiO_2 nanotube arrays measured under air mass 1.5 illumination (100 mW/cm^2) are shown in Figure 4 and Table 1. Four types of closed-ended freestanding TiO_2 nanotube array–based DSSCs were fabricated in order to assess the effect of each component on the energy conversion efficiency: closed-ended freestanding TiO_2 nanotube array–based DSSCs without Ag or large TiO_2 NPs (a), with Ag NPs (b), with large TiO_2 NPs (c), and with both Ag NPs and large TiO_2 NPs (d). The open-circuit voltage (V_{oc}), short-circuit current (J_{sc}), fill factor (*ff*), and energy conversion efficiency (η) values are shown in Table 1. The energy conversion efficiency of DSSCs based on closed-ended freestanding TiO_2 nanotube arrays lacking NPs was 5.86%. The energy conversion efficiencies of DSSCs based on closed-ended freestanding TiO_2 nanotube arrays with Ag NPs, with large TiO_2 NPs, and with both Ag NPs and large TiO_2 NPs were 6.40%, 6.24%, and 6.71%, respectively. The introduction of Ag NPs increased the energy conversion efficiency significantly, by 9.21% (actual efficiency change, from 5.86% to 6.40%) compared to closed-ended freestanding TiO_2 nanotube array–based DSSCs without Ag and large TiO_2 NPs, because of increased light harvesting by the plasmonic effect. The introduction of large TiO_2 NPs also increased the energy conversion efficiency significantly, by 6.48% (actual efficiency, from 5.86% to 6.24%), owing to increased light harvesting by the scattering effect. Moreover, the introduction of both Ag NPs and large TiO_2 NPs increased the energy conversion efficiency significantly, by 14.50% (actual efficiency, from 5.86% to 6.71%), because of increased light harvesting resulting from both the plasmonic and scattering effects.

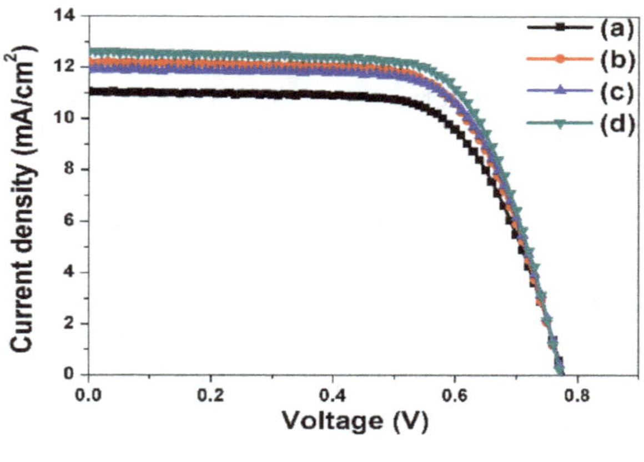

Figure 4. I–V curves of DSSC-based closed-ended freestanding TiO_2 nanotube arrays fabricated without NPs (**a**), with Ag NPs (**b**), with large TiO_2 NPs (**c**), and with Ag NPs and large TiO_2 NPs (**d**).

Table 1. Photovoltaic properties of dye-sensitized solar cells (DSSCs) based on closed-ended freestanding TiO_2 nanotube arrays.

DSSCs	J_{sc} (mA/cm^2)	V_{oc} (V)	*ff*	η (%)
(a) Closed-ended freestanding TiO_2 nanotube arrays without any NPs	11.05	0.78	0.68	5.86
(b) Closed-ended freestanding TiO_2 nanotube arrays with Ag NPs	12.22	0.77	0.68	6.40
(c) Closed-ended freestanding TiO_2 nanotube arrays with large TiO_2 NPs	11.90	0.76	0.69	6.24
(d) Closed-ended freestanding TiO_2 nanotube arrays with Ag NPs and large TiO_2 NPs	12.63	0.77	0.69	6.71

2.4. DSSCs with Open-Ended Freestanding TiO_2 Nanotube Arrays with Channels Containing Ag NPs and Large TiO_2 NPs

The photocurrent-voltage curves of DSSCs fabricated using open-ended freestanding TiO_2 nanotube arrays are shown in Figure 5 and Table 2; they are useful in assessing the effect of each component on the energy conversion efficiency. Four types of DSSCs based on open-ended freestanding TiO_2 nanotube arrays were fabricated: open-ended freestanding TiO_2 nanotube array–based DSSCs without Ag or large TiO_2 NPs (a), with Ag NPs (b), with large TiO_2 NPs (c), and with both Ag NPs and large TiO_2 NPs (d). The V_{oc}, J_{sc}, *ff*, and η values are summarized in Table 2. The energy conversion efficiency of DSSCs based on open-ended freestanding TiO_2 nanotube arrays lacking NPs was 6.12%. The energy conversion efficiencies of DSSCs based on open-ended freestanding TiO_2 nanotube arrays with Ag NPs, with large TiO_2 NPs, and with both Ag NPs and large TiO_2 NPs were 6.68%, 6.62%, and 7.05%, respectively. The introduction of Ag NPs, large TiO_2 NPs, and both increased the energy conversion efficiency by 9.15%, 8.17%, and 15.20%, respectively. Compared to closed-ended freestanding TiO_2 nanotube arrays, the energy conversion efficiency of DSSCs based on open-ended freestanding TiO_2 nanotube arrays was 5.07% (6.71%–7.05%) higher due to enhanced electron transport and electrolyte diffusion [33].

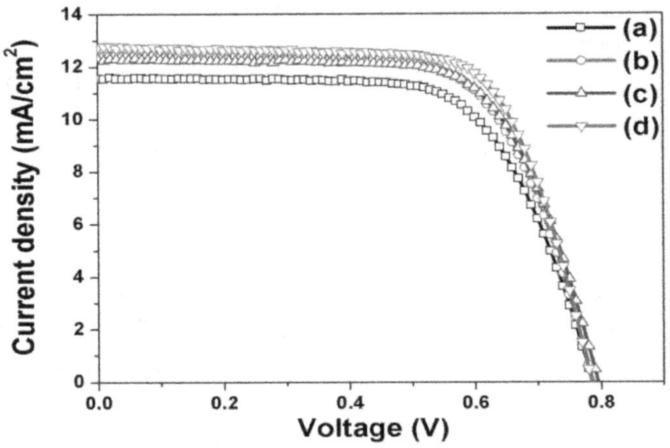

Figure 5. I–V curves of DSSCs based on open-ended freestanding TiO_2 nanotube arrays fabricated without NPs (**a**), with Ag NPs (**b**), with large TiO_2 NPs (**c**), and with both Ag NPs and large TiO_2 NPs (**d**).

Table 2. Photovoltaic properties of DSSCs based on open-ended freestanding TiO_2 nanotube arrays.

ADSSCs	J_{sc} (mA/cm^2)	V_{oc} (V)	*ff*	η (%)
(a) Open-ended freestanding TiO_2 nanotube arrays without any NPs	11.56	0.79	0.67	6.12
(b) Open-ended freestanding TiO_2 nanotube arrays with Ag NPs	12.45	0.79	0.68	6.68
(c) Open-ended freestanding TiO_2 nanotube arrays with large TiO_2 NPs	12.33	0.79	0.68	6.62
(d) Open-ended freestanding TiO_2 nanotube arrays with Ag NPs and large TiO_2 NPs	12.74	0.78	0.71	7.05

Although TiO_2 nanotube array-based DSSCs have great potential, as far as we know, the theoretical maximum improvement by Ag NPs or a TiO_2 scattering layer of TiO_2 nanotube-based DSSCs has not yet been reported. However, the open-ended TiO_2 nanotube-based devices exhibited an increase in one-sun efficiency from 5.3% to 9.1%, corresponding to a 70% increase, which is a much higher increase than we achieved [35]. We believe that there is ample room to improve efficiency by combining each parameter in an optimal condition based on theoretical studies.

3. Materials and Methods

3.1. Materials

Titanium plates (99.7% purity, 0.25 mm thickness, Alfa Aesar, Ward Hill, MA, USA), ammonium fluoride (NH_4F, Showa Chemical Industry Co., Beijing, China, 97.0%), ethylene glycol (Daejung Chemical, Siheung, Korea, 99%), hydrogen peroxide (H_2O_2, Daejung Chemical, Siheung, Korea, 30%), fluorine-doped tin oxide (FTO) glass (Pilkington, St. Helens, UK, TEC-A7), titanium diisopropoxide bis(acetylacetonate) solution (Aldrich, St. Louis, MS, USA, 75 wt % in isopropanol), n-butanol (Daejung Chemical, Siheung, Korea, 99%), TiO_2 paste (Ti-Nanoxide T/SP, Solaronix, Aubonne, Switzerland), scattering TiO_2 paste (18NR-AO, Dyesol, Queanbeyan, Australia), silver nitrate ($AgNO_3$, Aldrich, St. Louis, MS, USA, 99%), titanium chloride ($TiCl_4$, Aldrich, St. Louis, MS, USA, 0.09 M in 20% HCl), dye (cis-diisothiocyanato-bis(2,2′-bipyridyl-4,4′-dicarboxylato) ruthenium(II) bis(tetrabutylammonium) (N719, Solaronix, Aubonne, Switzerland), chloroplatinic acid hexahydrate ($H_2PtCl_6 \cdot 6H_2O$, Aldrich, St. Louis, MS, USA), 1-butyl-3-methyl-imidazolium iodide (BMII, Aldrich, St. Louis, MS, USA, 99%), iodine (I_2, Aldrich, St. Louis, MS, USA, 99%), guanidium thiocyanate (GSCN, Aldrich, St. Louis, MS, USA, 99%), 4-tertbutylpyridine (TBP, Aldrich, St. Louis, MS, USA, 96%), acetonitrile (CH_3CN, Aldrich, St. Louis, MS, USA, 99.8%), and valeronitrile ($CH_3(CH_2)_3CN$, Aldrich, St. Louis, MS, USA, 99.5%) were obtained from commercial manufacturers.

3.2. Preparation of Freestanding TiO_2 Nanotube Arrays

TiO_2 nanotube arrays were prepared by anodization of thin Ti plates. Ti anodization was carried out in an electrolyte composed of 0.8 wt % NH_4F and 2 vol % H_2O in ethylene glycol at 25 °C and at a constant voltage of 60 V direct current (DC) for 2 h. After being anodized, the Ti plates were annealed at 500 °C for 1 h under ambient conditions to improve the crystallinity of the TiO_2 nanotube arrays, and then a secondary anodization was conducted at a constant voltage of 30 V DC for 10 min. The Ti plates were immersed in 10% H_2O_2 solution for several hours in order to detach the TiO_2 nanotube arrays from the Ti plates and produce freestanding TiO_2 nanotube arrays. The bottom of the TiO_2 nanotube arrays was removed by ion milling with Ar^+ bombardment for several minutes in order to prepare open-ended freestanding TiO_2 nanotube arrays.

3.3. Preparation of Ag NPs in the Channel of Freestanding TiO_2 Nanotube Arrays

A TiO_2 blocking layer was formed on FTO glass by spin-coating with 5 wt % titanium di-isopropoxide bis(acetylacetonate) in butanol, followed by annealing at 500 °C for 1 h under ambient conditions to induce crystallinity. A TiO_2 paste was coated onto the TiO_2 blocking/FTO glass using the doctor blade method, and closed- and open-ended freestanding TiO_2 nanotube arrays were then introduced onto the TiO_2 paste. The substrate was annealed at 500 °C for 1 h under ambient conditions to enhance the adhesion between the TiO_2 NPs and freestanding TiO_2 nanotube arrays. The substrate was dipped in 0.3 mM $AgNO_3$ aqueous solution and exposed to 254 nm UV irradiation. The substrate was treated with 10 mM $TiCl_4$ solution at 50 °C for 30 min and then annealed at 500 °C for 1 h.

3.4. Fabrication of DSSCs with Freestanding TiO_2 Nanotube Arrays with Channels Containing Ag NPs

A substrate that consisted of freestanding TiO_2 nanotube arrays with channels containing Ag NPs was coated with ~400 nm TiO_2 NPs for scattering and then annealed at 500 °C for 1 h under ambient conditions to induce crystallinity and adhesion. The substrate was immersed in a dye solution at 50 °C for 8 h to function as a working electrode. The working electrode was sandwiched with a counter electrode, Pt on FTO glass, by a 60-µm-thick hot-melt sheet and filled with electrolyte solution composed of 0.7 M 1-butyl-3-methyl-imidazolium iodide (BMII), 0.03 M I_2, 0.1 M guanidium thiocyanate (GSCN), and 0.5 M 4-tertbutylpyridine (TBP) in a mixture of acetonitrile and valeronitrile (85:15, *v*/*v*).

3.5. Instruments

The morphology, thickness, size, and presence of Ag NPs in the channels of freestanding TiO_2 nanotube arrays were confirmed using a field emission scanning electron microscope (FE-SEM, JSM-6330F, JEOL Inc., Tokyo, Japan) and the high angular annular dark field (HAADF) technique with a scanning transmission electron microscope (TEM, JEM-2200FS, JEOL Inc., Tokyo, Japan). The current density−voltage ($J-V$) characteristics and the incident photon-to-current conversion efficiency (IPCE) of the DSSCs were measured using an electrometer (Keithley 2400, Keithley Instruments, Inc., Cleveland, OH, USA) under AM 1.5 illumination (100 mW/cm^2) provided by a solar simulator (1 kW xenon with AM 1.5 filter, PEC-L01, Peccell Technologies, Inc., Yokohama, Kanagawa, Japan), or using a solar cell IPCE measurement System(K3100, McScience Inc., Suwon, Korea) with reference to the calibrated diode.

4. Conclusions

In this study, we compared the natural consequences of altering three parameters, the plasmonic effect, the scattering effect, and open- *vs.* closed-ended freestanding TiO_2 nanotubes, as a basic means of exploring improvements in efficiency. We demonstrated that the plasmonic and scattering effects enhanced the energy conversion efficiency of freestanding TiO_2 nanotube arrays in DSSCs. Ag NPs were added to the channels of TiO_2 nanotube arrays by UV irradiation to induce a plasmonic effect, and large TiO_2 NPs were introduced to TiO_2 nanotube arrays to induce a scattering effect. The energy conversion efficiency of DSSCs with both Ag NPs and large TiO_2 NPs was higher than that of DSSCs without Ag NPs owing to the plasmonic effect, and it was higher than that of DSSCs without large TiO_2 NPs owing to the scattering effect. Compared to closed-ended freestanding TiO_2 nanotube arrays, open-ended freestanding TiO_2 nanotube arrays [40] exhibited enhanced energy conversion efficiency. We demonstrate that Ag NPs, TiO_2 NPs, and open-ended freestanding TiO_2 nanotube arrays enhanced the energy conversion efficiency; furthermore, the combination of all components exhibited the highest energy conversion efficiency. Our research suggests that the energy conversion efficiency of DSSCs is improved by both the plasmonic and scattering effects. This knowledge has applications in organic solar cells, hybrid solar cells, and perovskite solar cells.

Author Contributions: Won-Yeop Rho, Myeung-Hwan Chun, Jung Sang Suh and Bong-Hyun Jun designed the research and wrote the manuscript. Won-Yeop Rho, Ho-Sub Kim and Hyung-Mo Kim carried out experiments. All authors discussed the results and commented on the manuscript. Won-Yeop Rho and Bong-Hyun Jun guided all aspects of the work.

References

1. Oregan, B.; Gratzel, M. A low-cost, high-efficiency solar-cell based on dye-sensitized colloidal TiO_2 films. *Nature* **1991**, *353*, 737–740. [CrossRef]
2. Lim, J.; Kim, H.A.; Kim, B.H.; Han, C.H.; Jun, Y. Reversely fabricated dye-sensitized solar cells. *RSC Adv.* **2014**, *4*, 243–247. [CrossRef]
3. Jo, Y.; Yun, Y.J.; Khan, M.A.; Jun, Y. Densely packed setose ZnO nanorod arrays for dye sensitized solar cells. *Synth. Met.* **2014**, *198*, 137–141. [CrossRef]
4. Kim, J.; Lim, J.; Kim, M.; Lee, H.S.; Jun, Y.; Kim, D. Fabrication of carbon-coated silicon nanowires and their application in dye-sensitized solar cells. *ACS Appl. Mater. Interfaces* **2014**, *6*, 18788–18794. [CrossRef] [PubMed]
5. Ko, K.W.; Lee, M.; Sekhon, S.; Balasingam, S.K.; Han, C.H.; Jun, Y. Efficiency enhancement of dye-sensitized solar cells by the addition of an oxidizing agent to the TiO_2 paste. *Chem. Sus. Chem.* **2013**, *6*, 2117–2123. [CrossRef] [PubMed]
6. Balasingam, S.K.; Kang, M.G.; Jun, Y. Metal substrate based electrodes for flexible dye-sensitized solar cells: Fabrication methods, progress and challenges. *Chem. Commun.* **2013**, *49*, 11457–11475. [CrossRef] [PubMed]
7. Jung, C.-L.; Lim, J.; Park, J.-H.; Kim, K.-H.; Han, C.-H.; Jun, Y. High performance dye sensitized solar cells by adding titanate co-adsorbant. *RSC Adv.* **2013**, *3*, 20488–20491. [CrossRef]
8. Nath, N.C.D.; Ahammad, A.; Sarker, S.; Rahman, M.; Lim, S.-S.; Choi, W.-Y.; Lee, J.-J. Carbon nanotubes on fluorine-doped tin oxide for fabrication of dye-sensitized solar cells at low temperature condition. *J. Nanosci. Nanotechnol.* **2012**, *12*, 5373–5380. [CrossRef] [PubMed]
9. Deb Nath, N.C.; Lee, H.J.; Choi, W.-Y.; Lee, J.-J. Effects of phenylalkanoic acids as co-adsorbents on the performance of dye-sensitized solar cells. *J. Nanosci. Nanotechnol.* **2013**, *13*, 7880–7885. [CrossRef]
10. Grätzel, M. Dye-sensitized solar cells. *J. Photochem. Photobiol. C* **2003**, *4*, 145–153. [CrossRef]
11. Du, L.C.; Furube, A.; Hara, K.; Katoh, R.; Tachiya, M. Mechanism of particle size effect on electron injection efficiency in ruthenium dye-sensitized TiO_2 nanoparticle films. *J. Phys. Chem. C* **2010**, *114*, 8135–8143. [CrossRef]
12. Ahn, J.; Lee, K.C.; Kim, D.; Lee, C.; Lee, S.; Cho, D.W.; Kyung, S.; Im, C. Synthesis of novel ruthenium dyes with thiophene or thienothiophene substituted terpyridyl ligands and their characterization. *Mol. Cryst. Liquid Cryst.* **2013**, *581*, 45–51. [CrossRef]
13. Kwon, T.H.; Kim, K.; Park, S.H.; Annamalai, A.; Lee, M.J. Effect of seed particle size and ammonia concentration on the growth of ZnO nanowire arrays and their photoconversion efficiency. *Int. J. Nanotechnol.* **2013**, *10*, 681–691. [CrossRef]
14. Rho, W.-Y.; Kim, H.-S.; Lee, S.H.; Jung, S.; Suh, J.S.; Hahn, Y.-B.; Jun, B.-H. Front-illuminated dye-sensitized solar cells with Ag nanoparticle-functionalized freestanding TiO_2 nanotube arrays. *Chem. Phys. Lett.* **2014**, *614*, 78–81. [CrossRef]
15. Hwang, K.-J.; Cho, D.W.; Lee, J.-W.; Im, C. Preparation of nanoporous TiO_2 electrodes using different mesostructured silica templates and improvement of the photovoltaic properties of DSSCs. *New J. Chem.* **2012**, *36*, 2094–2100. [CrossRef]
16. Nath, N.C.D.; Kim, J.C.; Kim, K.P.; Yim, S.; Lee, J.-J. Deprotonation of N3 adsorbed on TiO_2 for high-performance dye-sensitized solar cells (DSSCs). *J. Mater. Chem. A* **2013**, *1*, 13439–13442. [CrossRef]
17. Nakade, S.; Saito, Y.; Kubo, W.; Kitamura, T.; Wada, Y.; Yanagida, S. Influence of TiO_2 nanoparticle size on electron diffusion and recombination in dye-sensitized TiO_2 solar cells. *J. Phys. Chem. B* **2003**, *107*, 8607–8611. [CrossRef]
18. Zhu, K.; Kopidakis, N.; Neale, N.R.; van de Lagemaat, J.; Frank, A.J. Influence of surface area on charge transport and recombination in dye-sensitized TiO_2 solar cells. *J. Phys. Chem. B* **2006**, *110*, 25174–25180. [CrossRef] [PubMed]

19. Chung, K.-H.; Rahman, M.M.; Son, H.-S.; Lee, J.-J. Development of well-aligned TiO_2 nanotube arrays to improve electron transport in dye-sensitized solar cells. *Int. J. Photoenergy* **2012**, *2012*. [CrossRef]
20. Lee, G.I.; Nath, N.C.D.; Sarker, S.; Shin, W.H.; Ahammad, A.S.; Kang, J.K.; Lee, J.-J. Fermi energy level tuning for high performance dye sensitized solar cells using sp 2 selective nitrogen-doped carbon nanotube channels. *Phys. Chem. Chem. Phys.* **2012**, *14*, 5255–5259. [CrossRef] [PubMed]
21. Nath, N.C.D.; Sarker, S.; Ahammad, A.S.; Lee, J.-J. Spatial arrangement of carbon nanotubes in TiO_2 photoelectrodes to enhance the efficiency of dye-sensitized solar cells. *Phys. Chem. Chem. Phys.* **2012**, *14*, 4333–4338. [CrossRef] [PubMed]
22. Yadav, S.K.; Madeshwaran, S.R.; Cho, J.W. Synthesis of a hybrid assembly composed of titanium dioxide nanoparticles and thin multi-walled carbon nanotubes using "click chemistry". *J. Colloid Interface Sci.* **2011**, *358*, 471–476. [CrossRef] [PubMed]
23. Bavykin, D.V.; Parmon, V.N.; Lapkin, A.A.; Walsh, F.C. The effect of hydrothermal conditions on the mesoporous structure of TiO_2 nanotubes. *J. Mater. Chem.* **2004**, *14*, 3370–3377. [CrossRef]
24. Macak, J.M.; Tsuchiya, H.; Schmuki, P. High-aspect-ratio TiO_2 nanotubes by anodization of titanium. *Angew. Chem. Int. Ed. Engl.* **2005**, *44*, 2100–2102. [CrossRef] [PubMed]
25. Law, M.; Greene, L.E.; Johnson, J.C.; Saykally, R.; Yang, P. Nanowire dye-sensitized solar cells. *Nat. Mater.* **2005**, *4*, 455–459. [CrossRef] [PubMed]
26. Mor, G.K.; Varghese, O.K.; Paulose, M.; Shankar, K.; Grimes, C.A. A review on highly ordered, vertically oriented TiO_2 nanotube arrays: Fabrication, material properties, and solar energy applications. *Sol. Energy Mater. Sol. Cells* **2006**, *90*, 2011–2075. [CrossRef]
27. Jennings, J.R.; Ghicov, A.; Peter, L.M.; Schmuki, P.; Walker, A.B. Dye-sensitized solar cells based on oriented TiO_2 nanotube arrays: Transport, trapping, and transfer of electrons. *J. Am. Chem. Soc.* **2008**, *130*, 13364–13372. [CrossRef] [PubMed]
28. Rho, W.-Y.; Jeon, H.; Kim, H.-S.; Chung, W.-J.; Suh, J.S.; Jun, B.-H. Recent progress in dye-sensitized solar cells for improving efficiency: TiO_2 nanotube arrays in active layer. *J. Nanomater.* **2015**, *2015*. [CrossRef]
29. Atwater, H.A.; Polman, A. Plasmonics for improved photovoltaic devices. *Nat. Mater.* **2010**, *9*, 865–865. [CrossRef]
30. Standridge, S.D.; Schatz, G.C.; Hupp, J.T. Distance dependence of plasmon-enhanced photocurrent in dye-sensitized solar cells. *J. Am. Chem. Soc.* **2009**, *131*, 8407–8409. [CrossRef] [PubMed]
31. Brown, M.D.; Suteewong, T.; Kumar, R.S.S.; D'Innocenzo, V.; Petrozza, A.; Lee, M.M.; Wiesner, U.; Snaith, H.J. Plasmonic dye-sensitized solar cells using core-shell metal-insulator nanoparticles. *Nano Lett.* **2011**, *11*, 438–445. [CrossRef] [PubMed]
32. Qi, J.F.; Dang, X.N.; Hammond, P.T.; Belcher, A.M. Highly efficient plasmon-enhanced dye-sensitized solar cells through metal@oxide core-shell nanostructure. *ACS Nano* **2011**, *5*, 7108–7116. [CrossRef] [PubMed]
33. Rho, C.; Min, J.H.; Suh, J.S. Barrier layer effect on the electron transport of the dye-sensitized solar cells based on TiO_2 nanotube arrays. *J. Phys. Chem. C* **2012**, *116*, 7213–7218. [CrossRef]
34. Rho, W.-Y.; Chun, M.-H.; Kim, H.-S.; Hahn, Y.-B.; Suh, J.S.; Jun, B.-H. Improved energy conversion efficiency of dye-sensitized solar cells fabricated using open-ended TiO_2 nanotube arrays with scattering layer. *Bull. Korean Chem. Soc.* **2014**, *35*. [CrossRef]
35. Lin, C.-J.; Yu, W.-Y.; Chien, S.-H. Transparent electrodes of ordered opened-end TiO_2-nanotube arrays for highly efficient dye-sensitized solar cells. *J. Mater. Chem.* **2010**, *20*, 1073–1077. [CrossRef]
36. Li, L.-L.; Chen, Y.-J.; Wu, H.-P.; Wang, N.S.; Diau, E.W.-G. Detachment and transfer of ordered TiO_2 nanotube arrays for front-illuminated dye-sensitized solar cells. *Energy Environ. Sci.* **2011**, *4*, 3420–3425. [CrossRef]
37. Hahm, E.; Jeong, D.; Cha, M.G.; Choi, J.M.; Pham, X.-H.; Kim, H.-M.; Kim, H.; Lee, Y.-S.; Jeong, D.H.; Jung, S. β-CD dimer-immobilized Ag assembly embedded silica nanoparticles for sensitive detection of polycyclic aromatic hydrocarbons. *Sci. Rep.* **2016**, *6*. [CrossRef] [PubMed]
38. Arockia Jency, D.; Umadevi, M.; Sathe, G. Sers detection of polychlorinated biphenyls using β-cyclodextrin functionalized gold nanoparticles on agriculture land soil. *J. Raman Spectrosc.* **2015**, *46*, 377–383. [CrossRef]
39. Xie, Y.; Wang, X.; Han, X.; Xue, X.; Ji, W.; Qi, Z.; Liu, J.; Zhao, B.; Ozaki, Y. Sensing of polycyclic aromatic hydrocarbons with cyclodextrin inclusion complexes on silver nanoparticles by surface-enhanced Raman scattering. *Analyst* **2010**, *135*, 1389–1394. [CrossRef] [PubMed]
40. Kim, H.-Y.; Rho, W.-Y.; Lee, H.Y.; Park, Y.S.; Suh, J.S. Aggregation effect of silver nanoparticles on the energy conversion efficiency of the surface plasmon-enhanced dye-sensitized solar cells. *Sol. Energy* **2014**, *109*, 61–69. [CrossRef]

Permissions

All chapters in this book were first published in MDPI; hereby published with permission under the Creative Commons Attribution License or equivalent. Every chapter published in this book has been scrutinized by our experts. Their significance has been extensively debated. The topics covered herein carry significant findings which will fuel the growth of the discipline. They may even be implemented as practical applications or may be referred to as a beginning point for another development.

The contributors of this book come from diverse backgrounds, making this book a truly international effort. This book will bring forth new frontiers with its revolutionizing research information and detailed analysis of the nascent developments around the world.

We would like to thank all the contributing authors for lending their expertise to make the book truly unique. They have played a crucial role in the development of this book. Without their invaluable contributions this book wouldn't have been possible. They have made vital efforts to compile up to date information on the varied aspects of this subject to make this book a valuable addition to the collection of many professionals and students.

This book was conceptualized with the vision of imparting up-to-date information and advanced data in this field. To ensure the same, a matchless editorial board was set up. Every individual on the board went through rigorous rounds of assessment to prove their worth. After which they invested a large part of their time researching and compiling the most relevant data for our readers.

The editorial board has been involved in producing this book since its inception. They have spent rigorous hours researching and exploring the diverse topics which have resulted in the successful publishing of this book. They have passed on their knowledge of decades through this book. To expedite this challenging task, the publisher supported the team at every step. A small team of assistant editors was also appointed to further simplify the editing procedure and attain best results for the readers.

Apart from the editorial board, the designing team has also invested a significant amount of their time in understanding the subject and creating the most relevant covers. They scrutinized every image to scout for the most suitable representation of the subject and create an appropriate cover for the book.

The publishing team has been an ardent support to the editorial, designing and production team. Their endless efforts to recruit the best for this project, has resulted in the accomplishment of this book. They are a veteran in the field of academics and their pool of knowledge is as vast as their experience in printing. Their expertise and guidance has proved useful at every step. Their uncompromising quality standards have made this book an exceptional effort. Their encouragement from time to time has been an inspiration for everyone.

The publisher and the editorial board hope that this book will prove to be a valuable piece of knowledge for researchers, students, practitioners and scholars across the globe.

List of Contributors

Chunze Yuan, Jing Huang, Zhijun Ning and Hans Ågren
Department of Theoretical Chemistry and Biology, School of Biotechnology, Royal Institute of Technology, 10691 Stockholm, Sweden

Lin Li and Licheng Sun
Center of Molecular Devices, Department of Chemistry, School of Chemical Science and Engineering, Royal Institute of Technology, 10044 Stockholm, Sweden

Zhen Lu, Wen Liu, Jingjing Li, Wanning Li and Feng Feng
College of Chemistry and Environmental Engineering, Shanxi Datong University, Datong 037009, China

Tao Fang, Jicheng Zhang and Wenhua Li
Beijing Key Laboratory of Energy Conversion and Storage Materials, College of Chemistry, Beijing Normal University, Beijing 100875, China

Qiang Zhang
School of Systems Engineering, Kochi University of Technology, Kami, Kochi 782-8502, Japan

Shengwen Hou
Center for Nanotechnology, Kochi University of Technology, Kami, Kochi 782-8502, Japan

Chaoyang Li
School of Systems Engineering, Kochi University of Technology, Kami, Kochi 782-8502, Japan
Center for Nanotechnology, Kochi University of Technology, Kami, Kochi 782-8502, Japan

Prathap Pathi
Ames Laboratory, Microelectronics Research Center, Iowa State University, Ames, IA 50011, USA
Silicon Solar Cell Division, CSIR-National Physical Laboratory, Dr. K.S. Krishnan Road, New Delhi-110012, India

Akshit Peer
Ames Laboratory, Microelectronics Research Center, Department of Electrical and Computer Engineering, Iowa State University, Ames, IA 50011, USA

Rana Biswas
Ames Laboratory, Microelectronics Research Center, Department of Physics and Astronomy, Department of Electrical and Computer Engineering, Iowa State University, Ames, IA 50011, USA

Michael Eck
Department of Microsystems Engineering (IMTEK), University of Freiburg, Georges-Köhler-Allee 103, 79110 Freiburg, Germany

Michael Krueger
Department of Physics, Carl von Ossietzky University of Oldenburg, Carl-von-Ossietzky-Straße 9-11, 26129 Oldenburg, Germany

Alvien Ghifari, Dang Xuan Long, Seonhyoung Kim, Brian Ma and Jongin Hong
Department of Chemistry, Chung-Ang University, 84 Heukseok-ro, Dongjak-gu, Seoul 06974, Korea

Jin Wang, Aurélie Habert and Nathalie Herlin-Boime
IRAMIS/NIMBE/LEDNA, UMR 3685, CEA Saclay, 91191 Gif sur Yvette, France

Kosti Tapio and Jussi Toppari
Nanoscience Center, Department of Physics, University of Jyväskylä, 40014 Jyväskylä, Finland

Sebastien Sorgues and Christophe Colbeau-Justin
Laboratoire de Chimie Physique, UMR8000, Université Paris-Sud, 91405 Orsay, France

Bernard Ratier and Johann Bouclé
XLIM UMR 7252, Université de Limoges/CNRS, 87060 Limoges Cedex, France

Monica Scarisoreanu
National Institute for Lasers Plasma and Radiation Physics, Bucharest, Romania

Yiming Bai, Lingling Yan, Lin Su and Nuofu Chen
State Key Laboratory of Alternate Electrical Power System with Renewable Energy Sources, North China Electric Power University, Beijing 102206, China

Jun Wang
Institute of Information Photonics and Optical Communications, Beijing University of Posts and Telecommunications, Beijing 100876, China

Zhigang Yin and Yuanyuan Liu
Key Laboratory of Semiconductor Materials Science, Institute of Semiconductors, Chinese Academy of Sciences, Beijing 100083, China

Tianyi Shen, Qiwen Tan, Zhenghong Dai, Nitin P. Padture and Domenico Pacifici
School of Engineering, Brown University, 184 Hope Street, Providence, RI 02912, USA

Jue Wei
Department of Electrical and Computer Engineering, University of Alberta, Edmonton, AB T6G 2V4, Canada
Key Laboratory of Coherent Light and Atomic and Molecular Spectroscopy of Ministry of Education, College of Physics, Jilin University, Changchun 130012, China

Qiuyang Xiong, Seyed Milad Mahpeykar and Xihua Wang
Department of Electrical and Computer Engineering, University of Alberta, Edmonton, AB T6G 2V4, Canada

Yunfei Shang
School of Chemical Engineering and Technology, Harbin Institute of Technology, Harbin 150001, China

Shuwei Hao and Chunhui Yang
School of Chemical Engineering and Technology, Harbin Institute of Technology, Harbin 150001, China
Harbin Huigong Technology Co., Ltd., Harbin 150001, China

Guanying Chen
School of Chemical Engineering and Technology, Harbin Institute of Technology, Harbin 150001, China
Institute for Lasers, Photonics, and Biophotonics, University at Buffalo, State University of New York, Buffalo, NY 14260, USA

Kodai Kishibe, Soichiro Hirata, Ryoichi Inoue, Tatsushi Yamashita and Katsuaki Tanabe
Department of Chemical Engineering, Kyoto University, Kyoto 615-8510, Japan

Tom Grace, LePing Yu, Christopher Gibson and Joseph Shapter
Centre for Nanoscale Science & Technology (CNST), Flinders University of South Australia, South Australia 5042, Australia

Daniel Tune
Centre for Nanoscale Science & Technology (CNST), Flinders University of South Australia, South Australia 5042, Australia
Institute of Nanotechnology, Karlsruhe Institute of Technology, 76021 Karlsruhe, Germany

Huda Alturaif and Zeid Al Othman
Advanced Material Research Chair, Chemistry Department, College of Science, King Saud University, Riyadh 11451, Saudi Arabia

Sofia Paulo
Institut Català d'Investigació Química (ICIQ), The Barcelona Institute of Science and Technology, Avda. Països Catalans 16, Tarragona 43007, Spain
Fundació Eurecat, Avda. Ernest Lluch 36, Mataró 08302, Spain

Emilio Palomares
Institut Català d'Investigació Química (ICIQ), The Barcelona Institute of Science and Technology, Avda. Països Catalans 16, Tarragona 43007, Spain
ICREA, Passeig Lluís Companys 23, Barcelona 08010, Spain

Eugenia Martinez-Ferrero
Fundació Eurecat, Avda. Ernest Lluch 36, Mataró 08302, Spain

Yanbin Wang, Changlong Zhuang, Yawen Fang, Huang Yu and Biaobing Wang
Jiangsu Key Laboratory of Environmentally Friendly Polymeric Materials, School of Materials Science and Engineering, Jiangsu Collaborative Innovation Center of Photovolatic Science and Engineering, Changzhou University, Changzhou 213164, Jiangsu, China

Hyung Do Kim and Hideo Ohkita
Department of Polymer Chemistry, Graduate School of Engineering, Kyoto University, Katsura, Nishikyo, Kyoto 615-8510, Japan

Yinan Zhang and Baohua Jia
Centre for Micro-Photonics, Faculty of Science, Engineering and Technology, Swinburne University of Technology, Hawthorn, Victoria 3122, Australia

Boyuan Cai
Institute of Photonics Technology, Jinan University, Guangzhou 510632, China

Il Ku Kim
National Photonics Semiconductor Lab, National Photonics Semiconductor Inc., Suwon-Si 17113, Korea

Jun Hyung Jo
School of Information and Communication Technology, Griffith University, Southport QLD 4222, Australia

Jung-Ho Yun
Nanomaterials Centre, School of Chemical Engineering, University of Queensland, Brisbane QLD 4072, Australia

Won-Yeop Rho
Department of Bioscience and Biotechnology, Konkuk University, Seoul 143-701, Korea
Department of Chemistry, Seoul National University, Seoul 151-747, Korea

Myeung-Hwan Chun, Ho-Sub Kim and Jung Sang Suh
Department of Chemistry, Seoul National University, Seoul 151-747, Korea

Hyung-Mo Kim and Bong-Hyun Jun
Department of Bioscience and Biotechnology, Konkuk University, Seoul 143-701, Korea

Index

A

Alloy, 1-3, 10, 71, 111, 144, 194

Aluminium Nanoparticles, 196, 205

Aluminum-doped Zinc Oxide, 26, 34, 180

Amorphous Silicon, 36, 46, 130, 139, 175, 205

B

Bulk Heterojunction, 23-24, 51-53, 60-62, 88, 133, 174, 182-183, 193-194, 213

Bulk-heterojunction Solar Cells, 51, 61, 193, 212

C

Cadmium-stearic Acid, 52-53, 60

Carbon Quantum Dots, 164-165, 168, 170, 180-182

Chemical Vapor Deposition, 26-27, 34, 152, 162

Chlorobenzene, 16-17, 20, 59-60, 186-187, 190-191, 208

Conjugated Polymers, 25, 51, 60-61, 184, 213

Conversion Efficiency, 1-2, 6, 10-11, 13-15, 18, 23, 26, 35, 50, 53-54, 60, 63, 65, 68, 78, 91-92, 94, 97, 99, 105, 111-112, 116, 118, 121, 125, 128, 130-134, 141-142, 145, 152, 157-158, 162, 184, 189, 193, 205, 209-210, 214-215, 218-219, 221, 223

Core-shell Nanorods, 26-32

Crystalline Silicon, 35, 37, 46, 49-50, 121-122, 128, 135, 148, 152

D

Dielectric Photonic Crystals, 35, 48

Diketopyrrolopyrrole, 14-15, 22-24, 194

Dye-sensitized Solar Cell, 26, 33-34, 63, 71, 88, 137, 139-141, 166

Dynamic Light Scattering, 54, 59-60, 62

E

Electrolyte, 1-2, 5-6, 8-9, 26, 29, 63-65, 68, 71-73, 77, 166, 172, 174, 177-178, 215, 219-221

Electrolyte Solution, 26, 221

Electron Injection, 1, 3, 5, 8, 12, 26, 34, 131, 172-173, 183, 222

Electron Injection Efficiency, 1, 34, 131, 222

Energy Conversion Efficiency, 134, 205, 214-215, 218-219, 221, 223

Energy Transfer, 123-126, 132, 138, 140, 173, 181-182, 184-185, 187-188, 190-191, 193, 195

Exciton Harvesting, 184-185, 192, 194-195

F

Fluorinated Phenyl, 14-15, 22

G

Graphene Quantum Dots, 165, 168, 171, 173, 181-183

H

Heterojunction Solar Cells, 2, 10, 12, 24, 51, 61-62, 88, 140, 151, 163, 174, 182, 193-194, 212-213

Hexadecylamine, 52-54, 60

Hexanoic Acid, 51-54, 58, 60

Hole Transport Layer, 172, 180, 182

Hybrid Solar Cells, 51, 54, 56, 58-62, 73, 182-183, 221

I

Indium Tin Oxide, 60, 120, 208

L

Ligand Sphere, 51-52, 54, 56, 59

Light Incoupling, 196-199, 201-204

Light Trapping, 35-41, 43, 45, 47, 49-50, 97-98, 110, 112-116, 118-119, 196, 204-205

Light-trapping Architecture, 36, 42

M

Mass Production, 14, 45, 135

Metal-oxide Semiconductor, 196-197

Mist Chemical Vapor, 26-27, 34

Mist Chemical Vapor Deposition, 26-27, 34

N

Nano-scale Phase, 14-15, 19, 22

Nanocrystals, 11-13, 51-53, 60-62, 73-74, 84, 89, 125-126, 128-130, 136-140, 178

Nanoparticles, 33, 36, 61, 63, 65-66, 68, 71, 73, 84-86, 88-92, 95, 97-99, 110-111, 129, 132-133, 136-139, 141-142, 150, 164, 168, 177, 196-197, 204-206, 214, 216, 223

Nanorod, 27, 30, 34, 60-61, 136, 176, 222

O

Open Circuit Voltage, 14-15, 32, 79, 93-94, 152, 157-158

Organic Solar Cell, 14, 49, 58, 133, 166-167, 194, 207

P

Parallel Resistance, 56, 60

Photoactive Layer, 51, 173, 184-185, 207-208, 210-212

Photocurrent, 1, 3, 6-10, 12-13, 34, 65, 78-79, 84, 86, 88-89, 91-92, 94, 96-97, 130, 133-134, 148, 162, 185, 189, 191, 202-205, 218-219, 223

Photoelectrode, 10, 26, 140

Photoluminescence, 52-53, 60, 62, 78, 87, 129, 137-138, 171, 180, 183, 187

Photonic Crystal, 36-38, 40-46, 50, 149

Photosensitizers, 1, 3-4, 10

Photovoltaic Performance, 1-2, 6-8, 10, 14, 18-19, 22-24, 33, 68, 73, 77, 111, 119, 137, 141, 162, 183, 185, 189

Polydimethylsiloxane, 45-46, 118

Polymer Solar Cells, 23-24, 61-62, 98, 141-142, 150, 175, 181, 183-186, 189, 191-195, 213

Post-synthetic Treatment, 51-52, 54-55, 60

Power Conversion Efficiency, 11, 13-15, 18, 23, 26, 35, 53-54, 60, 65, 68, 78, 99, 105, 112, 116, 118, 130-133, 141, 152, 157-158, 162, 184, 189, 193, 209-210

Q

Quantum Dot-sensitized Solar Cells, 1, 11-13, 181

Quantum Dots, 1, 10-13, 51-53, 60-62, 97, 114, 119, 122, 126, 134, 136, 142, 151, 164-165, 168, 170-171, 173, 178, 180-183

Quantum Efficiency, 10, 38, 78, 93-94, 96-97, 121, 125, 128, 130, 134, 184, 187, 189, 191, 193, 202, 208, 210, 213

Quantum Yield, 53, 60, 142, 168

S

Series Resistance, 1, 56-57, 60, 176, 178

Short-circuit Current Density, 1, 7, 10, 39, 55-56, 60, 79, 93-94, 116, 157, 184-185, 189

Small Molecule, 14-15, 18, 20-21, 23-24, 166-167, 175, 178, 180, 193-194, 207, 209-210, 213

Solar Absorption, 35, 119

Solar Cell Performance, 1, 13, 33, 51, 54, 56, 58, 60, 62, 97, 108, 157, 196, 208

Solar Cell Technology, 35, 49, 135

Solar Cells, 1-4, 6-14, 18, 23-26, 32-37, 48-63, 65, 71-74, 77-78, 84, 86-89, 91-93, 96-99, 110-113, 115-116, 118-122, 124-125, 127-131, 133-136, 139-144, 148-152, 156, 160-168, 171-178, 180-186, 189, 191-207, 209, 212-214, 216, 219, 221-223

Solar Panels, 35, 43, 46, 151

Solution Process, 14, 50

Surface Energy, 184-185, 187, 190, 192-193

Surface Plasmon Resonances, 109, 196-197, 205

T

Ternary Alloy, 1, 10

Ternary Blend Solar Cells, 184-185, 189, 194

Thermal Annealing, 51, 55-56, 59, 74, 98

Titanium Dioxide, 11, 26-27, 33, 37-38, 72, 88, 223

Trioctylphosphine Oxide, 52-54, 60

Trioctylphosphine-selenid, 52-53, 60

Triphenyl Amine, 14-15

Triphenylamine, 15, 24

U

Ultraviolet-visible, 16-17, 52-53, 60, 94, 217-218

V

Vapor Deposition, 26-27, 34, 152, 162

Z

Zinc Oxide, 26-27, 33-34, 49, 180

Zinc Oxide Nanorods, 26, 34

Zinc Oxide-titanium Dioxide, 26-27

www.ingramcontent.com/pod-product-compliance
Lightning Source LLC
LaVergne TN
LVHW082216200926
848484LV00008B/2266